Foundations of Safety Science

Foundations of Safety Science

Science
A Century of Understanding Accidents and Disasters

Sidney Dekker

CRC Press
Taylor & Francis Group
Boca Raton London New York

CRC Press is an imprint of the
Taylor & Francis Group, an **informa** business

CRC Press
Taylor & Francis Group
6000 Broken Sound Parkway NW, Suite 300
Boca Raton, FL 33487-2742

International Standard Book Number-13: 978-1-138-48178-7 (Paperback)
International Standard Book Number-13: 978-1-138-48177-0 (Hardback)

Library of Congress Cataloging-in-Publication Data

Names: Dekker, Sidney, author.
Title: Foundations of safety science : a century of understanding accidents and disasters / authored by Sidney Dekker.
Description: Boca Raton : Taylor & Francis, CRC Press, 2019. | Includes index.
Identifiers: LCCN 2018057030 | ISBN 9781138481787 (pbk. : alk. paper) | ISBN 9781138481770 (hardback : alk. paper) | ISBN 9781351059794 (e-book)
Subjects: LCSH: Industrial safety—History. | Industrial accidents—History.
Classification: LCC T55 .D399 2019 | DDC 363.11—dc23
LC record available at https://lccn.loc.gov/2018057030

Visit the Taylor & Francis Web site at
http://www.taylorandfrancis.com

and the CRC Press Web site at
http://www.crcpress.com

Contents

Preface

I wrote this book foremost for safety practitioners and students. Confession: I never was a safety practitioner. Nor am I one now. For sure, I talk about safety a lot. And I write a lot about it too, or about topics affiliated with it. But I was never educated as a safety practitioner—I was educated in psychology and engineering, and trained how to fly a big jet. Well, check that: out of curiosity, I did once take a weeklong course that awarded me a Certificate IV in Occupational Health and Safety. But to say that it either educated or qualified me would be a stretch.

Over the past decades, I have worked increasingly with those who *have* been trained as safety practitioners—in a range of industries. Much of the education they have gone through was organized around applicable laws, regulations, policies, best practices, methods, and techniques, often driven by peer-to-peer influence—inspirations from what others in other organizations have done—and hand-me-down knowledge.

And actually, not all safety practitioners were educated as safety practitioners. In fact, many safety practitioners have backgrounds in operations, in HR, in engineering or chemistry or a mechanical trade or psychology, or something else altogether.

Whether it is a background in safety practice or something else, they are all great ways to get into, as David Provan would say, the safety of work and the work of safety. What I have learned, though, is that all could benefit from a more solid grounding in the foundations of the science of safety (such as it is, I hear Erik Hollnagel justifiably say (Hollnagel, 2014)).

I have found that without that grounding, it is easy to reinvent the wheel and happily embrace an idea or slogan simply because it is shiny and seemingly new. Without that grounding, it is tempting to apply a putative solution (such as putting a barrier in place) to a problem that is not only immune to the solution, but may well bite back by spawning more problems than the safety practitioner bargained for. Without that grounding, it is seductive to fall for expensive solutions (enterprise-wide introduction of risk matrices, hazard awareness campaigns, spoggles for supposed eye protection) and force-feed them into the organization, even when they are based on a particular conceptualization of danger that is not applicable at all.

I have chosen an episodic approach to organizing this book. That is, I have divided it up into time slices. Every chapter is founded on the ideas of a particular era—each roughly a decade from the past century. It then explores how these have influenced our thinking in safety in other decades or ever since. Of course, the lines and categories of what belongs to which decade, or what inspired what exactly, can always be debated, as it should. They are not in this book to radiate an impression of linear, historical truth. Rather, they are a way for me to organize the ideas, and for you to start thinking with them.

How are today's 'hearts and minds' programs, for example, linked to a late-19th-century definition of human factors as people's moral and mental deficits? What do Heinrich's 'unsafe acts' from the 1930s have in common with the Swiss Cheese

Model of the early 1990s? Why was the reinvention of Human Factors in the 1940s such an important event in the development of safety thinking? What makes many of our current systems so complex and impervious to Tayloristic safety interventions? I have tried to review the theoretical origins of major schools of safety thinking, and to trace the heritage of, and interlinkages between, the ideas that make up safety science today.

I spend some time exploring questions like these in historical and theoretical depth. Yet I hopefully move fast enough to keep you captivated. If all is well, the book should offer you a comprehensive overview of theoretical foundations of safety science, showing interlinkages and cross-connections between them. To make it useable as a textbook, I have provided chapter overviews at the beginning of each chapter and study questions at the end of each chapter. References are organized per chapter also, so that you can easily find what you might have become inspired to look for from that particular decade or approach.

Chapter 1 introduces safety science as the interdisciplinary study of accidents and accident prevention. It contains theories inspired by engineering, physical sciences, epidemiology, sociology, psychology, anthropology, and more. The chapter explains how most of the theories that guide current safety practices were developed during the 20th century. The same principles of scientific experimentation, theorizing, and logical reasoning—which had increasingly shown their value during the 18th and 19th centuries—could be brought to bear on the problem of safety. This accompanied a shift toward believing that the causes of accidents could be scientifically understood, and that there was a moral responsibility to engineer or organize preventative measures. It laid the basis for the emergence of new institutions for the creation and maintenance of safety rules and practice, including regulators, inspectorates, and investigative bodies which directly represented the government. Other institutions represented the common interests of employers or workers (standards bodies, professional associations), and still others combined the two, as in the case of government-mandated private insurance schemes. The chapter also describes how the first political concern for safety grew from the mines, factories, railroads, and steamships of the Industrial Revolution. The interplay between concerned citizens and scientists, government regulators, and insurers would continue to drive the creation of safety theories and their practical applications into the 20th and 21st centuries.

Chapter 2 discusses scientific management and its implications for how we look at the relationship between work and rules even today. The approach was labeled and exemplified by Frederick Taylor and the Gilbreths, who applied a scientific method to determine the most efficient way to perform a specific task. Workers needed to comply with the one best way to do the task in order to maximize efficiency (and safety). Taylorism accelerated and solidified the division between those who plan, manage, and supervise the work, and those who execute it. Thinking and problem-solving was heavily concentrated in the former, while the latter merely needed to do what they were instructed. It lies at the basis of the belief that control over workers and operators is possible through better compliance with procedures; through better hierarchical order and the imposition of rules and obedience. Taylor's ideas have left deep traces in our thinking about the safety of work: workers need to be told what to do. They need to be supervised and monitored, and their work has to be specified in great detail. Autonomy and initiative are undesirable. Worker departures from the one best method can be seen as a 'violation,' which requires sanctions or reminders, or the rewriting of the rule (but that should not be done by the worker). They can also be seen as local, adaptive resilience necessary to close the gap between how work is imagined and how it actually gets done. However, adherence to rules can indeed make it impossible to get the work done, lead to blindness to new situations and the squashing of innovation, resentment at the loss of freedom and discretion, and the growth of bureaucratic and supervisory control to impose compliance.

Chapter 3 relates the approach taken by accident-proneness as one of the earliest attempts to scientifically investigate the 'human element' in safety. It took the emerging sciences of psychology and eugenics, and applied them to explain patterns in industrial data. Accident-proneness built on two patterns that were hard to dispute: (1) Human performance is variable, in ways that can be relevant for accidents, and (2) Some people are involved in more accidents than others, in ways that are unlikely to arise purely by chance. In doing so, it provided an object lesson in the limitations and moral perils of treating human capacities in a reductionist, 'scientific' way. And, as the chapter explains, accident-proneness is ultimately difficult to prove. Safety outcomes are never purely due to specific individual attributes. If the same activities take place in different contexts (busy times versus low workload, or night versus day), even the same individual may well have different safety outcomes. It can never be determined with certainty how much of the outcome is attributable to the individual, and how much to context.

Chapter 4 discusses Heinrich's ideas and behavior-based safety. Heinrich reestablished the idea that many accidents and injuries are preventable. He used a metaphor of a row of dominoes to explain how distant causes lead to injuries. Ancestry and the social environment give rise to character flaws in a worker, such as bad temper, ignorance, or carelessness. Character flaws give rise to unsafe conditions, mechanical hazards, and unsafe acts. These factors in turn lead to accidents, which lead to injuries and fatalities. Like a row of falling dominoes, Heinrich suggested that the sequence could be interrupted by removing the right factor in the sequence. Heinrich's opinion on the best point for intervention shifted throughout his career. Early on, he placed a strong emphasis on improving the physical conditions and physical safeguards of work. Later he placed increasing emphasis on eliminating unsafe acts by workers. He advocated creating an environment where even small undesirable acts are not tolerated. It was mostly through these later, human-error-focused ideas that Heinrich influenced the theory and practice of safety. Behavior-based safety is one of the most visible expressions of it, with us to this day. Three key ideas of Heinrich's have influenced safety practices (and even some theories) for decades: (1) Injuries are the result of linear, single causation; (2) there is a fixed ratio between accidents (or simply "occurrences"), minor injuries and major injuries; and (3) worker unsafe acts are responsible for 88% of industrial accidents. All three have been proven false.

Chapter 5 explains how the human became increasingly acknowledged to be a recipient of safety trouble—trouble that was created upstream and then handed down to them by their tools, technologies, organizations, working environments, or tasks. Human factors was a field that grew from the insights of engineering psychologists in the 1940s. Confronted with the role of the human in increasingly complex technological systems, it represented an important hinge in this thinking about the relationship between humans, systems, and safety. Systems and technologies were considered malleable, and they should be adapted to human strengths and limitations. Indeed, individual differences were less important than devising technologies and systems that would resist or tolerate the actions of individuals, independent of their differences. Safety problems had to be addressed by controlling the technology. The approach of human factors led to a rekindling of our interest in mental and social phenomena. These became important for understanding how best to design and engineer technologies that fit the strengths and limitations of human perception, memory, attention, collaboration, communication, and decision-making. A few decades later, the field departed from an overly technicalized, individualist, laboratory task-based and mentalist information processing paradigm and took the study of cognition 'into the wild' to understand people's collaborative sensemaking in their interaction with actual complex, safety-critical technologies. It led to an entirely new take on human factors, in the field known as cognitive systems engineering.

Chapter 6 details the contributions of system safety. It was driven by the commitment that safety should get built into the system from the very beginning. And once a system was in operation, system safety specified the requirements for its effective and safe management. This required system safety to recognize the technical, human, and environmental contributors to the creation and erosion of safety, and to map and resolve (to the extent possible) the conflicts and trade-offs between safety and other factors in the design and operation of the system. To do so, systems engineering for safety involves standardized process steps, with many variations in the detail of technique applied at each step. The steps are (semi-) formal modeling of the system under development, analysis of draft system designs, and analysis of the final design to demonstrate safety and to inform post-design safety efforts. From a design perspective, systems can be unsafe through requirements error (designing the wrong system), or implementation error (designing the system wrong). The aim is to prevent foreseeable events and minimize the consequences of unforeseen ones. The increasing complexity of automation and computerization (particularly when added to legacy systems can make this very difficult. System safety, through its formal language and techniques, has defined safety as freedom from unwanted events, and protection against unwanted outcomes. As systems have become more complex and anticipating all pathways to failure becomes virtually impossible, an emphasis is shifting to assuring the presence of capacity to handle unforeseen events, rather than assuring the absence of failure modes.

Chapter 7 recounts how safety was taken out of the engineering space and expert-driven language of system safety by high-visibility disasters and a wave of emancipatory movements in the 1960s and 1970s. The size and complexity of many of society's safety-critical systems were becoming apparent to many—and in certain cases alarmingly so. Large disasters with socio-technical systems, and many near-disasters, brought safety and accidents to center stage. This greater visibility helped give rise to two decades of productive scholarship, and set the stage for a lot of the conversation about safety, accidents, and disasters we are having to this day.

Accidents were increasingly understood as social and organizational phenomena, rather than just as engineering problems. Man-made disaster theory was the first to theorize this, closely followed by high reliability theory and normal accident theory. Disasters and accidents are preceded by sometimes lengthy periods of gradually increasing risk, according to man-made disasters theory. This buildup of risk goes unnoticed or unrecognized. Turner referred to this as the incubation period. During this period, he suggested, latent problems and events accumulate which are culturally taken for granted or go unnoticed because of a collective failure of organizational intelligence. The accumulation of these events can produce a gradual drift toward failure.

Chapter 8 compares two (North American) approaches that emerged from the greater social preoccupation with accidents and safety in the 1970s and 1980s. As high-visibility accidents and disasters put the limits of safety engineering and risk management on display, questions arose: Was there a limit to the complexities we could handle? Were there things that we perhaps should not do, or build at all? Normal accident theory (NAT) suggested that some accidents are 'normal'—and thus in a sense predictable—because they can be traced to the interactive complexity and coupling of the systems we design, build, and operate. Interactive complexity and tight coupling built into the very structure of these systems will generate certain accidents, the theory said, independent of how much risk management we do. Yet there are interactively complex and tightly coupled systems that do not generate accidents, or that have not yet. So are there characteristics of 'high reliability' that can somehow be distilled from the things that these systems are doing? This is what high reliability organizations (HROs), also known as high reliability theory (HRT), suggested. A comparison of the two approaches serves as an introduction to the debate that was triggered by the publication in the early 1990s of *The Limits of Safety*. Both theoretical schools have had considerable influence on what has happened in safety science since.

Chapter 9 discusses how the Swiss cheese model became an important icon of the idea that problems experienced at the sharp end (or front line) of an organization are not created there, but are inherited from imperfect upstream processes and parts. By this time, a strong consensus had already formed: human performance at the sharp end is shaped by local workplace conditions and distal organizational factors. The Swiss cheese model is a defenses-in-depth or barrier model of risk, which suggests (as did Heinrich) that risk should be seen as energy that needs to be contained or channeled or stopped. The chapter explains why this makes it difficult for Swiss cheese to be a true 'systemic' model, since it is not capable of explaining or portraying the complex emergence of organizational decisions, the erosion of defenses, drift and a normalization of deviance. Swiss cheese conceptually aligns with safety management systems. These direct safety efforts and regulation at the administrative end of an organization, where assurance that safety is under control is sought in management systems, accountabilities, processes and data. The gradual shift to 'back-of-house,' to organizational and administrative assurance of safety, had been long in the making. It is now intuitive that all work is shaped by the engineered context and workplace conditions and upstream organizational factors. If we want to understand or change anything, then that is where we need to look. This trend has also given rise to large safety bureaucracies and cultures of compliance. It has left us with a safety profession that broadly lacks purpose and vision in many industries, and with regulators whose roles have in certain cases been hollowed out and minimized.

Chapter 10 continues the focus on the sorts of things that can be found and fixed in an organization before they can create or contribute to an accident. Encouraging organizations to build a 'good' safety culture is the logical continuation of this trend. Safety culture has given organizations an aspiration, getting leaders and others to think about what they want to have rather than what they want to avoid. Researchers and practitioners became concerned with specifying what is necessary inside an organization and its people to enhance safety. A functionalist approach to safety culture sees and measures it as something that an organization 'has.' A culture can be taken apart, redesigned, and formed. Management can 'work' on parts of that culture (e.g., hazard reporting, procedural compliance). It assumes that values drive people's attitudes and beliefs, which in turn determine their behavior. The interpretivist, or qualitative, approach defines culture as something that an organization 'does.' It considers culture as a bottom-up, complex, emergent phenomenon; greater than the sum of its parts; resistant to reductionist analysis, measurement, and engineering. For this reason, it cannot be trained and injected into individual minds. Critiques of 'safety culture,' discussed in the chapter, have been targeted at (1) the normative idea that some cultures are 'better' than others, (2) the implication that cultures are consistent and coherent rather than full of conflict and contradiction, (3) the avoidance of any mention of power in most models of safety culture, (4) the methodological individualism that sees culture as the aggregate of measurable individual attitudes, beliefs, and values, (5) the lack of usefulness of the concept in safety regulation and investigations, (6) the lack of predictive value and (7) the fact that the 'container term' of 'safety culture' tries to say so much that it ends up saying very little.

Chapter 11 covers the field of resilience engineering. Resilience engineering is about identifying and then enhancing the positive capabilities of people and organizations that allow them to adapt effectively and safely under varying circumstances. Resilience is not about reducing negatives (incidents, errors, violations). Resilience engineering wants to understand and enhance how people themselves build, or engineer, adaptive capacities into their system, so that systems keep functioning under varying circumstances and conditions of imperfect knowledge. How do they create safety—by developing capacities that help them anticipate and absorb pressures, variations and disruptions? Resilience engineering is inspired by a range of fields beyond traditional safety disciplines, such as physical, organizational, psychological, and ecological sciences. The organic systems studied in these fields are effective (or not) at adjusting when they recognize a shortfall in their adaptive capacity—which is key to the creation of resilience (or its disruption). Although this is not unique to this theory, the chapter finishes by concluding that resilience engineering, too, appears vulnerable to three analytical traps: a reductionist, a moral, and a normative one.

Finally, the book's postscript reflects on a pattern that all approaches over the past century have seemingly adopted. From an innovation that typically targets the system in which people work, almost every approach seems to end up reverting, one way or another, to the people who work in that system. At the heart of this pattern is a dialectic—system or person, upstream or downstream, organization or individual? The future for safety science may well lie in our ability to break out of this dialectic and see people *in* systems, rather than people versus systems.

The way the book covers certain topics may not be how others remember them or how they would have approached them. I know. This book is my take on the foundations of safety science—as imaginative or ill-informed as it might be. But by laying out my understanding of the foundations of safety science, I hopefully invite others to bounce their own views and interpretations off it. I have picked out what I could from my understanding of the history of our science, and then deformed it into the linear confines of a book.

Note that this is not a book of safety techniques or methods. There are other books that tell you about those. This, instead, is a book about the ideas, the foundations, from which those methods and techniques stem. Take ICAM (Incident Cause Analysis Method) as an incident investigation method. It is based on the 1990s idea of Swiss cheese, of accident trajectories that penetrate barriers or imperfect defenses in depth, which in turn is based in part on 1930s linear domino models. For many safety practitioners, ICAM is simply a method that their organization or industry was already using. They have adopted it or adapted along with it—no further questions asked. After reading this book, it is my hope that methods like that become much more: a set of assumptions and premises that influence what can be done with the method, and what it will not be capable of; a set of possibilities and limitations that are imported from down the decades.

I have had some help in writing this book. My colleagues Drew Rae of Griffith University, Verena Zimmerman of the Technische Universität Darmstad in Germany, and Johan Bergström of Lunds Universitet in Sweden have all done some portions of writing—in chapters 1, 6, 8, and 11. I am grateful for their help, and they are listed as co-authors of the respective chapters. Jop Havinga has helped with writing up some of the case studies. The cartoonish figures you will see throughout the book are all mine, so I am to blame for any offense they might cause.

The main purpose of this book is to increase the literacy of safety practitioners and of those who want to become safety practitioners. Those who do safety work can use it to build up their fluency in the ideas, concepts, and theories on which much of what they do is based. Others may want to use it as a reference, as an occasional guide, or a one-volume library of ideas and concepts and theories in safety.

In a previous book, *The Safety Anarchist* (Dekker, 2018), I used the cover of the title to pillory some of the more superficial, ill-conceived safety interventions I have come across. Many readers have told me they recognized themselves, or their colleagues, or their organizations in it. Some of them shook their heads in disdain and resignation.

Indeed, I know organizations that have developed safety into something akin to a 'mystery religion,' replete with its sacred texts, saintly thought leaders, holy relics, sacred places (think of ten golden rules, high-viz vests and altar-like safety notice boards), rituals (the prayer-like 'safety share'), and idolatry: the giving of ultimate meaning and allegiance to something that should be serving only an intermediate purpose (like Zero Harm, which is at best a moral mobilizer, an inspiration or aspiration, but is idolized as a promised land that will deliver mankind from suffering).

It is my hope that *Foundations of Safety Science* will allow them, and others, to recognize the foundations they stand on, and thereby become firmer in standing for something. Because if you don't stand for something, you'll fall for anything.

REFERENCES

Dekker, S. W. A. (2018). *The safety anarchist: Relying on human expertise and innovation, reducing bureaucracy and compliance.* London, UK: Routledge.
Hollnagel, E. (2014). Is safety a subject for science? *Safety Science, 67,* 21–24.

Author

Sidney Dekker, PhD (Ohio State University, USA, 1996) is professor and director of the Safety Science Innovation Lab at Griffith University in Brisbane, Australia, and professor in the Faculty of Aerospace Engineering at Delft University in the Netherlands. He has been flying the Boeing 737 as airline pilot on the side. He is a film maker ('Safety Differently,' 2017; 'Just Culture,' 2018) and author of, most recently: *The Safety Anarchist* (2018); *The End of Heaven* (2017); *Just Culture* (2016); *Safety Differently* (2015); *The Field Guide to Understanding 'Human Error'* (2014); *Second Victim* (2013); *Drift into Failure* (2012); *Patient Safety* (2011), and co-author of *Behind Human Error* (2010). More information is available at sidneydekker.com

1 The 1900s and Onward
Beginnings

Drew Rae and Sidney Dekker

KEY POINTS

- Safety science is the interdisciplinary study of accidents and accident prevention. It contains theories inspired by engineering, physical sciences, epidemiology, sociology, psychology, anthropology, and more.
- Most of the theories that guide current safety practices were developed during the 20th century. The same principles of scientific experimentation, theorizing, and logical reasoning—which had increasingly shown their value during the 18th and 19th centuries—could be brought to bear on the problem of safety.
- This accompanied a shift toward believing that the causes of accidents could be scientifically understood, and that there was a moral responsibility to engineer or organize preventative measures.
- It laid the basis for the emergence of new institutions for the creation and maintenance of safety rules and practice, including regulators, inspectorates, and investigative bodies that directly represented the government. Other institutions represented the common interests of employers or workers (standard bodies, professional associations), and still others combined the two, as in the case of government-mandated private insurance schemes.
- The first political concern for safety grew from the mines, factories, railroads, and steamships of the Industrial Revolution. The interplay between concerned citizens and scientists, government regulators, and insurers would continue to drive the creation of safety theories and their practical applications into the 20th and 21st centuries.

1.1 INTRODUCTION

Safety Science is the interdisciplinary study of accidents and accident prevention.

- As a *social science* discipline, Safety Science describes how society makes sense of and responds to the possibility of accidents.
- As a *psychology* discipline, Safety Science examines how humans behave as individuals, teams, and organizations during the incubation and aftermath of accidents.

- As a discipline in *population health*, Safety Science describes trends and patterns in the occurrence of accidents.
- As a multitude of *physical sciences* disciplines, it describes the physical processes by which accidents occur.
- As an *engineering* discipline, it seeks to identify and suggest practices and other interventions that can reduce the likelihood and consequences of accidents.

Most of the theories that guide current safety practices were developed during the 20th century, which is why this book concentrates on this time period. In this first chapter, however, we briefly consider the social and intellectual roots from which these theories have grown. These roots were planted in the 18th and 19th centuries: the historical germination and appearance of "modernity." It is not surprising that much of the origins of safety thinking can be found there. One of the defining characteristics of modernity was a widespread faith in human-engineered progress. People began to believe that the same principles of scientific experimentation and logical reasoning—which had harnessed lightning, predicted the movement of the planets, and supported the invention of steam engines—could be brought to bear on problems such as poverty, famine, and war. Under modernity, accidents came to be seen as problems that were caused by human failings, and that were fixable by human efforts. Through engineering, social, and legislative efforts, humans were held to have the capacity—and the responsibility—to make their world safer.

This chapter traces two themes that dominated safety in the 19th century, and still shape the way we see safety today.

1. A shift toward believing that the causes of accidents could be scientifically studied, and that there was a moral responsibility to engineer or organize preventative measures.
2. The emergence of new institutions for the creation and maintenance of safety rules and practice. Some of these institutions were regulators who directly represented the government. Other institutions represented the common interests of employers or workers, and still others combined the two, as in the case of government-mandated private insurance schemes.

Together, these changes created a social system in which governments had the right and responsibility to create or endorse rules for safe design and operation, and to enforce those rules through inspection, licensing, or punishment. They also created an individual right, if not to be protected from harm at work, then at least to be compensated and taken care of when injury occurred. This right in turn placed an obligation on employers to fund insurance schemes for workers' compensation.

1.2 SAFETY AND RISK: DIVINE OR HUMAN?

The history of safety thought can be charted by examining the flurry of social, political, and intellectual activities following major accidents or 'crises' arising from large numbers of smaller accidents. From newspaper articles, letters, official reports, and academic papers, we can see how people at the time tried to make sense of catastrophic events. How people explained accidents determined the types of actions they took to prevent future accidents:

- An act of divine retribution demanded repentance and prayer;
- A chance event beyond human control created a need for insurance;
- An engineering failure suggested engineered solutions.

Throughout every country and industry, almost every change to safety law or practice can be linked to one or more accidents.

To find an exact beginning for Safety Science, it would be necessary to artificially divide history into an era when accidents were seen as divine or random acts, versus an era when accidents were seen as preventable or insurable. There is indeed broad support for the gradual shift from divine to engineering views of risk and accidents during the modern era, with an acceleration toward the end of the 19th century (Green, 1997). Yet in a sense these worldviews have always been contested, with movement back and forth and a gradual shift in dominance from one view to the other.

Early hints of what we might term modern thinking, for example, can be seen in ancient texts. Some trace safety back to the Code of Hammurabi (Hollnagel, 2009; Noy, 2009), a set of 282 Babylonian laws recorded around 1754 BCE. Five of these laws dealt with shoddy construction, promising severe retribution for those found at fault (The Avalon Project, n.d.):

229 If a builder builds a house for someone, and does not construct it properly, and the house which he built falls in and kills its owner, then that builder shall be put to death.
230 If it kills the son of the owner, the son of that builder shall be put to death.
231 If it kills a slave of the owner, then he shall pay, slave for slave, to the owner of the house.
232 If it ruins goods, he shall make compensation for all that has been ruined, and inasmuch as he did not construct properly this house which he built and it fell, he shall re-erect the house from his own means.
233 If a builder builds a house for someone, even though he has not yet completed it; if then the walls seem toppling, the builder must make the walls solid from his own means.

Parts of the Pentateuch referenced safety and negligence. Deuteronomy (22:8 NIV), written around 1400 BCE, includes the safety-through-design rule, with the understanding that flat roofs were (and are) used for more than just covering a house. At night, they offer escape from the heat and are even slept on:

When you build a new house, make a parapet around your roof so that you may not bring the guilt of bloodshed on your house if someone falls from the roof.

Exodus (21:28–29), written between 600 and 400 BCE, draws a distinction between accident and negligence:

> If a bull gores a man or woman to death, the bull is to be stoned to death, and its meat must not be eaten. But the owner of the bull will not be held responsible.
>
> If, however, the bull has had the habit of goring and the owner has been warned but has not kept it penned up and it kills a man or woman, the bull is to be stoned and its owner also is to be put to death.

These passages hint at a world where the future can be predicted by studying the past, and where individuals hold responsibility for the safety of those around them. There are numerous examples of court rulings that upheld human responsibility for injury and property damage, but these notions stood alongside the idea that the divine acted in daily life, and that sickness and misfortune were the wages of sin (Hall, 1993; Loimer & Guarnieri, 1996).

The Lawyer's Logic, endorsed by the Bishop of London, opined that the use of profane words like 'fortune,' or 'chance,' or 'haphazard' was evidence that people did not understand the "first cause: God's providence" (Loimer & Guarnieri, 1996, p. 105). In 1615, a boy of seven or eight "was drowned in Goodmans ffeilds in a Pond, playing with other Boyes there and swymming" (Forbes, 1979). The boy's death was ruled the result of a "Visitation of God," or "Act of God" (Burnham, 2009, p. 7).

A few decades after the drowning of the boy at Goodman's, however, we can find stirrings of a modern view. Consider the *Bills of Mortality*, published in London in 1647, for example. Over 13,000 "premature" deaths were recorded that year, with epidemic diseases (smallpox, measles, bubonic plague, malaria, tuberculosis, and enteric diseases) as their overwhelming cause (Loimer & Guarnieri, 1996). But 27 additional deaths resulted from "accidents." Drowning killed another 47 and burning killed 3 (today these categories might be labeled as accidents too).

Reflecting on the *Bills*, an early amateur demographer by the name of John Graunt noted that deaths from accidents were 'chronical.' Their number, he observed in his 1662 *Natural and Political Observations Made upon the Bills of Mortality*, was rather constant from year to year, like that of homicide or suicide. This was in sharp contrast with the bursts of deaths from epidemic diseases. It suggested a different etiology, or set of causes. Though a devout man, Graunt connected these deaths to people's occupations and suggested divine intervention might say

> …nothing of the numbers of those that have been drowned, killed by falls from scaffolds, or by carts running over them, etc. because [this] depends upon the casual trade, and employment of men.
>
> **Loimer and Guarnieri (1996, p. 102)**

The naturalist, or human-made view of accidents, however, would have another couple of centuries of contest with divine views ahead of it. Like in many other spheres of life (everything from the divine right of kings to an understanding of pathogens), these were heady and unsettled centuries where secular interpretations of how to understand, govern and order life slowly gained ground. The Scientific Revolution, the Enlightenment, and the Industrial Revolution would all contribute, as would the many developments and disasters during the 20th century.

1.3 MODERNITY AND HUMANKIND'S CONTROL OF NATURE

Francis Bacon was arguably the most influential scientist never to make any original scientific discoveries. In his 1620 masterpiece, *The New Organon*, he explicitly set out to tear down the reputations of the ancient philosophers whose writings—along with biblical studies—formed the basis of medieval scholarship. Bacon believed that reliance on logical thought was a certain pathway to self-deception:

> The mental operation which follows the act of sense I for the most part reject; and instead of it I open and lay out a new and certain path for the mind to proceed in, starting directly from the simple sensuous perception. The necessity of this was felt, no doubt, by those who attributed so much importance to logic, showing thereby that they were in search of helps for the understanding, and had no confidence in the native and spontaneous process of the mind. But this remedy comes too late to do any good, when the mind is already, through the daily intercourse and conversation of life, occupied with unsound doctrines and beset on all sides by vain imaginations. And therefore that art of logic, coming (as I said) too late to the rescue, and no way able to set matters right again, has had the effect of fixing errors rather than disclosing truth. There remains but one course for the recovery of a sound and healthy condition—namely, that the entire work of the understanding be commenced afresh, and the mind itself be from the very outset not left to take its own course, but guided at every step; and the business be done as if by machinery.
>
> **Bacon (1620, p. 8)**

Bacon's approach was based on a form of inductive reasoning, and bears little resemblance to the hypothetico-deductive 'scientific method' taught today. But his core argument that logic and argument led to self-deception, and that experiment and observation guided by strict methods led to fundamental truths, formed the heart of a new empirical approach to science.

Bacon's work was (and in the history of philosophy, still is) contrasted with René Descartes. On the surface, Bacon and Descartes held very different ideas about science. Bacon argued that human reason was too fallible to result in reliable knowledge; Descartes believed that reason could be used to extend knowledge beyond direct observation, so long as strict rules of logic were followed. Bacon and Descartes agreed on something more important and long-lasting than either of their methods. They were both spiritual men who believed in God, but they were engaged in the same grand project of developing systems of scientific enquiry to understand nature—and by understanding, to control it.

In 1637, Descartes published his *Discourse on the Method of Rightly Conducting One's Reason and of Seeking Truth in the Sciences*. Observing the tribulations of Galileo, he was originally not eager to have his own work printed and distributed, but reasoned:

> But as soon as I had acquired some general notions respecting physics, and beginning to make trial of them in various particular difficulties, had observed how far they can

carry us, and how much they differ from the principles that have been employed up to the present time, I believed that I could not keep them concealed without sinning grievously against the law by which we are bound to promote, as far as in us lies, the general good of mankind. For by them I perceived it to be possible to arrive at knowledge highly useful in life; and in room of the speculative philosophy usually taught in the schools, to discover a practical, by means of which, knowing the force and action of fire, water, air the stars, the heavens, and all the other bodies that surround us, as distinctly as we know the various crafts of our artisans, we might also apply them in the same way to all the uses to which they are adapted, and thus render ourselves the lords and possessors of nature. And this is a result to be desired, not only in order to the invention of an infinity of arts ... but also and especially for the preservation of health.

Descartes (1637, p. 27)

Like Bacon, Descartes believed that the application of appropriate methods of reasoning provided humanity not only with reliable knowledge, but with the power to shape the world. A century later, in 1755, the infamous All Saints' Day earthquake hit Lisbon, the capital of Portugal. With its epicenter in the Atlantic Ocean about 200 km from shore, the quake toppled buildings, ripped open fissures up to 5 m wide throughout the city center and triggered fires that raged for days (Kozák & Čermák, 2010; Kozak & James, 1998). Some inhabitants fled out to open spaces, including the harbor and docks, where they were promptly washed away by a tsunami. Tens of thousands of people were killed or injured, not only in Lisbon but in many coastal communities, and 85% of Lisbon houses were destroyed or heavily damaged (Aguirre, 2012; Kozák & Čermák, 2010).

The earthquake is seen by many as a drastic turning point in Western thinking about the origin of accidents and disasters (de Almeida, 2009; Dynes, 1999; Larsen, 2006). Portugal was a devout country, and a staunch supporter of the Pope in Rome. But this disaster struck on an important religious holiday. It destroyed most of Lisbon's churches and convents, while sparing the city's red-light district. Was there still some divine purpose behind the quake? This was in the middle of the Enlightenment—the cultural and intellectual movement that reached its high point in Europe in the 18th century. Enlightenment thinkers questioned the supremacy of church and crown, particularly of their imposition of hand-me-down knowledge and ideas. Inspired by earlier thinkers, such as Descartes, they believed in social progress, in the liberating possibilities of individual reason and scientific knowledge. God, they decided, had nothing to do with the causes of an earthquake. It had no sacred purpose. Appealing to divine intervention was no substitute for the human responsibility to understand and mitigate disasters like this (Dekker, 2017; Voltaire, 1759/2005).

1.4 MODERNITY AND SAFETY ENGINEERING

The Scientific Revolution of the 17th century provided the scaffold for the Industrial Revolution of the 18th century. The Industrial Revolution made commonplace the key conditions for disaster—concentration of people into a small area, and the potential for an uncontrolled release of energy. Nothing symbolized the promise and potential of the Industrial Revolution more than the coal-fired steam engine.

The first practical steam engine was developed for pumping water from mines. Like most major technologies, it was a progressive development of ideas rather than a single discovery by a single inventor. Early designs, credited to the military engineer Thomas Savery, had no moving parts other than the valve gear, and used the expansion and condensation of steam to create a vacuum (Hills, 1993). The French physicist Denis Papin contributed the idea of a piston, and the preacher/inventor Thomas Newcomen added a system of self-operating valves to create a repeating, two-stroke action. This basic design was refined and improved by civil engineer John Smeaton, who provided methods for analyzing the power and efficiency of various designs, driving further improvements (Dickinson, 2011). Between 1776 and 1788, James Watt and his business partner Matthew Boulton filed a series of patents that together earned Watt recognition as the "inventor of the steam engine." Watt's improvements dramatically increased the power, efficiency, and controllability of engines, and provided new gear mechanisms for driving machinery (Museum of Applied Arts & Sciences, n.d.).

Powerful and efficient steam engines created the possibility of steam-powered rail transport—including Trevithick's 'Coalbrookdale Locomotive' and 'Catch me who can,' Hedley's 'Puffing Billy,' and Stephenson's 'The Rocket.' The first fully steam-powered railway ran from Liverpool to Manchester, and commenced operations on September 15, 1830. The opening festivities included a derailment, a collision, and a widely reported fatality (Garfield, 2002). A Member of Parliament, William Huskisson, was struck and killed by The Rocket.

Present at the event was Charles Babbage, grandfather of the modern computer. Babbage had read the works of Bacon, Descartes, and Hume, and held, for his time, an extreme position on the determinability of the universe. For Babbage, the universe operated according to invariable laws. This, of course, is the ideal universe of Isaac Newton (Dekker, 2015). In such a universe, the apparent operation of chance and miracles were evidence that humanity's current knowledge of the laws of the universe was incomplete. By understanding the laws completely, and the current state of every atom in the world perfectly, Babbage believed that it was possible to work out what the world looked like at any time in the past or future.

> No motion impressed by natural causes, or by human agency, is ever obliterated. The ripple on the ocean's surface caused by a gentle breeze, or the still water which marks the more immediate track of a ponderous vessel gliding with scarcely expanded sails over its bosom, are equally indelible. The momentary waves raised by the passing breeze, apparently born but to die on the spot which saw their birth, leave behind them an endless progeny, which, reviving with diminished energy in other seas, visiting a thousand shores, reflected from each and perhaps again partially concentrated, will pursue their ceaseless course till ocean be itself annihilated.
>
> **Babbage (1841, p. 114)**

Babbage's determinism was not just philosophical musing. After the accident he witnessed, Babbage was inspired to devise a mechanism for mechanically recording the speed, force, and balance of any locomotive. This would provide an "unerring record of facts, the incorruptible witnesses of the immediate antecedents of any catastrophe" (Babbage, 1864, p. 334). Babbage believed that complete knowledge of the physical circumstances of an accident, free from the distortions and biases of human perception and recollection, would allow the causes of any catastrophe to be fully understood.

Babbage represented an increasingly prevalent worldview that saw accidents, rather than as the result of divine or diabolic action in the world, as something that was human in origin (Green, 2003), and controllable through human efforts and ingenuity.

Throughout his life, Babbage invented and tested a number of mechanical inventions for rail safety. His recording mechanism was used in experiments to determine the optimum gauge for railway tracks, but was never used as an accident analysis device. Similar ideas, though, would eventually form the basis of the "black box" central to many aviation accident investigations.

1.5 THE RISE OF SAFETY INSTITUTIONS

1.5.1 THE POLITICS OF SAFETY

It is perhaps not surprising that the first political concern for safety grew from the mines, factories, railroads, and steamships of the Industrial Revolution. These had congregated around the places where fuel for steam engines was plentiful: the coal deposits of the German Ruhr, U.S. Appalachia and the Midwest, and the British Northeast. Farm work was also still quite plentiful in the 19th century, but factories and mines concentrated labor in large masses, and exposed workers to higher temperatures and faster machinery, as well as the risk of explosions and chronic workplace illnesses. As the Industrial Revolution took hold, injury and fatality rates escalated in the late 1700s and expanded rapidly in the 1800s:

> The coal-powered economy brought to bear much more energy than existing technologies could easily control. Many jobs consequently became exceedingly hazardous. By the early 20th century, tens of thousands of workers were dying every year on the railroads, in factories, and especially in coal mines, including many boys and adolescents. For each laborer killed directly, several were maimed, and several more found their lives shortened by coal dust, lead, and other poisons.
>
> **Andrews (2010, p. 213)**

Railroad, textile, and mining industries began recording work-related injuries in the early 1800s (Loimer & Guarnieri, 1996). The kinds of accidents, as well as their number, changed with the mass concentration of humans and machines. The visibility of the accidents increased too. A fatal incident in farmer's field may have affected only the immediate stakeholders. An incident on a factory floor, and a subsequent funeral in a company town, would have been noticed by far more people.

> Accidents changed as humans had more and more interaction with technology, particularly machines and powered devices. By the twentieth century, 'the accident' had become a social institution, part of the world of modernity.
>
> **Burnham (2009, p. 7)**

With it came a greater push to do something about them, often as part of a package of demands for broader reforms.

An illustrative example of an early Industrial Revolution accident was the Felling Pit Disaster of 1812. The Felling Colliery, just south of Newcastle upon Tyne, had a design common at the time. The working shaft, called "John Pit," provided access for the miners (Coal Mining History Resource Centre, 1999). The up-shaft, called 'William Pit,' was 550 yards away on the surface. A furnace at the top of William Pit was kept constantly burning to ensure a flow of air down John Pit and up William Pit. Underground there was a maze of tunnels. Trap doors and stoppings below ground directed the flow of air in a single path through the maze, to prevent any buildup of noxious or explosive gasses. The underground layout was constantly reconfigured as work progressed, and leaving a door incorrectly open or closed could lead to a pocket of stagnant air. These pockets were sometimes discovered and removed by ventilation workers, or 'wastemen,' and sometimes accidentally ignited by miners using candles for illumination. Small explosions caused burn injuries, but seldom fatalities. Larger explosions could damage the ventilation system and disperse clouds of coal dust, leading to even larger explosions and ongoing fires.

> ...the whole mine is instantly illuminated with the most brilliant lightning - the expanded fluid drives before it a roaring whirlwind of flaming air, which tears up everything in its progress, scorching some of the miners to a cinder, burying others under enormous heaps of ruins shaken from the roof, and, thundering to the shafts, wastes its volcanic fury in a discharge of thick clouds of coal dust, stones, timber, and not unfrequently limbs of men and horses.

Coal Mining History Resource Centre (1999, p. 4)

On May 25, 1812, an explosion was heard and felt in the nearby villages, followed by an inverted cone of coal dust that fell like snow (Coal Mining History Resource Centre, 1999). Thirty-three workers and two bodies were quickly recovered from the mine, leaving 87 men and boys unaccounted for. Several efforts were made to reenter the pits, but the rescuers were forced back by choking gas. Another explosion during one of the rescues caused the mine managers to prohibit further attempts. The community waited for weeks, while rumors circulated of voices heard underground and men who had escaped after the rescue was abandoned. Eventually, through a combination of starving the fire and re-ventilation, the mine was rendered sufficiently safe to remove the bodies, and a parade of coffins was filled with unidentifiable scorched and decayed remains (Coal Mining History Resource Centre, 1999).

Early mine disaster accident analysis was an amateur undertaking. Committees of concerned citizens—not just mine managers and engineers but nobles, clergy, doctors, schoolteachers and other 'learned men'—would meet together to discuss the causes and treatment of the accidents. Some of the speakers at these meetings had no knowledge of mines or vested interest in particular solutions, just a moral outrage at the harm that had been caused. Others were seeking to justify their own past actions, to build political careers, or to promote their scientific ideas.

After the Felling Pit Disaster, the local parish priest, Reverend Hodgson, produced a pamphlet containing an account of the accident (Coal Mining History

Resource Centre, 1999). This pamphlet was republished in the Proceedings of the Royal Philosophical Society, which in turn led a London barrister to publish a proposal for setting up a "Society for the Prevention of Accidents in Coal Mines." This group came to be known as the Sunderland Society. It included among its members George Stephenson (later famous for The Rocket locomotive) and William Clanny, an inventor. The Sunderland Society aimed to establish prizes for the invention of safer schemes for ventilating and lighting mines. Their meetings and correspondence focused on demonstrations of mine safety technology—in particular, the society became a forum for debate about who should be credited with the invention of the safety lamp. The society's efforts contributed to the widespread use of three different lamp designs: the Davy Lamp (invented by Sir Humphrey Davy), Clanny's lamp, and Stephenson's "Geordie Lamp."

The adoption of new safety technology was not as effective as the Sunderland Society hoped. In fact, misplaced confidence in the lamps had encouraged more aggressive mining practices (Fynes, 1873). This led to an increase rather than a decrease in coal-mine explosions. It is an early example of what would become known as risk homeostasis almost a century later. The idea of risk homeostasis is that added protections can rapidly get converted into productive gain, leaving the system as exposed to risk as it was before, or even more so.

The St. Hilda Colliery was less than ten miles from Felling. In 1839, 51 miners were killed in a single explosion in the West working in the mine (Blower, 2014). Coal-mining communities had been told that safety lamps would make mines safe, but the procession of bodies recovered from the pit shattered that belief. A new society, the "South Shields Committee" was formed (Durham County Advertiser, 1839), with a similar mix of clergy, nobility, businessmen, and engineers.

The South Shields Committee embodied the promise and the curse of modernity. On the one hand, frustration with the failure of simple technical solutions led to deep curiosity and wide-ranging inquiry. The committee studied government reports, wrote to scientific men, conducted experiments, and held public hearings. They produced recommendations including improved ventilation design, education of mine officers, and restrictions on the use of child labor (Mills, 2016). On the other hand, their most influential finding was the need to embed their own scientific judgment in formal power structures. In the minds of the committee safety could only be assured if the implementation of their other recommendations was overseen by an independent and knowledgeable authority—a government inspector of mines.

1.5.2 INSPECTORS AND INVESTIGATORS

During the remainder of the 19th century, committees of 'learned men' gave way to increasingly powerful government bodies charged with approval and oversight of dangerous industries. The community could feel safe knowing that their representatives—not the commercially compromised industry owners—were ultimately responsible for safety. In every industry, the same pattern was followed. One or more major accidents would create an intolerable situation (Pellew, 1974). A private or government-sponsored committee would begin examining the technical causes of the accidents, and then broaden their recommendations to address a perceived need for regulation. An inspectorate would be established, which would use any further accidents as proof of the need for greater power.

The Factories Act of 1833 and the Coal Mines Inspection Act of 1850 established inspectorates for their respective industries (Hutchins & Harrison, 1911; Parliament, n.d.). The inspectors were originally charged with overseeing working hours and living conditions, particularly for child workers, but their duties gradually extended to the enforcement of safety practices and the investigation of accidents.

The U.K. Rail Regulation Act of 1840 was established because it was considered expedient for the safety of the public to provide for the due supervision of railways (His/Her Majesty's Government, 1840). This Act created the new occupation of 'Inspecting Officer.' Inspecting Officers, also known as Railway Inspectors, were primarily charged with ensuring that new railways were constructed to an adequate standard, and were fitted with safety features such as signals and level-crossing barriers. In 1871, Inspectors were given further powers to investigate and report on the causes of major accidents (His/Her Majesty's Government, 1871).

The United States had their own share of railroad accidents, but it was steam-powered paddleboats, rather than rail locomotives, that were the symbols of modernity and calamity. Steamboats and their Masters were celebrities (Lloyd, 1856). Fast voyages were reported in newspapers, and crowds gathered on riverbanks to watch impromptu races. Steamboats also were prone to boiler explosions.

STEAMBOAT MOSELLE

The Steamboat Moselle was "regarded as the very paragon of western steamboats; she was perfect in form and construction, elegant and superb in all her equipments, and enjoyed a reputation for speed which admitted of no rivalship" (Lloyd, 1856, p. 26). On April 25, 1838, Moselle was carrying 280 passengers from Cincinnati to St. Louis, the return leg of a journey for which she had just broken the speed record by several hours. Moselle's Master, Captain Perrin, declared an intention to overtake a rival steamer that had already started the trip. Moselle cast off from the Cincinnati docks and raced one mile up the river, then pulled in to collect a final few passengers. As she was being pushed off again, all four boilers exploded simultaneously. Debris and bodies were said to be thrown onto both banks of the quarter-mile wide river.

A public meeting was held the next day, presided over by the Mayor of Cincinnati. The meeting resolved to raise before Congress the matter of what was seen as a great and increasing carelessness in the navigation of steam vessels (Lloyd, 1856). Two months later, the Steamboat Act of 1838 introduced judge-appointed safety inspectors (U.S. Congress, 1838). Steamboat accidents continued, and so revisions to the Act in 1852 and 1871 increased the power of the Inspectors and established a permanent Steamboat Inspection Service charged with licensing not only the boats themselves, but masters, pilots, and engineers (The U.S. National Archives and Records Administration, 2016; U.S. Congress, 1852, 1871).

In a trend that continues to the present day, industry sectors developed at different rates and in different directions. Every step forward involved one or more major accidents creating moral outrage followed by government action, usually in the form of regulations overseen by an inspectorate. In some cases, the safety rules were included in the legislation, but usually the inspectorates wrote their own regulations—these inspectorates were quasi-scientific institutions populated by technical experts, and constantly seeking out and sponsoring technical solutions to safety problems. Often the inspectorates were instrumental in the next iteration of their own establishing Acts, lobbying for increased powers and wider responsibilities. Inspectorates were established for gunpowder (and later in the century, chemical explosives) factories, "locomotives on highways," and agricultural machinery.

1.5.3 Standards and Professional Associations

Initial reassurance that safety was in the hands of responsible, independent, and scientific government agencies gave way to the realization that accidents were still happening even on the inspectors' watch. It was no consolation to the families of those who died on the Sultana, the Annie Faxon, and the Senator to know that those steamboats had passed inspections. Nor was it reassuring to London commuters to know that the same Railway Inspectorate responsible for their safety had approved the Tay Bridge less than 2 years before it collapsed, dropping an entire train into the Firth of Tay and killing everyone on board. An inspectorate is—at best—only as effective for safety as the rules it enforces. Who determines what specific set of rules is 'safe' for each industry or application?

The Sunderland Society and the South Shields Committee were temporary organizations, but many other learned societies formed in the late 1700s and early 1800s became entrenched institutions. Several of these institutions began establishing standards for the certification of equipment and competence. Notable examples include the U.K. Institution of Civil Engineers (ICE) and Institution of Mechanical Engineers (IMechE), the German Technischer Überwachungsverein (TUV), and the American Society of Mechanical Engineers (ASME).

The ICE began in 1818 as a knowledge-sharing organization (Buchanan, 1985; Institution of Civil Engineers, n.d.). Military engineers had formal schooling and a professional community, but nonmilitary (civil) engineers learned their trade through informal pupillages and apprenticeships. Concerns for increasing the competency of the civil engineering community gradually built into a formal system of examinations and professional qualifications. The IMechE split from ICE in 1847 (Institution of Mechanical Engineers, n.d.), but followed a similar historical path, beginning as an informal knowledge-sharing organization and maturing into a professional licensing body.

TUV and ASME developed independently, but had common origins as independent boiler inspection organizations (TÜV SÜD, n.d.). TUV (originally DUV) was formed in 1866 in response to a brewery boiler explosion. The establishment of the association was encouraged by both industry and government, who were keen to show that they took the problem of inadequate boiler maintenance seriously. DUV was, from the start, a commercial enterprise, conducting inspections as a paid service.

ASME was founded in 1880 (The American Society of Mechanical Engineers, n.d.), and was originally an information-sharing professional body similar to ICE and IMechE. However, the ASME's early activities concentrated on standardizing equipment parts rather than formalizing professional training. This included the publication of "codes"—minimum requirements for safe equipment design and inspection.

THE GROVER SHOE FACTORY DISASTER

A key point in ASME's development was the Grover Shoe Factory disaster of 1905 (The American Society of Mechanical Engineers, n.d.). At the time Massachusetts was a heavily industrialized state, but the prevailing political climate discouraged government interference in business practices. Safety was ensured through private enterprise; factory owners purchased insurance, and insurers arranged for safety inspections. R.B. Grover operated a four-story wooden shoe factory in Brockton, Massachusetts, with around 400 employees. Like most large buildings, the factory was heated by an industrial boiler. On March 20, the usual boiler was not working, so a much older backup was connected and started. The subsequent explosion collapsed the building, trapping workers and breaking open gas pipes. Fifty-eight workers were crushed or burned.

Since the boiler had been inspected and passed by the insurance company (The Taney County Republican, 1905), ASME was able to use the public and political interest generated by the accident to push for government oversight of an inspection regime. At the urging of ASME, a new law was written incorporating and expanding the ASME Boiler Testing Code, originally established in 1884. The code was rewritten as the Boiler and Pressure Vessel Code, and issued in 1914 as a 114-page volume. This document was adopted by other states, and became the basis of boiler regulation across the United States (Canonico, 2011).

1.5.4 Insurers, the State, and Workers' Compensation

Committees of 'learned men,' government inspectors, and professional associates were all very effective at creating rules, but they had limited power to ensure that those rules were followed. The main tool at their disposal was the withholding of a license to operate—a threat that gave no incentive for businesses to cooperate in lifting safety standards. A much more effective mechanism was found in government regulation of accident insurance.

The insurance industry was a byproduct of the increase in international trade during the 1600s and 1700s. A merchant ship filled with cargo represented a substantial investment of capital, with the potential for both high profit and catastrophic loss. Merchants could share the risk by taking out loans that were only repayable if the voyage was successful. In 14th-century Italy, this evolved into the practice of insurance, spreading risk from a single merchant to a group of underwriters, in exchange for a premium. Insurance required underwriters who could match premiums to risk, and merchants who trusted the underwriters to pay out in the event of a disaster, and so it quickly became a business in its own right, with expert brokers to match merchants and underwriters, to draft contracts, and to recommend premiums.

Along with insurance came the need to translate informal qualitative information about risk into quantitative estimates. An insurance syndicate or company that routinely set premiums too low would pay out more than their income. If they set premiums too high, then they would lose business to underwriters who offered lower premiums. The Dutch mathematician, Christiaan Huygens, published a treatise on probability theory in 1657. This included the concept of "expected value," weighting a future reward (or loss) based on its likelihood of occurrence. The work of Huygens and others became the foundation of an insurance market based on the mathematical calculation of risk.

The new insurance industry also needed a way to protect its investments from fraud and "moral hazard." Fraud included practices such as overinsuring a cargo and then deliberately sinking the ship, and could be combated through government regulation of insurance contracts, including the establishment of courts to arbitrate disputes. Moral hazard is where a ship owner took greater risks in the knowledge that they were covered by insurance. This required insurers to start thinking of risk not as a fixed uncertainty, but as an emergent property of human decision making. Risk could not only be calculated, but it could also be controlled.

EDWARD LLOYD'S COFFEE HOUSE

One of the centers of the insurance trade was Edward Lloyd's coffee house, in London (Palmer, 2007). Lloyd's provided a place for merchants to share information and to negotiate insurance. In 1764, an action was brought in the Court of Exchequer concerning a ship, the Mills Frigate, insured by underwriters operating out of Lloyds. The Mills Frigate was of French construction, considered inferior to English ships due to the use of iron bolts, which could become rusted and were not readily able to be inspected. The underwriters payed for the ship to be inspected before departure, but by the time it arrived at its destination in the Leeward Islands, it was in "weak, leaky and distressed condition." The underwriters were asked and agreed to renew the insurance for the return voyage. The Mills Frigate set off back to England, but started taking on water, and returned to St. Christopher's, where it was judged unseaworthy. The underwriters refused to pay out on the grounds that an insurance contract could only be valid for a ship that was seaworthy when it left court.

The Mills Frigate case provided legal recognition of the idea that insurance created obligations on the insured. Underwriters covered the remaining risk after reasonable precautions had been taken, but they were not a substitute for mitigating risk by choosing seaworthy vessels, competent captains, and appropriate routes and cargo. Quite the opposite—insurers began to take active measures to help or demand risk reduction. For example, seaworthiness inspections were often arranged and paid for by the underwriters.

The state became involved around the same time, but it was a slow start. With a new understanding of the values and capacities of individual citizens that had both produced and emerged from the Enlightenment, came a new realization of the role of the state. Louis XIV, king of France from 1643 to 1715, had once proclaimed that *l'état, c'est moi* (the state, that's me). But in modernity, the state was no longer just the sovereign's. It was a collective production. It was made by, and of, the people. As such, the state also had collective responsibilities for people. Naturally, it should develop the capabilities to live up to those responsibilities. Rationality, standardization, and the application of state power led to a greater uniformity of laws, language, and units of measurement during the 18th and 19th centuries. It eventually led to public education, universal suffrage, and many other innovations, which we have come to associate with a modern state. The state had started to care, even about people's safety:

> The idea that one of the central purposes of the state was the improvement of all the members of society—their health, skills and education, longevity, productivity, morals and family life—was quite novel. There was, of course, a direct connection between the old conception of the state and this new one. A state that improved its population's skills, vigor, civic morals, and work habits would increase its tax base and field better armies; it was a policy that any enlightened sovereign must pursue. And yet, in the nineteenth century, the welfare of the population came increasingly to be seen, not merely as a means to national strength, but as an end in itself.
>
> **Scott (1998, p. 91)**

In 1884, the Imperial Senate of a newly united Germany passed the *Accident Insurances Act*. Otto von Bismarck, Emperor at the time, was under pressure to demonstrate the government's willingness to recognize the hardships suffered by workers. Work days lasted anywhere from 12 to 14 h, there was work on Sundays, and the conditions of work were harmful to health and life. The numbers of injuries and deaths, even by the standards of the day, were indeed rather shocking, said to reach 'crisis proportions.' Between 1885 and 1908, two million industrial accidents occurred in Germany alone (Cooter & Luckin, 1997).

Bismarck's main concern was the socialist threat that German workers would be attracted to various left-wing parties, including the Social Democrats. To build a broad enough political base, Bismarck had to ally himself with Catholics. They made up a third of the population and had earlier been sidelined and discriminated against. At the same time, he had to assuage rightist National Liberal concerns about state socialism. Compromises were needed to pass the Act. While previous acts that covered worker illness were funded by the state, the entire expense for the Accident Act was to be underwritten by employers. The risk of a work accident was considered an enterprise risk, and thus the cost for insuring against it should be borne by the enterprise.

Bismarck gave employers the power to administer the program, arranging for it to be managed by a cooperative, *Der Arbeitgeberverband in den beruflichen Korporationen*. The cooperative established insurance offices at the imperial and state level, and agreed to pay medical treatment and pensions of up to two-thirds of the earned wages if the worker was fully disabled. This arrangement was *unabhängig von der Schuldfrage,* or independent of whom was to be blame for the accident (similar to modern "no fault" insurance).

Before the passage of the Act, workers who suffered accidents (or their surviving dependents) were left to fend for themselves. In Germany, the 1884 Act covered 4.7 million workers. It must have made a difference for the aftermath (if not prevention) of accidents.

The initiative, however, was met with substantial criticism from within the German empire. Social Democrats and the nascent labor movement did not want the state to use the Act as a substitute for more structural industrial relations reforms, including a reduction in work hours and an increase in job security. But Bismarck refused categorically. He had run out of political room for maneuver: Liberals kept agitating against a creeping state socialism. A young emperor Wilhelm II was seen as more "worker friendly": after Bismarck's dismissal in 1890, he spoke out in favor of health and safety measures, limiting work hours to 11 a day, regulating Sunday rest, and restricting child labor. It would take until the massive unrest of post-World War I Germany in 1918–1919, however, for trade unions to be recognized and work to be reduced to 8 h a day. The 1884 Accident Insurances Act is still in effect today, though with many amendments and partial repeals.

The German approach would later be implemented in other European countries as well, with local variations. The Netherlands established its Occupational Injuries Act in 1901 and gave employers the choice of whether to pay for their premiums themselves or outsource payments to a private supplier. Influential employers put together a cooperative like the German one, which introduced competition into the injury insurance market.

In England, employers could not be held responsible for injuries to their workers until the Employers Liability Act of 1880. In 1887, the German model was adopted as the Workmen's Compensation Act 1887.

The widespread introduction of inspectorates and mandatory workers compensation schemes was an indication of how far society had shifted from the start of the 19th century. In part, the governmental, bureaucratic, or organizational programs to improve safety—or at least to deal with the consequences of accidents—were responses to the demands of industry and technology. The historical examples show, though, that for something to be done about safety there has to be a moral or, more critically, a *political* imperative to act. And there must be political room for maneuver to act within. This is as true for the chief executive of a resources company today as it was for Bismarck in 1884. What amounts to a 'safety problem' is in part historically and politically constructed. It is not just a self-evident issue to be managed. It arises from what certain people find important, or—if you are more cynical—what certain people need to be seen to find important, and can afford to find important, at a given time (Cooter & Luckin, 1997).

A TALE OF TWO BRIDGES

The shift in safety thinking during the 19th century can be seen by comparing how society responded to two disasters of similar magnitude at different points in time.

On May 2, 1845, the Bure Suspension Bridge at Great Yarmouth collapsed under the weight of several hundred spectators gathered to watch a circus publicity stunt (Barber, 2013). There were 79 fatalities, including 59 children.

On December 28, 1879, the Tay Bridge collapsed during a storm (Rolt, 1966). A passenger train was crossing the bridge at the time, and fell into the Firth of Tay, killing approximately 75 people.

The physical causes of the Catastrophe at Great Yarmouth and the Tay Bridge Disaster were very similar. In each case, the bridge had physical manufacturing defects that significantly reduced the load it was capable of carrying. The Bure Suspension bridge had an incomplete weld (Thorold, 1845), and could not hold the weight of spectators all on one side of the bridge. The Tay Bridge had poorly manufactured bracing ties and cast iron lugs, and could not withstand the lateral force of the wind (Rolt, 1966). In each case, the bridge design allowed insufficient "safety margin"—allowance for imperfection or extreme events.

However, the reaction to the collapses was completely different. A single engineer was sent by the Home Secretary to investigate the catastrophe at Yarmouth. He spoke to the owner, architect and manufacturer, and inspected the bridge. The local inquest took a single day, and the engineer's report was sent to the government on June 9.

After the Tay Bridge collapsed, the Board of Trade established a Court of Inquiry. The Inquiry collected documents, examined witnesses on at least three separate occasions, and commissioned expert examination of the evidence. The final report is 50 pages long (Board of Trade, 1880). It contains long discussions of what could or should have been done to ensure better design and quality of the bridge. On numerous occasions, the report tries to identify standards for good practice, and to hold individuals to account for failing to meet those standards.

> It is said that engineers are always liable to be deceived by the borers, and that therefore Sir Thomas Bouch could not be held to blame on that account. But that argument does not satisfy me. I should have thought that, if engineers are liable to be deceived by borers, it is all the more important that before designing a bridge, they should satisfy themselves, beyond a doubt, of the accuracy of the borings
>
> I think also that it is not sufficient to say that the supervision at the Wormit Foundry, and in the subsequent maintenance of the bridge was insufficient, without saying in what that insufficiency consisted, and who was to blame for it.

Board of Trade (1880, p. 48)

Between the Catastrophe at Great Yarmouth and the Tay Bridge Disaster, the social response to accidents had been transformed by modernity. Accidents now had causes that could be uncovered by scientific investigation, and prevented through human endeavors.

1.6   SAFETY SCIENCE AND THE ROLE OF THE HUMAN

Safety Science in the 19th century focused primarily on the design and maintenance of equipment and facilities. The new institutions—learned associations, inspectorates, standard bodies, and insurers—worked to create and enforce rules for safe design. Safety could be verified at any point in time by inspecting the equipment to check that it matched the design rules. Progress in safety came from expanding the existing rules to cover more things that could go wrong, and failure in safety came either from gaps in the rules, or failure to enforce them.

Where though, did humans fit into the picture? 19th century safety initiatives had a lot to say about managers and professionals, but very little to say about workers. Engineers, inspectors, and investigators of 19th century thought about human fallibility, but only to the extent that they thought it was the responsibility of designers to eliminate, as much as possible, the possibility of catastrophic human failure. Frontline workers, for the most part, were seen as powerless victims of their work environment.

As Safety Science developed into the 20th century, it brought with it the attitudes and institutions of the 19th century. The promise of modernity—that humans could apply scientific principles to control the world they lived in—continued at the heart of safety. The interplay between concerned citizens, government regulators, and insurers continued to drive the creation and enforcement of safety rules.

The big changes to come centered on the roles of humans and organizations. Already in the Tay Bridge Inquiry, investigators showed signs of thinking of human rules in the same way that they thought of scientific principles. As a matter of abstract philosophy, Charles Babbage was mostly right. Examination of the current state of the world, combined with knowledge of the laws that govern it, allows us to project backwards in time to uncover the causes of accidents. In practice, that only works for simple mechanical systems.

STUDY QUESTIONS

1. How can you see the 1755 Lisbon earthquake and the Scottish Tay Bridge Disaster of 1879 as representing turning points to modernity with respect to our views on accidents and safety?
2. What were the reasons for private corporations (and insurers) to get involved in safety?
3. When did the state begin to play a role in assuring the safety of workers, and why were states later to get in on this than private corporations?
4. Can you name institutions in your own industry that have their origins in changing ideas about safety in previous centuries?
5. In what sense was concern for safety in the 19th century born out of politics? Do you have examples from our current era that demonstrate this as well, even if on a smaller scale?
6. In what sense was Charles Babbage right with his assertion about using the current state of the world to understand the causes of accidents, and where did this go wrong?

7. Can you consider Safety Science to still be a 'modernist' science even today, in the sense that it might hold on to the aspiration to make the world a better place?

REFERENCES AND FURTHER READING

Aguirre, B. E. (2012). Better disaster statistics: The Lisbon earthquake. *Journal of Interdisciplinary History, 43*(1), 27–42. doi:10.1162/JINH_a_00337

Andrews, T. G. (2010). *Killing for coal: America's deadliest labor war.* Cambridge, MA: Harvard University Press.

Babbage, C. (1841). *The ninth Bridgewater treatise: A fragment.* London, UK: John Murray.

Babbage, C. (1864). *Passages from the life of a philosopher.* London, UK: Longman, Green, Longman, Roberts, & Green.

Bacon, F. (1620). *Novum organum scientiarum.* London, UK: McMillan & Co.

Barber, M. (2013). Great Yarmouth suspension bridge disaster memorial unveiled. *BBC News.* Retrieved from http://www.bbc.com/news/uk-england-norfolk-24240357

Blower, J. (2014). Pit disaster touched every family in town. *The Shields Gazette.* Retrieved from https://www.shieldsgazette.com/lifestyle/nostalgia/pit-disaster-touched-every-family-in-town-1-6617622

Board of Trade. (1880). *Tay bridge disaster: Report of the court of inquiry, and report of Mr. Rothery, upon the circumstances attending the fall of a portion of the Tay bridge on the 28th December 1879.* London, UK: His/Her Majesty's Stationery Office.

Buchanan, R. A. (1985). Institutional proliferation in the British engineering profession, 1847–1914. *The Economic History Review, 38*(1), 42–60. doi:10.1111/j.1468-0289.1985.tb00357.x

Burnham, J. C. (2009). *Accident prone: A history of technology, psychology and misfits of the machine age.* Chicago, IL: University of Chicago Press.

Canonico, D. (2011). The history of ASME's boiler and pressure vessel code. Retrieved from https://www.asme.org/engineering-topics/articles/boilers/the-history-of-asmes-boiler-and-pressure

Coal Mining History Resource Centre. (1999). Felling Colliery 1812: An account of the accident. Retrieved from www.mineaccidents.com.au/mine-accident/230/1812-felling-colliery

Cooter, R., & Luckin, B. (Eds.). (1997). *Accidents in history: Injuries, fatalities and social relations.* Amsterdam, The Netherlands: Editions Rodopi.

de Almeida, A. B. (2009). The 1755 Lisbon earthquake and the genesis of the risk management concept. In L. Mendes-Victor, C. S. Oliveira, J. Azevedo, & A. Ribeiro (Eds.), *The 1755 Lisbon earthquake: Revisited* (pp. 147–165). Heidelberg, Germany: Springer.

Dekker, S. W. A. (2015). *Safety differently: Human factors for a new era.* Boca Raton, FL: CRC Press/Taylor & Francis Group.

Dekker, S. W. A. (2017). *The end of heaven: Disaster and suffering in a scientific age.* London, UK: Routledge.

Descartes, R. (1637). *Discourse de la methode pour bien conduire la raison et chercher la verite dans les sciences.* Leiden, The Netherlands: Jan Maire.

Dickinson, H. W. (2011). *A short history of the steam engine.* Cambridge, UK: Cambridge University Press.

Durham County Advertiser. (1839). The Rev. James Carr in the chair. Durham County Advertiser.

Dynes, R. R. (1999). The dialogue between Voltaire and Rousseau on the Lisbon earthquake: The emergence of a social science view. Retrieved from http://udspace.udel.edu/handle/19716/435

Forbes, T. R. (1979). By what disease or casualty: The changing face of death in London. *Health, Medicine and Mortality in the Sixteenth Century,* 117–139. doi:10.1093/jhmas/XXXI.4.395

Fynes, R. (1873). *The miners of Northumberland and Durham: A history of their social and political progress*. Leiden, The Netherlands: John Robinson.

Garfield, S. (2002). From triumph to tragedy. *The Guardian*. Retrieved from https://www.theguardian.com/education/2002/oct/13/highereducation.news

Green, J. (1997). *Risk and misfortune: The social construction of accidents*. London, UK: Routledge.

Green, J. (2003). The ultimate challenge for risk technologies: Controlling the accidental. In J. Summerton & B. Berner (Eds.), *Constructing risk and safety in technological practice* (pp. 29–42). London, UK: Routledge.

Hall, C. G. (1993). An unsearchable providence: The lawyer's concept of act of God. *Oxford Journal of Legal Studies, 13*(2), 227–248. doi:10.1093/ojls/13.2.227

Hills, R. L. (1993). *Power from steam: A history of the stationary steam engine*. Cambridge, UK: Cambridge University Press.

His/Her Majesty's Government. (1840). *An act for regulating railways*. London, UK: His/Her Majesty's Stationery Office.

His/Her Majesty's Government. (1871). *An act to amend the law respecting the inspection and regulation of railways*. London, UK: His/Her Majesty's Stationery Office.

Hollnagel, E. (2009). *The ETTO principle: Efficiency-thoroughness trade-off: Why things that go right sometimes go wrong*. London, UK: CRC Press.

Hutchins, B. L., & Harrison, A. (1911). *A history of factory legislation*. London, UK: P. S. King & Son.

Institution of Civil Engineers. (n.d.). Our history. Retrieved from https://www.ice.org.uk/about-ice/our-history

Institution of Mechanical Engineers. (n.d.). History of the Institution of Mechanical Engineers. Retrieved from https://www.imeche.org/about-us/imeche-engineering-history/institution-and-engineering-history

Kozák, J., & Čermák, V. (2010). *The illustrated history of natural disasters*. Dordrecht, The Netherlands: Springer.

Kozak, J. T., & James, C. D. (1998). Historical depictions of the 1755 Lisbon earthquake. *National Information Service for Earthquake Engineering*. Retrieved from https://web.archive.org/web/20100209031537/http://nisee.berkeley.edu:80/lisbon

Larsen, S. E. (2006). The Lisbon earthquake and the scientific turn in Kant's philosophy. *European Review, 14*(3), 359–367. doi:10.1017/S1062798706000366

Lloyd, J. T. (1856). *Lloyd's steamboat directory, and disasters of the western waters*. Mississippi River: J.T. Lloyd & Company.

Loimer, H., & Guarnieri, M. (1996). Accidents and acts of god: A history of the terms. *American Journal of Public Health, 86*(1), 101–107.

Mills, C. (2016). *Regulating health and safety in the British mining industries, 1800–1914*. London, UK: Routledge.

Museum of Applied Arts & Sciences. (n.d.). Boulton and Watt engine. Retrieved from https://maas.museum/event/the-boulton-and-watt-engine/

Noy, Y. I. (2009). *Occupational safety: Charting the future*. Paper presented at the Contemporary Ergonomics 2009: Proceedings of the International Conference on Contemporary Ergonomics, London, UK.

Palmer, S. (2007). Lloyd, Edward. *Oxford Dictionary of National Biography*. Retrieved from http://www.oxforddnb.com/view/10.1093/ref:odnb/9780198614128.001.0001/odnb-9780198614128-e-16829

Parliament. (n.d.). Coal mines. Retrieved from https://www.parliament.uk/about/living-heritage/transformingsociety/livinglearning/19thcentury/overview/coalmines/

Pellew, J. H. (1974). The home office and the explosives act of 1875. *Victorian Studies, 18*(2), 175–194.

Rolt, L. T. C. (1966). *Red for danger*. Devonshire, UK: David & Charles.

Scott, J. C. (1998). *Seeing like a State: How certain schemes to improve the human condition have failed.* New Haven, CT: Yale University Press.

The American Society of Mechanical Engineers. (n.d.). Engineering history. Retrieved from https://www.asme.org/about-asme/engineering-history

The Avalon Project. (n.d.). The code of Hammurabi. Retrieved from http://avalon.law.yale.edu/ancient/hamframe.asp

The Taney County Republican. (1905). *Shoe factory horror: Nearly one hundred killed by a boiler explosion.* Missouri, TX: The Taney County Republican. Retrieved from http://chroniclingamerica.loc.gov/lccn/sn89067390/1905-04-06/ed-1/seq-2/

The U.S. National Archives and Records Administration. (2016). Records of the Bureau of marine inspection and navigation. Retrieved from https://www.archives.gov/research/guide-fed-records/groups/041.html#41.2

Thorold, W. (1845). An account of the failure of the suspension bridge at great Yarmouth. *Minutes of the Proceedings of the Institution of Civil Engineers, 4*(1845), 291–293. doi:10.1680/imotp.1845.24454

TÜV SÜD. (n.d.). Our history: From inspectorate to service provider. Retrieved from https://www.tuev-sued.de/company/tuev-sued-group/history

U.S. Congress. (1838). *An Act to provide for the better security of the lives of passengers on board of vessels propelled in whole or in part by steam.*

U.S. Congress. (1852). *An Act to amend an Act entitled "An Act further to provide for the better security of the lives of passengers on board of vessels propelled in whole or in part by steam," and for other purposes.*

U.S. Congress. (1871). *An Act to provide for the better security of life on board of vessels propelled in whole or in part by steam, and for other purposes.*

Voltaire. (1759/2005). *Candide: Or, optimism.* London, UK: Penguin Classics.

2 The 1910s and Onward
Taylor and Proceduralization

KEY POINTS

- Whereas the early history of safety was preoccupied with engineering design, by 1920, safety management was dominated by measures to control worker behavior. Companies started paying bonuses to foremen and workers in the belief that cash incentive schemes reduced injury rates.
- Scientific management, labeled and exemplified by Frederick Taylor and the Gilbreths, is key to this development. It applied the scientific method to determine the most efficient way to perform a specific task. Workers needed to comply with the one best way to do the task in order to maximize efficiency (and safety).
- Taylorism accelerated and solidified the division between those who plan, manage, and supervise the work, and those who execute it. Thinking and problem-solving was heavily concentrated in the former, while the latter merely needed to do what they were instructed.
- This lies at the basis of the belief that control over workers and operators is possible through better compliance with procedures; through better hierarchical order and the imposition of rules and obedience.
- Rules and procedures, following on Taylor, embody the wisdom and guidance of others and previous lessons. They save time and effort and provide clarity of tasks and responsibilities. They impose order, uniformity, and predictability. They are the basis for risk management and control.
- Taylor's ideas have left deep traces in our thinking about the safety of work: Workers need to be told what to do. They need to be supervised and monitored, and their work has to be specified in great detail. Autonomy and initiative are undesirable.
- Worker departure from the one best method can be seen as a 'violation,' which requires sanctions or reminders, or the rewriting of the rule (but that should not be done by the worker). They can also be seen as local, adaptive resilience necessary to close the gap between how work is imagined and how it actually gets done.
- Adherence to rules can indeed make it impossible to get the work done, lead to blindness to new situations and the squashing of innovation, resentment at the loss of freedom and discretion, and the growth of bureaucratic and supervisory control to impose compliance.

2.1   INTRODUCTION

The early history of safety was preoccupied with engineering design. Up until the start of the 20th century, insurers covered their risk by insisting on physical inspections, accident reports discussed mainly the design and maintenance of equipment, and professional bodies focused on the competency of engineers or standardized engineering controls. A mere twenty years later, safety management was full of measures to control worker behavior. It was common practice for companies to have safety committees made up of selected foremen who wore badges to indicate their special status as safety supervisors. Companies were paying bonuses to foremen and workers in the belief that cash incentive schemes reduced injury rates.

How did workplace safety shift from a movement to protect workers from unscrupulous management, to a system protecting management from the irresponsibility of workers? In order for worker unsafe acts to even be an available explanation, there needed to be a default assumption that work happened according to set rules or procedures that were 'safe.' To understand the origin and logic of procedures, at least on a large scale, we have to start in a place that is not safety related. The ideas that were generated there, however, have had a significant impact on the link between procedures, compliance, and safety, even today.

This chapter examines the contributions of *scientific management* labeled and exemplified by Frederick Taylor (1911). Even though he never collated his principles in any consistent or coherent way, Taylor expressed his ideas as such:

1. Do not allow people to work by heuristic (rule of thumb), habit, or common sense. Such human judgment cannot be trusted, because it is plagued by laxity, ambiguity, and unnecessary complexity. Instead use the scientific method to study work and determine the most efficient way to perform a specific task;
2. Do not just assign workers to any job. Instead, match workers to their jobs based on capability and motivation, and train them to work at maximum efficiency;
3. Monitor worker performance, and provide instructions and supervision to ensure that they are complying with the most efficient way of working. Affairs of workers are best guided by experts;
4. Allocate the work between managers and workers so that the managers spend their time planning and training. This allows the workers to perform their assigned tasks most efficiently.

The categories and distinctions suggested by Taylor's ideas form the basis for understanding modern debates in safety. His assumptions about workers and their limited capacity for insight, autonomy, and innovation do too.

2.2   THE INTERSECTION OF SCIENCE, MANAGEMENT, AND SAFETY

2.2.1   FOUNDATIONS OF PROCEDURES AND SAFETY

We need to pick up the story of the industrializing world around the 1910s. This is where we can find one of the most important foundations of enduring beliefs about procedures and safety. The Industrial Revolution had fundamentally been changing the economics of labor. In preindustrial labor relations, people with capital generally purchased a worker's actual labor. This could, for instance, be measured in the output of that labor. Think of harvesting: laborers might be paid per bushel or other measure of their productivity. People also bought products of labor directly at an agreed price (a painting, a printing press). In an industrialized society, it no longer worked that way. Capital purchased a worker's labor *time*, or *potential* labor, as opposed to products or actual work. It became natural, in such an arrangement, to pursue strategies that regulated the laborer's productive processes. The point was to derive as much work, and thus value, as possible from a given amount of purchased labor time (Newlan, 1990). In this arrangement:

> Managers did not make use of social science out of a sense of social responsibility, but out of a recognized need to attach age-old problems of costs and worker loyalty with new weapons designed to fit the needs and problems of the twentieth century.

Baritz (1960, p. 196)

Meeting the needs and problems of the 20th century was a new type of management: scientific management. Its best-known proponent was Frederick Taylor. In testimony before a Special House Committee of the U.S. Congress in 1912, Taylor expressed that

> ...true Scientific Management requires a mental revolution on the parts both of management and of workers: ... the interest of both and of society in the long run call for ever greater output of want-satisfying commodities. Output requires expenditure of human and material energies; therefore both workers and management should join in the search for discovery of the laws of least waste.

Taylor (1912, p. xiii)

In an industrialized economy full of *potential* labor, pursuing least waste made good sense. It was, in a way, getting the greatest 'bang for the buck.' Let's look in greater detail how scientific management aimed to accomplish this.

2.2.2 TAYLOR AND TIME STUDIES

Frederick Winslow Taylor (1856–1915) was born into a wealthy and highly educated family (Kanigel, 2005). After securing a place to study law at Harvard, he changed course abruptly and apprenticed as a pattern-maker and machinist. Through a combination of family connections and his own efforts, he made his way up the manufacturing work hierarchy—laborer, clerk, leading hand, foreman, and, after earning a degree through night-study, chief engineer.

Throughout his career, Taylor demonstrated an almost fanatical drive for the scientific attainment of efficiency. This is exemplified in his 1906 paper "On the Art of Cutting Metals." The paper details 16,000 experiments over 26 years, all to determine the optimum relationship between the type of metal being cut, the size and shape of the cut, the shape of the cutting tool, and other variables of the cutting process. Taylor collaborated with mathematicians, professors, and expert metallur-gists in the design, conduct, and analysis of his experiments, and kept abreast of literature from around the world on industrial experimental design. The result of his work was a set of practices and formulas that could supposedly give a manufacturer an edge in efficiency over their competitors.

> The three great questions, as to shape of tools, speed and feed ... are daily answered for all the men in each shop far better by our one trained mechanic with the aid of his slide rule than they were formerly by the many machinists, each one of whom ran his own machine, etc., to suit his foreman or himself.
>
> **Taylor (1906, p. 46)**

It is unsurprising that Taylor, who put so much effort into determining the most time-efficient method of cutting metal, would experience moral outrage at the prob-lem of 'soldiering'—workers deliberately working slower than necessary. Having eliminated, to his own satisfaction, the problem that the workmen did not necessarily know the fastest way to complete each job, he sought explanations in the systems of work and incentives.

> The greater part of the systematic soldiering, however, is done by the men with the deliberate object of keeping their employers ignorant of how fast work can be done. So universal is soldiering for this purpose, or under any of the ordinary systems of com-pensating labor, who does not devote a considerable part of his time to studying just how slowly he can work and still convince his employer that he is going at a good pace.
>
> **Taylor (1911, p. 21)**

Taylor did not, as some critics of his work assume, consider soldiering to be the fault of the workers. Rather, he was most offended by 'unscientific' management prac-tices that created perverse incentives. For example, as he explained in his 1903 work 'Shop Management,' piece-work (paying workers based on how many parts they produced, rather than how much time they worked for) might seem to incentivize workers to be more productive. However, when workers worked faster, this caused managers to reduce the payment per piece to keep wage expenses constant.

> After a workman has had the price per piece of the work he is doing lowered two or three times as a result of his having worked harder and increased his output, he is likely to entirely lose sight of his employer's side of the case and to become imbued with a grim determination to have no more cuts if soldiering can prevent it.
>
> **Taylor (1903, p. 1351)**

Taylor believed that work incentives did not need to be a trade-off between the interests of employers and workers. Instead, he reasoned that there was an optimum balance whereby workers would receive the highest possible wages and employers the highest possible profits. To reach this point it was necessary to know exactly how long a task should take, and then offer stable incentives to work at that speed. Offering incentives that were too high or too low would result in soldiering or poor work. Offering variable incentives would also lead workers to adjust their speed to induce the employers to adjust the incentives. The solution was to find out how long a task actually took to complete, when workers were applying their best efforts. A high wage would be paid to workers who met productivity targets, and a lower wage to everyone else.

The crucial part of Taylor's system, in his own eyes and in the legacy of Taylorism that followed, was a reimagination of the role of management. Taylor said "Almost every act of the workman should be preceded by one or more preparatory acts of the management which enable him to do his work better and quicker than he otherwise could" (1911, p. 26). As was repeated 90 years later, "a worker should not be left to his own unaided devices" (Geller, 2001, p. 26). Supervisory control emerged as a new class of work. Supervisors set the initial conditions for, received information about the details of activities on their shop floor, and accordingly fine-tuned and adjusted the work assigned:

> But little can be accomplished with these laws unless the old-style shop foreman and shop superintendent have been done away with, and functional foremanship has been substituted ... in fact the correct use of slide rules involves the substitution of our whole task system of management for the old style of management.
>
> *(1911, p. 26)*

Taylor is generally regarded as a great believer in his own ideas, and in the moral superiority of those ideas. Greater organizational performance and cost efficiency were key, of course. Yet for him, these ideas stood to benefit not just managers, factory owners, corporations, or the nation, but also individual workers. Through doing a simple and tightly defined task, and doing it well, a worker could become what Taylor advertised in 1911 as 'a high-class man.' This sounded seductively like climbing the societal ladder, but it simply meant that this worker could earn more than his peers (Evans & Holmes, 2013). One problem of the 'science' in the scientific method was that results were driven both by worker selection, as well as by task design and control. This could of course lead to confounds, or a difficulty with determining what exactly was responsible for any increase in efficiency. It also opened the door to a favorable presentation of the results in order to promote the method (as can be seen in the following chapters as well).

2.2.3   THE GILBRETHS AND MOTION STUDIES

There are supposedly 18 basic motions that make up all human movement, according to the work of Dr. Lillian Gilbreth (1878–1972) and her husband Frank Gilbreth (1868–1924) (Barnes & Mundel, 1939). By decomposing work into the smallest elements, studying them, and then recombining them into standardized practices, the Gilbreths sought to make work easier and more productive.

The individual contributions of Lillian Gilbreth and Frank Gilbreth cannot easily be separated, since they worked and published together for most of their careers. Lillian Gilbreth trained in literature, education, and psychology. She completed two PhD dissertations. Her first, published as *The Psychology of Management* was disqualified due to residency requirements (Lancaster, 2004), but secured her place as the first industrial and organizational psychologist. Frank Gilbreth worked his way from bricklayer's helper to chief superintendent and management engineer (Ricci, 2012).

The Gilbreths were concerned with 'fatigue' and 'waste.' Fatigue was not used in the modern sense of the word—it referred to overexertion due to inefficient ways of working. The Gilbreth's observed that when groups of workers left a factory at the end of the day, having performed the same amount of work, some were clearly more tired than others. This led them to the belief that there was an ideal way to perform every task, and that scientific study could determine that 'one best way.'

> The waste in work not done, or in work done with the wrong method, is a serious economic waste. The waste in unnecessary fatigue is not only an economic waste, it is a waste of life, and it calls for immediate attention from every one of us, whether interested in the individual, the group, or the economic prosperity of our country.
>
> **Gilbreth and Gilbreth (1919, p. 13)**

The Gilbreth method, a motion study, was to decompose a task into its smallest possible elements, and to measure the time taken for each element. The task could then be reconstructed using the most efficient set of elements, and with each element performed in the least wasteful fashion. The Gilbreths' applied this method originally to highly repetitive tasks such as bricklaying, and then in World War I to the assembly and disassembly of weapons (Gilbreth, 1911).

2.2.4 Differences and Similarities between Time and Motion Studies

Taylor and the Gilbreths were aware of and influenced by each other's work. There was some animosity between them—and in particular between their followers. This was mainly concerned with claims about who should get credit for what, and assertions that the other 'camp' was more interested in saving money for managers than in supporting the well-being of workers.

Taylor and the Gilbreths agreed on several fundamental principles of management:

1. For each task, there is one best method which is derived with the scientific method by managers and planners. Consistent and uniform compliance of workers with this one best method will lead to predictable and efficient results.
2. All work can and should be systematically observed, measured, tabulated, and analyzed. The aim is to identify the optimum sequence and pace, as well as to eliminate all the unnecessary elements and movements. Work can be reduced to its absolute minimum required components and movements.
3. There needs to be a clear functional division between managers who manage and workers who carry out the work. Job design and job execution need to be separated. Managers and planners are to use their knowledge to plan, coordinate, and control how work is done. Workers just need to *do* the work.

The reverberations of these principles are still felt today. Whenever problems in safety or efficiency arise, it is still quite natural to reach for Tayloristic solutions. We might tell people to follow a procedure more carefully, to comply better to the 'one best method' that had been determined for their task. We might choose to increase the number of occupational health and safety experts who plan and monitor work, for example. Or we might consider ways of automating parts of the work: if machines do it, then human errors will no longer be possible. Indeed, Taylorism gave rise to the idea of the redundancy of parts. Because scientific management chunked tasks up into their most elemental parts, those could simply become seen as interchangeable parts of an industrial process—whether done by human or machine. Whichever did the task more reliably and efficiently should get to do it.

The idea is still with us in what is known as function allocation or MABA/MABA (Men-Are-Better-At/Machines-Are-Better-At) lists (Dekker & Woods, 2002). The basis for this is Taylor's redundancy-of-parts idea: by conceiving work in terms of the most reducible component tasks, it became possible to think about automating parts of those component tasks. If a machine could do that task more reliably and more efficiently than a human, the task should be automated. Today this is known as the substitution myth—the false idea that humans and machines are interchangeable. This myth, however, has a powerful hold on many ideas (popular as well as scientific) in human factors and safety. More will be said about it in the chapter on automation.

The separation of planning from execution also represented a major departure from the craftsman phase that had preceded mass industrialization. With it went the image of humans as innovative, creative, and intrinsically motivated. Taylorism became the embodiment of ideas from earlier in the Industrial Revolution: that of humans as machines, as "locomotives capable of working" (Billings, 1997, p. 58). Taylor was said to be genuinely surprised at the bottom-up resistance that his ideas created (Derksen, 2014). He had not foreseen (or had perhaps no way to see or interpret) how his work dehumanized labor, how the eradication of worker initiative, of local expertise and craftsmanship not only brought efficiency, but also hollowed out the meaning of work. The immense personnel turnover in major factories like Ford's bewildered him. Did he not bring those people a crisper, cleaner, more predictable world? Henry Ford had to go to great lengths to 'bribe' his workers to stay, making the radical move of offering a $5 a day wage (Cwiek, 2014). This had the added benefit that workers like his could now afford to buy the cars they produced, thus increasing the size of Ford's market dramatically. Ford's initiative helped convert what had been a society of producers (scratching out a living on farms and in workshops) into one of consumers.

Where Taylor and the Gilbreths' differed were in their approach to how workers should be influenced to perform a task in the best way. The Gilbreths' motion studies often resulted in physical reconfiguration of the workplace to support efficient task execution—the forerunner to modern ergonomics. In fact, on several occasions they returned to places they had previously worked with, only to find that the workers were no longer following the 'correct' motions. However, the efficiency gains had not been lost. It suggested that physical changes in the workplace had led to the improvements. The improvements were not so much due to training the workers in specific motions. This theme would reappear in the following decades, particularly in the realization that behavior-based safety (targeting behavior to change behavior) is nowhere near as effective as changing the environment in which that behavior takes place (see Chapters 4 and 5). If you want to change behavior, in other words, change the circumstances and conditions that help produce the behavior. Target *those*—not the behavior itself.

2.2.5 IMPLICATIONS FOR SAFETY SCIENCE

What are the implications of scientific management for safety? To see how Taylorism is an important foundation for Safety Science, consider the following three intertwined issues:

1. Should work be planned and designed from the top-down, or do the best way(s) of working emerge from those who actually do the work?
2. Is it possible or desirable to identify one best method for that work, and write it down as a procedure?
3. Is strict compliance with that one best method a guarantee of safety?

Taylor's and the Gilbreths' position on the first two of these questions should be clear. According to scientific management, work is to be planned and designed from the top-down, by people who do not do that work themselves. And yes, there is one best method for every particular job. Strict compliance with that one best method is a guarantee of reliability and efficiency. Here is how Taylor put it, referring to the transformation brought on by his ideas as the split between an 'old type of management' and 'scientific management.' He warned that relying on the initiative of workmen themselves would get a manager nowhere. His scientific methods, instead, would generate such initiative (if that would still be the right word to use) with absolute uniformity. And it would give managers new powers and authorities:

> Under the old type of management success depends almost entirely upon getting the 'initiative' of the workmen, and it is indeed a rare case in which this initiative is really attained. Under scientific management, the 'initiative' of the workmen (that is their hard work, their good-will and their ingenuity) is obtained with absolute uniformity and to a greater extent than is possible under the old system; and in addition to improvement on the part of the men, the managers assume new burdens, new duties and new responsibilities never dreamed of in the past.
>
> **Taylor (1911, p. 31)**

So did workers have nothing to say about how they were to work under scientific management? Or whether they should comply with the one best method under all circumstances? Taylor wrote that he believed in the "intimate cooperation of management with the workmen, so that they together do the work in accordance with the scientific laws which have been developed" (Taylor, 1911, p. 115). In reality, this cooperation was never very intimate, nor actually cooperative. The cooperation, such as it was, was one-directional and very specific. Managers would expect workers to carry out their tasks exactly as instructed and without complaint or deviation.

In return, workers could expect a fair reward for the work they put in. The Tayloristic approach relied on "a purely instrumental, low-involvement employment relationship of pay in return for obedience and effort" (Hales, 2013, p. 19). Think about the legacy of Taylor as you proceed further into this chapter. How much consultation with those who do the safety-critical work is done in your world? Who makes the procedures? Do the people who actually do the work get to sign off or veto procedures that they will have to work with? (Figure 2.1).

In 1964, long after Taylor, the sociologist Daniel Katz summarized the fundamental weakness of Taylorism. Katz wrote:

> Three basic types of behavior are essential for a functioning organization: (1) People must be induced to enter and remain within the system, (2) They must carry out their role assignments in a dependable fashion, (3) There must be innovative and spontaneous activity in achieving organizational objectives which go beyond the role specifications.

> **Katz (1964, p. 131)**

Taylor was concerned only with the second of these types of behavior, and only then to the extent that 'carrying out a role assignment' meant following instructions to the letter. Ignoring the first and third requirements led to serious problems even within the workplaces Taylor was trying to improve. There is some historical evidence to suggest that one of Taylor's most publicized experiments, involving loading pig-iron onto rail cars, did not in fact happen as he later reported it, because the original workers went on strike rather than participate in the experiment (Wrege & Perroni, 1974).

FIGURE 2.1 'Workers are dumb, managers and planners are smart.' The Tayloristic division of those who think, plan, and supervise from those who simply need to carry out the one best method given to them. (Drawing by author.)

2.3 PROCEDURES, SAFETY RULES, AND "VIOLATIONS"

2.3.1 The Relationship between Safety and Rules

A rule or procedure, said Leplat (1998, p. 190),

> ...defines the characteristics an action must fulfill: it says what must be done (or not done) and the conditions to do this. The canonical form of the rule is: 'if (such conditions are fulfilled), then (do such an action).' Conditions may be more or less explicit in the rule. Often, only the conditions specifically connected with the rule are made explicit.

For the purposes of this chapter, let's use the definition for safety rules and procedures supplied by Hale and Swuste. They suggested that a safety rule or procedure:

> is a defined state of a system, or a defined way of behaving in response to a predicted situation, established before the event and imposed upon and accepted by those operating the system as a way of improving safety or achieving a required level of safety.
>
> *(1998, p. 164)*

In addition to these, they noted, there are safety rules that define broader goals to be achieved (such as duty of care requirements) and safety rules that define the way in which decisions about a course of action must be made (i.e., the specification of a decision-making or governance process). Compliance with safety rules and procedures is commonly seen as part of an organization's overall risk management (Hopkins, 2011). If organizations need to demonstrate, for example, to their regulator, that they have processes and controls in place to manage their risks, then rules and compliance with those rules can play an important role. This is particularly the case when an industry has arrived at standards of good practice that they believe should be adopted by all players, or when a mishap exposes deficiencies that need fixing.

But people do not always follow safety rules and procedures. You can easily observe this when watching people at work, and indeed experience this when you yourself do such work. Managers, supervisors, and regulators (or anybody else responsible for safe outcomes of work) often consider deviations from procedures or safety rules to be a large practical problem. As an example, a:

> seminal study of safety rules in the Dutch railways showed that only 3% of workers surveyed used the rules often, and almost 50% never; 47% found them to be not always realistic, 29% thought they were used only to point the finger of blame, 95% thought that, if you kept to the rules, the work could never be completed in time, 79% that there were too many rules, 70% that they were too complicated and 77% that they were sometimes contradictory.
>
> **Hale and Borys (2013, p. 208)**

In hindsight, after a mishap, rule violations seem to play a dominant causal role. If only people had followed the procedure! Studies keep returning the basic finding that procedure violations precede accidents. For example, an analysis carried out for an aircraft manufacturer identified "pilot deviation from basic operational procedure" as primary factor in almost 100 accidents (Lautman & Gallimore, 1987, p. 2). The title of their paper was "Control of the crew caused accident." Crews, or humans, in other words cause accidents. Control over crews is possible through better compliance with procedures, better hierarchical order, and imposition of rules and obedience. In that kind of thinking, procedural violations easily get overestimated for their role in the sequence of events. But consider this for a moment:

- An ideological problem with such work is that it was, in this case, sponsored by a party (the aircraft manufacturer) which stands to benefit from results that show that crew actions crashed their airplanes—not their own engineering work.
- And a methodological problem with such work as empirical evidence is that it selects its cases on the dependent variable (the accident). Procedural violations go on the whole time, and most do not result in accidents. So a study that selects on the dependent variable cannot count as proof of any causal link between procedure violations and bad outcomes.

Hindsight always turns complex, tangled histories laced with uncertainty and pressure, into neat, linear anecdotes with obvious choices. That kind of analysis tends to highlight the places where people could have followed a rule or procedure, but did not:

> The interpretation of what happened may be distorted by naturalistic biases to overestimate the possible causal role of unofficial action or procedural violation... While it is possible to show that violations of procedures are involved in many safety events, many violations of procedures are not, and indeed some violations (strictly interpreted) appear to represent more effective ways of working.
>
> **McDonald, Corrigan, and Ward (2002, p. 3)**

As you will see later on, what look like "violations" from the outside and hindsight are often actions that make sense given the pressures and trade-offs that exist on the inside of real work. Finding procedure violations as causes or contributors to mishaps, in other words, says more about us and the biases we introduce when looking back onto a sequence of events, than it does about people who were doing actual work at the time.

If procedure violations are construed to be such a large ingredient of mishaps, however, then it can be tempting, in the wake of failure, to introduce even more procedures or to change existing ones or to enforce stricter compliance. For example, shortly after a fatal shoot down of two U.S. Black Hawk helicopters over Northern Iraq by U.S. fighter jets, "higher headquarters in Europe dispatched a sweeping set of rules in documents several inches thick to 'absolutely guarantee' that whatever caused this tragedy would never happen again" (Snook, 2000, p. 201). It is a common, but not typically satisfactory reaction. Introducing more procedures does not necessarily avoid the next incident. And exhortations to follow rules more carefully do not necessarily increase compliance or enhance safety. In the end, a mismatch between procedures and practice is not unique to accident sequences. Not following procedures does not necessarily lead to trouble, and safe outcomes may be preceded by just as (relatively) many procedural deviations as accidents are. There is, in other words, a deep and perpetual duality when it comes to procedures and compliance. As Bieder and Bourrier observed:

> Sociologists and ergonomists have long demonstrated that procedures and rules are constraints at one level, but they do offer protection against arbitrary orders, deviant shop practices, poor rules of thumb and inconsistent management. It is often necessary to go beyond rules to achieve goals and missions, yet they also serve as comforting guidelines when the course of action is not easy to find or too controversial. Finally, they are tangible embodiment of successive lessons.
>
> *(2013, p. 3)*

When rules are 'violated,' are these bad people ignoring the rules? Or are these bad rules, ill-matched to the demands of real work? To be sure, procedures, with the aim of standardization, can play an important role in shaping safe practice. Commercial aviation is often held up as prime example of the powerful effect of standardization on safety. But there is a deeper, more complex dynamic where real practice is continually adrift from official written guidance, settling at times, unsettled and shifting at others. There is a deeper, more complex interplay whereby practice sometimes precedes and defines the rules rather than being defined by them. In those cases, is a violation an expression of defiance, or an expression of compliance—people following "practical rules" rather than official, impractical ones? These possibilities lie between two opposing models of what procedures mean, and what they in turn mean for safety. These models of procedures guide how organizations think about making progress on safety. Let us look at the two bookends that are implied in this in more detail, and study some of the foundational literature underneath both (in addition to what has been discussed in Section 2.1 of this chapter). We will call them Model 1 and Model 2, respectively.

2.3.2 Model 1 and the Scientific Management Legacy

Model 1 comes from the rationalist, top-down approach of scientific management. In this model, procedures are rules to follow. Work is designed from the top-down, by people who are not likely going to do the work themselves. Strict compliance with the best way to do the job is expected of operational people, and it is expected to lead to reliable outcomes. In this model, safety rules and procedures are about the specification, communication, and control of safe behavior in dangerous situations (Hale, 1990). Hale and Borys (2013) offer a comprehensive overview of Model 1:

- Procedures are the embodiment of the one best way to carry out activities, and they cover all known contingencies;

- Procedures are devised by experts to guard against the errors and mistakes of fallible human operators at the sharp end, who are more limited than the experts in their competence and experience, and/or in the time necessary, in the heat of operations, to work out that one best way;
- Procedures must be devised in advance, based on task and risk analyses. Once devised, they are "carved in stone," communicated to and imposed on the operators or workforce by management;
- Procedures are documented in manuals or databases. These are made available to workers and incorporated in training;
- Procedures are seen as essentially static and in a sense linear. Devising them is a one-off exercise, and they are to be modified only when the activity changes substantially. This is not imagined to be very often;
- Violations (intentional deviations) and errors (unintentional deviations) are seen as essentially negative actions made with free choice, either by operators who think they know better or as ill-informed actions that need to be understood, countered, and suppressed;
- Procedures should be enforced by suitable means. Activities to ensure compliance are needed to overcome the human tendency to make errors and deviate from the rules, either intentionally or unintentionally.

The first model supports the notion that not following procedures can lead to unsafe situations. We can summarize the premises as follows (Dekker, 2003):

- Procedures represent the best thought-out, and thus the safest way to carry out a job.
- Procedure-following is mostly simple IF-THEN rule-based mental activity: IF this situation occurs, THEN this algorithm (e.g., checklist) applies.
- Safety results from people following procedures.
- For progress on safety, organizations must invest in people's knowledge of procedures and ensure that procedures are followed.

To those who accept it, Model 1 offers a number of advantages. Compliance with safety rules and procedures not only helps assure safe outcomes, but can also minimize the effects of human variability and maximize predictability. Aviation and many other industries have taken this on board: procedures represent and allow a routinization that makes it possible to conduct safety-critical work with perfect strangers. Procedures are a substitute for knowing coworkers. In aviation, for example, the actions of a copilot are predictable not because the copilot is known (in fact, you may never have flown with her or him) but because the procedures make the copilot predictable. Without such standardization it would be impossible to cooperate safely and smoothly with unknown people. Model 1 also suggests that safety rules and procedures save work and make life easier for people. Consider these examples of decisions where safety rules and procedures, according to Model 1, come in handy (Hopkins, 2011, p. 111):

- Should I wear a hard hat on a production site, to reduce the risk of being injured by falling objects, or not?

- Is it too hot to work in the normal way, or not?
- Am I too fatigued to fly this aircraft, or not?
- Should I stop a process now because of the risks involved, or not?
- Should I authorize this expenditure now as a means of reducing risk, or not?
- We already have several safeguards in place. Do we need one more, or not?

There is no point, such authors believe, in leaving it up to individual workers to set the parameters for making such decisions. Because all these things can, to a greater or lesser extent, be specified in advance. Hard hats are required before entering a particular area, for example, and you put up signs that say so. Exceeding certain temperature will lead to work being cancelled. A particular kind and number of safeguards is necessary to commence a particular job. Of course, there are judgment calls (e.g., about fatigue, about risks involved), but these can be guided by the rules and procedures as well. Model 1 does not believe that giving workers discretion outside of such rules is smart or safe.

REDUCING AMBIGUITY AND WORKAROUNDS TO PREVENT MEDICAL ERROR

In an effort to improve patient care, a study from the Harvard Business School targets ambiguity and workarounds, because they are the putative sources of medical error (Spear & Schmidhofer, 2005). Ambiguity and workarounds would have raised red flags for Taylor as well, because they indicate that the work is not well designed, and that compliance with the best method of working is not complete. In the words of the paper:

Error-prone organizations tolerate ambiguity, a lack of clarity about what is expected to happen when work proceeds. Therefore, defining what constitutes a problem is difficult for several aspects of work. It is not perfectly clear (1) what the workgroup is trying to achieve; (2) who is responsible for what tasks; (3) how to exchange information, materials, or services; or (4) exactly how to perform tasks. Moreover, even when recognized, problems are worked around; people improvise to "get the job done," even when indicators suggest something amiss (p. 627).

In one example offered by the authors, a nurse mistakenly used insulin rather than heparin to flush the arterial line of a patient (Spear & Schmidhofer, 2005). This caused severe hypoglycemia, seizures, coma, and, ultimately, death. In another example, unreliable processes for identifying patients, assuring consent, and exchanging information led to a Mrs. Morris being mistaken for a Mrs. Morrison. The patient was subjected to an unnecessary and potentially dangerous electrophysiological examination. If the work processes had been highly specified and complied with, then these things would not have happened. But aside of their potential to prevent such tragic outcomes, the insights generated by this study are consistent with Gilbreth's and Taylor's position against waste and inefficiency. You can also recognize the commitment to time-and-motion studies, and the tabulating and counting necessary to identify and target such inefficiency:

In one hospital, on each shift nurses averaged 23 searches for keys to the narcotics cabinet; this wasted 49 min per shift and delayed analgesia to patients. Rather than tolerate continued searches, administrators tested assigning numbered keys at the start of each shift, with safeguards to prevent loss or misuse. This procedure nearly eliminated searches for keys and saved 2895 nurse-hours yearly in a 350-bed hospital (p. 629).

The authors concluded that by meticulous specification of who should supply what goods, information, or services, to whom, in what fashion, and when, problems could be identified even before they produced adverse events (Spear & Schmidhofer, 2005). With this sort of system, they believed, the consequences of problems would not propagate. Investigations would result in design changes that could reduce the likelihood of recurrence. This would work even in a complex and unpredictable system like healthcare, they said, because you can start simple. Toyota, the car manufacturer that pioneered much of the total quality movement that embodies such principles, did it too. Most patients actually require many of the same things, or what they require can be delivered to them in standardized ways. Linearity and regularity are possible and desirable. You can recognize much of Model 1 in this proposal.

2.3.3   "Violations" as a Preoccupation of Model 1

"Violations" are a preoccupation of Model 1. For this model, violations undermine the order and uniformity that procedures are supposed to bring. To be sure, "violations" are one particular way to describe the behavior that does not conform to the procedure. It is one way to describe the gap between how work is imagined and how work is actually done. "Violation" has a judgmental and moralizing tone. After a maintenance-related mishap with an aircraft in Europe, for example, investigators for the regulator found that "the engineers who carried out the flap change demonstrated a willingness to work around difficulties without reference to the design authority, including situation where compliance with the maintenance manual could not be achieved" (JAA, 2001, p. 21). "The engineers demonstrated a willingness"—such terminology implies intention and volition (the engineers had a free choice either to comply or not to) and full rationality (they knew what they were doing): they violated willingly. Violators, in Model 1, are thought to be wrong, because rules and procedures prescribe the best, safest way to do a job. This is independent of who does that job. Rules and procedures are for everyone.

WHY DO PEOPLE VIOLATE ACCORDING TO MODEL 1?

From reviewing the literature, Hale and Borys (2013) were able to find four groups of reasons that Model 1 uses to explain procedure violations. These are (1) individual factors, (2) hardware or activity factors, (3) organizational or safety climate factors, and (4) rule-related factors. Recognize that this is the way that Model 1 looks at noncompliant behavior. Its premises (see above) drive and restrict what the possible interpretations are. You will see later that different ways of constructing "non-compliance" are possible too. This both requires and represents a different way of looking at work. But first have a look at what Model 1 understands the reasons for violations to be:

Individual factors
Attitude to and habits of noncompliance
Intention to comply or violate
Previous accident involvement
Worker level of training, experience, or knowledge
Worker sees ways to make shortcuts or workarounds
High value on macho, exciting, quick, efficient, or productive way to work
Self-image and status among peers
Fatigue
Perception of low risk or consequence
High self-efficacy, independence
Does things on the fly or haphazardly
Other personal characteristics such as gender and age

Hardware or activity factors
Unfamiliarity with the design
Complicated, difficult, or changed design
Design or layout makes violation necessary to achieve the goal
Use of incorrect materials
Compensate for poor equipment

Organizational or safety climate factors
Management turns a blind eye
Management is inconsistent in sanctioning people for violations
Supervisors do not participate in work
Poor group cohesion and social control
Failures of site or shop organization
Conflicts between trades or management and trades
Norm that says that skilled person can violate with impunity
Time pressure
Conflicting demands, especially with productivity

Subjective group norm to violate
Lack of trust

Rule-related factors
Procedure is difficult to understand
Procedure is difficult to comply with
Violation is needed to get the job done
Procedure is outdated
Procedures conflict with each other, no priorities are given
Procedure is seen as not appropriate for the organization
Too many procedures and rules

Model 1 can only offer limited remedies against violations. For sure, the model allows some space to reconsider procedures that are outdated or incompatible with the job or the design of the equipment. But procedures (and those who write them) are judged to be in principle "right." The word "workaround" has the same implication, of course. It can only make sense if there is *something* that someone is actually "working around." That something is the standard, the rule, the procedure, the protocol, the correct way of doing things, which is somehow seen as an obstacle. This means that a violator, in principle, is wrong; the violator needs correcting (sanctioning, training, reminding, observing, monitoring). This means that people, in Model 1, are a problematic resource that needs to be controlled, just like they were for Taylor's scientific management. This notion has driven the growth of behavior-based safety programs too (see Chapter 4).

The kind of control and correction that is necessary and possible depends, according to Model 1, on what kind of rule or procedure people are violating, and what kind of violation people are committing. Hale and Swuste (1998) offered the following distinctions among different types of rules and procedures:

- *Performance goals.* These say only what needs to be achieved. They do not say how it should be done. Performance goals may be expressed in everything from numerical risk targets to expressions of demonstrating a duty of care. So-called performance-based regulation is based on this. The regulator does not specify how an organization needs to achieve goals, only that it should (e.g., have processes in place that monitor the controls for certain hazards). To be sure, performance-based regulation does not necessarily reduce the total number of safety rules and procedures. In order to show that (and how) it achieves the goal set by the regulator, an organization typically generates internal rules and procedures. Rather than a reduction in the total number, this shifts where specific rules and procedures come from and who administers them.
- *Process rules.* These define the process by which somebody or some organization will operate. It still leaves considerable freedom about what that operation will be. For example, an airline may be told to have a process in place for specified accountable managers to meet at particular

intervals to drive the development of acceptable flight standards. The process in this is specified; the product (what those flight standards are exactly) is not.

- *Action rules.* These specify actions in terms of 'If – Then' statements. They say exactly how people need to behave in response to situations, cues or indications, or how equipment needs to be designed or tested (e.g., wearing a seat belt when in a moving car, checking that various equipment is working before taking off in an aircraft, counting surgical items before closing an incision).

In his review on the application of safety rules, Leplat (1998) reminded us that all of those goals, rules, and procedures have typically been established progressively, with successive additions and revisions. Also, they have been established by different authorities both inside and outside the organization, which can lead to problems of overlap, contradiction, and inconsistency (see also Amalberti, 2001). Leplat showed that safety rules and procedures find their way into a variety of documents:

- Documents that do not specifically concern safety or take safety into account;
- Documents that contain a separate part that concerns safety;
- Documents devoted only to safety;
- Documents that integrate safety aspects in their description of equipment design and the different work phases related to it.

The source and specification of the rule or procedure can help workers understand the implications of violating it. And it can help supervisors, managers, and regulators determine how pressing it is to do something about it. Increasingly, rule and procedural documents have been written to manage corporate liabilities (by delegating those liabilities down to the people doing the work), insurance demands, and sometimes even training requirements (Dekker, 2014). A cynical reading is that rules and procedures are written not necessarily in the expectation that they will be followed to the letter or that they will help make work safe. Rather, they allow stakeholders to claim that they have provided all reasonable guidance, cautions, and warnings that can be expected of them, and that the rest is up to the worker to whom the work and rule is delegated (Hale & Swuste, 1998). The flip side has been observed too. Workers may hold off on a particular task until they have it committed to paper— demanding a traceable validation from the organizational hierarchy before engaging with the work. This can make for overspecified, ill-directed, and generally large documents and bureaucratic processes surrounding them. These may have lost their direct connection to the work-as-done (Bieder & Bourrier, 2013).

The various sources for rules and procedures, and the varying level of specification in them, also produce different kinds of violations. Research based on Model 1 has made the following distinctions of the different types of violations (Hale & Borys, 2013; Reason, 1990):

- *Routine violations.* These have become normal and accepted ways of behaving in and by the person's peer group (and often their supervisors). Routine violations are often linked to rules that are seen as overly restrictive or outdated, and where there is a lack of supervisory monitoring or discipline.
- *Situational violations.* These occur in response to specific situations where the rule does not appear to work or be relevant, and which are typically accepted by supervision in the name of efficiency or production.
- *Exceptional violations.* These are responses to situations that are seldom encountered, where the consequences of the violation may not be clear or thought through.
- *Optimizing violations.* These are violations committed to solve trade-offs between safety and other objectives (e.g., production pressure, effort) or to explore the boundaries of system operation and find new solutions to these trade-offs.

The first two of these and to an extent the last one as well, implicate management and supervisors in the violation. Model 1 understands that even if workers ultimately need to be encouraged not to violate rules and procedures, their violations are sometimes aided and abetted by their superiors and their organization. As an example, Hale and Borys (2013) reported a study of mining supervisors who were reluctant to discipline workers for venturing into an area of an underground mine whose roof was not supported. About 86% of the supervisors would simply tell the worker not to do it again, but not take any other action against the worker. About the same percentage of supervisors did suggest that they would take the violations up with their superiors if it happened again, but only half of the workers said they believed their supervisors would actually do so. The reasons were consistent with Model 1 (see the box on violations above): production pressures, supervisors breaking rules themselves and not wanting to set a double standard, a danger that was judged only as slight, a lack of backup by superiors, and fear of hostile reactions from workers or their union. Organizational complicity in rule- and procedural breaches, then, are understood by Model 1 as violations on part of supervisors and managers. This does not challenge or change the model or its assumptions. Rather, it just widens the model's coverage to include supervisors and managers as well—they, too, should know better, stick with the one best method, and follow and enforce the rules (Table 2.1).

Even within Model 1, there is a realization that demands for strict compliance can be a double-edged sword. Once a safety rule or procedure has been 'carved in stone,' change, adaptation, and innovation may become more difficult. In order to innovate, after all, you *have* to violate; you *have* to be noncompliant. Those who believe in Model 1, however, argue that evolving insights about the best method, and the need to learn and adapt, can all be accommodated by that model (Hopkins, 2011). The basics laid down by scientific management do not change: rules and procedures are made and enforced by those who do not do the work themselves. They represent best practice, and compliance with the one best method should be demanded of those who do the work. If, despite this, discoveries and evolving insight suggest that procedures and

TABLE 2.1
Advantages and Disadvantages of Model 1 According to Hale and Swuste (1998)

Model 1 Advantages	Model 1 Disadvantages
For the rule follower	**For the rule follower**
Saves time and effort, no need to reinvent the wheel	Blindness to new situations
Clarity of tasks and responsibilities	Resentment at loss of freedom and control
For the rule imposer	**For the rule imposer**
Increases predictability and uniformity	Represses bottom-up innovation
Clarity of responsibilities	Little learning or adaptation in the system
Basis for control and assessment	Higher demands on management of rule compliance and control

rules be changed, then this is not something that workers should engage in themselves without any supervision or approval. Instead, proper organizational risk management channels need to be followed:

> Once 'surprised' [by a situation where the rule or procedure didn't fit], leaders treat discrepancies as something that is not 'normal' and should be investigated immediately. This approach contains problems, generates knowledge, and leads to improvements.

> **Spear and Schmidhofer (2005, p. 628)**

This process will then produce a new 'one best method' for how to do the work—from the top-down. Such a commitment to Model 1 means that:

> ...front-line workers should comply as strictly as possible with (good) rules to prevent them from indulging in casual non-compliance or risk-taking... Risk assessment is not for them to indulge in. If they meet a rule that cannot be complied with, they should appeal to management to authorise a work-around rather than devising one themselves. [Even if] rules can never be complete and compliance should never be blind, the solution is a modification to model 1, not its scrapping.

> **Hale and Borys (2013, p. 214)**

In the spirit of scientific management, Model 1 assumes that order and stability in operational systems are achieved rationally, mechanistically, and that control is implemented vertically (e.g., through task analyses that produce prescriptions of work-to-be-carried out). As you will see in Chapter 5, the strong influence that information-processing psychology has had on human factors and Safety Science has reinforced the idea of procedures as IF-THEN rule following. Procedures are akin to a program in a computer which in turn serves as input signals to the human information processor. The algorithm specified by the procedure becomes the software on which the human processor runs. But is it that simple? Applying procedures in practice perhaps requires more than rote behavior. It requires additional cognitive work. This brings us to the second model of procedures and safety.

2.4 MODEL 2: APPLYING PROCEDURES AS SUBSTANTIVE COGNITIVE ACTIVITY

2.4.1 PROCEDURES AND THE COMPLEXITY OF WORK

Procedures are seen as an investment in safety—but it turns out that they are not always. Procedures are thought to be required to achieve safe practice—yet they are not always necessary nor likely ever sufficient for creating safety. Procedures spell out how to do the job safely—yet following all the procedures can lead to an inability to get the job done. According to the scientific management tradition, managers are supposed to make the procedures, and workers are supposed to follow them. But Martin (2001) showed that managers are just as frequent violators of rules as the workforce. What are some of further observations of contemporary research about the application of procedures in practice?

- Operational work takes place in a context of limited resources and multiple goals and pressures. Procedures assume that there is time to do them, certainty (of what the situation is), and sufficient information available (e.g., about whether tasks are accomplished according to the procedure). This already keeps rules at a distance from actual tasks, because real work seldom meets those criteria. Work-to-rule strikes show how it can be impossible to follow the rules and get the job done at the same time. In nuclear power plants, operators were once accused of "malicious compliance": staying so true to the rules and procedures for running a nuclear plant that production came to a complete standstill (Morgan, 1998).
- Some of the safest complex, dynamic work not only occurs despite the procedures—such as aircraft line maintenance—but without procedures altogether. Rochlin, LaPorte, and Roberts (1987, p. 79), commenting on the introduction of even heavier and capable aircraft onto naval aircraft carriers, noted that "there were no books on the integration of this new hardware into existing routines and no other place to practice it but at sea... Moreover, little of the process was written down, so that the ship in operation is the only reliable manual." Work is "neither standardized across ships nor, in fact, written down systematically and formally anywhere." Yet naval aircraft carriers, with inherent high-risk operations, have a remarkable safety record, like other so-called high reliability organizations.
- Procedure-following can also be antithetical to safety. In the 1949, U.S. Mann Gulch disaster, wildland firefighters, who perished were the ones sticking to the organizational mandate to carry their tools everywhere (Weick, 1993). Many of those who survived the Piper Alpha drilling platform fire in 1988, which killed 167 workers, were those who violated the safety rules and jumped off the platform into the sea (McGinty, 2008). In this case, as in others, people faced the choice to either following the procedure or surviving (Dekker, 2001).
- There is always a distance between a written rule and an actual task. Documentation cannot present any close relationship to situated action

because of the unlimited uncertainty and ambiguity involved in the activity (Loukopoulos, Dismukes, & Barshi, 2009). Especially where normal work mirrors the uncertainty and criticality of emergencies, rules emerge from practice and experience rather than preceding it. Procedures, in other words, end up following work instead of specifying action beforehand.

Safety Science has been limited in its ability to trace and model such coevolution of human and system of work and rules. When nothing seems to be going on in an operation, and managers and others might believe that all is calm and well because people are following the rules and procedures, then this is because people adapt smoothly and sometimes invisibly. When nothing is going on, in other words, *a lot* is going on. Model 1 cannot remotely describe what actually goes on in practice, or how rule-based or procedural activities are in fact practiced (Hale & Borys, 2013). In a cockpit, for example, "the analysis of what is implicit in the prescriptions shows that the implementation of pilot procedures is far from being a work of pure and simple execution of items read on the checklist" (LePlat, 1998, p. 196). The complexities and difficulties, the interpretations and adaptations—these are all hidden from managerial oversight because of the successful results and because operational people are so practiced and fluid at it. Paradoxically, this can sponsor a belief in the validity and value of Model 1 and the imposition of a mechanistic, static prescription of one best method from the top-down (Amalberti, 2013). This can lead to what McDonald and colleagues called a "job perception gap."

THE 'JOB PERCEPTION GAP'

Aviation line maintenance is emblematic: A 'job perception gap' exists (McDonald et al., 2002), where supervisors are convinced that safety and success result from mechanics following procedures—a sign-off means that applicable procedures were followed. But mechanics may encounter problems for which the right tools or parts are not at hand; the aircraft may be parked far away from base. Or there may be too little time: aircraft with a considerable number of problems may have to be turned around for the next flight within half an hour. Mechanics consequently see success as the result of their evolved skills at adapting, inventing, compromising, and improvising in the face of local pressures and challenges on the line—a sign-off means the job was accomplished in spite of resource limitations, organizational dilemmas, and pressures. Those mechanics, who are most adept are valued for their productive capacity even by higher organizational levels. Unacknowledged by those levels, though, are the vast informal work systems that develop so mechanics can get work done, advance their skills at improvising and satisficing, impart them to one another and condense them in unofficial, self-made documentation

(McDonald et al., 2002). Seen from the outside, a defining characteristic of such informal work systems would be routine nonconformity. But from the inside, the same behavior is a mark of expertise, fueled by professional and inter-peer pride. And of course, informal work systems emerge and thrive in the first place because procedures are inadequate to cope with local challenges and surprises, and because procedures' conception of work collides with the scarcity, pressure, and multiple goals of real work.

Studies over the past decades give ample testimony to the limits and shortcomings of Model 1:

> [One team] describes a case study in which workers often ignored the normal accounting procedures associated with a manufacturing process in order to meet the contingencies presented by rush orders. [Others] observed how methodologies imposed on design teams are "made to work" in practice by those using them. Such *making to work* involved shortcutting steps in the sequential process and patching up the product afterwards. [Yet others] describe the dilemma faced by staff in a radiography department who have concerns both for following procedures and for the care of their patients. Their analysis demonstrates a subtle decision process in which at times procedures are followed and at others *workarounds* are used in order to resolve the conflicting concerns. These authors do not interpret such behaviors as bad practice or procedural violation. Instead they argue that flexibility in the light of local contingencies in in the nature of human work, and it is in in the detail of this kind of engagement with methods, procedures and other formal artefacts of work that the intelligence of human-involved work resides.
>
> **Wright and McCarthy (2003, p. 681)**

These studies have focused on the expertise and sometimes 'craftmanship' necessary to translate and apply rules and procedures in practice. They typically highlight (Hale & Borys, 2013):

- *The value of discretion.* This de-emphasizes formal rules, because people who made the rules cannot possibly foresee all situations. And even if such foresight is possible, then they might still not be in the best position to determine how to control or manage those situations;
- *Tacit knowledge.* This is knowledge that is not explicit and that even the practitioner may have difficulty expressing. Such knowledge, however, is rich, learned by situated experience, exposure, and repetition more than by formal training. It also comes from interaction with experts who have done the work for a long time, mapping the boundaries of safe practice;
- *Personal accountability for outcomes.* When discretion and tacit knowledge are relied on to get the work done, more so than a blind application of procedures and rules, practitioners typically take more ownership and responsibility for outcomes (Dekker, 2013).

After his studies of pilots and seafarers, ethnographer Ed Hutchins (1995) pointed out that a procedure is not just an externalized cognitive task (i.e., a task that has been transplanted from the head into the world, for example, onto a checklist). Rather, following a procedure itself requires cognitive tasks that are not specified in the procedure. Transforming the written procedure into activity requires cognitive work. This is so because procedures are incomplete specifications of action: they contain abstract descriptions of objects and actions that relate only loosely to particular objects and actions that are encountered in the actual situation. People at work must interpret procedures with respect to a collection of actions and circumstances that the procedures themselves can never fully specify.

LUBRICATING A JACKSCREW

Take as an example the lubrication of the jackscrew on MD-80s—something that was done incompletely and at increasingly greater intervals before the crash of Alaska 261. You can read more about this accident in the book *Drift into Failure* (Dekker, 2011). This is part of the written procedure that describes how the lubrication work should be done (NTSB, 2002, pp. 29–30):

A. Open access doors 6307, 6308, 6306, and 6309

B. Lube per the following…

3. JACKSCREW

Apply light coat of grease to threads, then operate mechanism through full range of travel to distribute lubricant over length of jackscrew.

C. Close doors 6307, 6308, 6306 and 6309

This leaves a lot to the imagination, or to the mechanic's initiative. How much is a "light" coat? Do you do apply the grease with a brush (if a "light coat" is what you need), or do you pump it onto the parts directly with the grease gun? How often should the mechanism (jackscrew plus nut) be operated through its full range of travel during the lubrication procedure? None of this is specified in the written guidance. It is little wonder that: "Investigators observed that different methods were used by maintenance personnel to accomplish certain steps in the lubrication procedure, including the manner in which grease was applied to the acme nut fitting and the acme screw and the number of times the trim system was cycled to distribute the grease immediately after its application" (NTSB, 2002, p. 116).

In addition, actually carrying out the work is difficult enough. The access panels of the horizontal stabilizer were just large enough to allow a hand through, which would then block the view of anything that went on inside. As a mechanic you can either look at what you have to do or what you have just done, or actually do it. You cannot do both at the same time, since the access doors are too small. This makes judgments about how well the work is being done rather difficult. The investigation discovered as much when they interviewed the mechanic responsible for the last lubrication of the accident airplane: "When asked how he determined whether the lubrication was being accomplished properly and when to stop pumping the grease gun, the mechanic responded, 'I don't'" (NTSB, 2002, p. 31).

The time the lubrication procedure took was also unclear, as there was ambiguity about which steps were "included" in the procedure. Where does the procedure begin and where does it end? After access has been created to the area, or before? And is closing the panels part of it as well, as far as time estimates are concerned? Having heard that the entire lubrication process takes "a couple of hours," investigators learned from the mechanic of the accident airplane that:

> ...the lubrication task took 'roughly...probably an hour' to accomplish. It was not entirely clear from his testimony whether he was including removal of the access panels in his estimate. When asked whether his 1-hour estimate included gaining access to the area, he replied, 'No, that would probably take a little— well, you've got probably a dozen screws to take out of the one panel, so that's—I wouldn't think any more than an hour.' The questioner then stated, 'including access?', and the mechanic responded, 'Yeah'.
>
> **NTSB (2002), p. 32**

As the procedure for lubricating the MD-80 jackscrew above indicates, formal documentation can neither be relied on, nor is normally available in a way which supports a close relationship to action. Sociologist Carol Heimer makes a distinction between *universalistic* and *particularistic* rules: universalistic rules are very general prescriptions (e.g., "Apply light coat of grease to threads"), but remain at a distance from their actual application (Dekker, 2005). In fact, all universalistic rules or general prescriptions develop into particularistic rules as experience accumulates. With experience, people encounter the conditions under which universalistic rules need to be applied. They become increasingly able to specify those conditions and indeed the rules themselves. As a result, universalistic rules assume appropriate local expressions through practice. This actually goes so far that Becker (2005) shows logically that precise actions can never be deduced by looking at the rules for it, and that, conversely, observations of actions can never be specify, by induction, the rules that drive those actions. Quoting Wittgenstein, Becker drives the point home (italics in the original from Wittgenstein): "However many rules you give me, I give you a rule that justifies *my* use of your rule" (p. 253).

2.4.2 PROCEDURES AS RESOURCES FOR ACTION

In the Anglophone tradition, such insights are usually traced back to the work of Lucy Suchman, who studied repairmen at Xerox (Suchman, 1987). Her work is contrasted against the idea of humans as pure information processors who can follow a procedure like a computer program:

> What Suchman and others achieved by their critique of procedure following is an exposure of the limits of a purely information processing account of procedure. Suchman

turned our attention away from the idea that procedures are rather like programs in a computer that the operator can follow without intelligence. Instead, she turned our attention to the idea that it is only through the intelligent, pragmatic sensemaking of the operator that the procedure can be made to work in the diverse situations in which it is required to work. For Suchman, meaning and sensemaking, not information processing, are at the heart of procedure following. Procedures are not context-free algorithms that can be followed by an unintelligent agent; they are resources for action that have to be made sense of afresh in each new situation.

Wright and McCarthy (2003, p. 681)

Procedures, then, are not the work itself. Work, especially that in complex, dynamic workplaces, often requires subtle, local judgments with regard to timing of subtasks, relevance, importance, prioritization, and so forth. For example, there is no technical reason why a before-landing checklist in a commercial aircraft could not be automated. The kinds of items on such a checklist (e.g., hydraulic pumps OFF, gear down, flaps selected) are mostly mechanical and could be activated on the basis of predetermined logic without having to rely on, or constantly remind, a human to do so. Yet no before-landing checklist is fully automated today. The reason is that approaches for landing differ—they can differ in terms of timing, workload, priorities, and so forth. Indeed, the reason is that the checklist is not the job itself. The checklist is, to repeat Suchman (1987), a resource for action; it is one way for people to help structure activities across roughly similar yet subtly different situations. Variability in this is inevitable. Circumstances change or are not as was foreseen by those who designed the procedures.

The gap between procedures and practice is not constant. After the creation of new work (e.g., through the introduction of new technology), time can go by before applied practice stabilizes, likely at a distance from the rules as written for the system on the shelf. Social Science has characterized this migration from tightly coupled rules to more loosely coupled practice variously as 'fine-tuning' (Starbuck & Milliken, 1988) or practical drift' (Snook, 2000). Through this shift, applied practice becomes the pragmatic imperative; it settles into a system as normative. Deviance (from the original rules) becomes normalized; nonconformity becomes routine (Vaughan, 1996). The literature has identified important ingredients in the normalization of deviance, which can help organizations understand the nature of the gap between procedures and practice:

- Rules that are overdesigned (written for tightly coupled situations, for the worst case) do not match actual work most of the time.
- Emphasis on local efficiency or cost effectiveness pushes operational people to achieve or prioritize one goal or a limited set of goals (e.g., customer service, punctuality, capacity utilization). Such goals are typically easily measurable (e.g., customer satisfaction, on-time performance) and thus easier to prioritize on a daily basis than safety (particularly when everything keeps going right).
- Past success is taken as guarantee of future safety. Each operational success achieved at incremental distances from the formal, original rules can establish a new norm. From here a subsequent departure is once again only

a small incremental step. From the outside, such fine-tuning constitutes incremental experimentation in uncontrolled settings—on the inside, incremental nonconformity is an adaptive response to scarce resources, multiple goals, and often competition.

- Departures from the routine become routine. Seen from the inside of people's own work, violations become compliant behavior. They are compliant with the emerging, local ways to accommodate multiple goals important to the organization (maximizing capacity utilization but doing so safely; meeting technical requirements, but also deadlines). They are compliant, also, with a complex of peer pressures and professional expectations in which unofficial action yields better, quicker ways to do the job; in which unofficial action is a sign of competence and expertise; where unofficial action can override or outsmart hierarchical control and compensate for higher level organizational deficiencies or ignorance.

Although a gap between procedures and practice always exists, there are different interpretations of what this gap means and what to do about it. Some in safety may still see the gap between procedures and practice in Tayloristic terms. For example, as insufficiently stringent supervision or as a sign of complacency— operators' self-satisfaction with how safe their practice is or their system is or a lack of discipline. Psychologists may see routine nonconformity as expressing a fundamental tension between multiple goals (production and safety) that pull workers in opposite directions: getting the job done but also staying safe. Others highlight the disconnect that exists between distant supervision or preparation of the work (as laid down in formal rules) on the one hand, and local, situated action on the other. Sociologists may see in the gap a political lever applied on management by the work floor, overriding or outsmarting hierarchical control and compensating for higher level organizational deficiencies or ignorance. To the ethnographer, routine nonconformity would be interesting not just because of what it says about the work or the work context, but because of what it says about what the work means to the operator.

The distance between procedures and practice can create widely divergent images of work. Is routine nonconformity an expression of elitist operators who consider themselves to be above the law, of people who demonstrate a willingness to ignore the rules? Work in that case is about individual choices, supposedly informed choices between doing that work well or badly, between following the rules or not. Or is routine nonconformity a systematic by-product of the social organization of work, where it emerges from the interactions between organizational environment (scarcity and competition), internalized pressures, and the underspecified nature of written guidance? In that case, work is seen as fundamentally contextualized, constrained by environmental uncertainty and organizational characteristics, and influenced only to a small extent by individual choice. People's ability to balance these various pressures and influences on procedure following depends in large part on their history and experience. And, as Wright and McCarthy (2003) pointed out, there are not always good or established ways in which this experience can be given a legitimate voice in the design of procedures.

2.4.3 WORK-AS-IMAGINED VERSUS WORK-AS-DONE

The examples and studies above show that there is always a gap between how work is imagined and how it is actually done. One can see this gap in either of two ways:

- As *non-compliance*, or as a 'violation.' If that is the position you take, then there is only one thing you can do: try to close the gap by having people comply (by whatever means: sanctioning them, putting up posters, or adopting other behavioral interventions). The gap is not inevitable, it is just evidence of workers who need to try harder, or managers and designers who need to tweak the procedure or rule more so that people can actually comply.
- As the local, adaptive resilience necessary to close the distance between a static, linear procedure, and a dynamic, complex world. If this is the position you take (and it is the position of Model 2), then trying to close the gap by telling people to try harder makes no sense. The gap is inevitable, because it is embedded in the unbridgeable difference between a rule and the world. The solution, instead, is to better understand work: what it takes to get the job done safely.

The first to clearly articulate the difference between work-as-imagined and work-as-done was the Francophone tradition of work studies (see also Chapter 5), which made a distinction between *tâche* and *activité* (De Keyser, Decortis, & Van Daele, 1988). *Tâche* (or task) represents work-as-imagined, whereas *activité* means work-as-done; the actual activity. As Erik Hollnagel explained:

> Work-As-Imagined tends to see the action in terms of the task alone and cannot see the way in which the process of carrying out the task is actually shaped by the constantly changing conditions of work and world. An analogy, although inadequate, is the difference between a journey as actually carried out on the ground and as seen on a map. The latter inevitably smooths over the myriad decisions made in regard to imprecise information and changing conditions in the environment... Although the map is very useful, it provides little insight into how ad hoc decisions presented by changing conditions can be resolved (and of course, each resolved decision changes the conditions once more). As the journey becomes more complicated, the map increasingly conceals what is actually needed to make the journey.
>
> *(2014, p. 122)*

In a study of electronics repairmen, Rasmussen and Jensen (1974), found how very little information was actually available to those who designed the system or wrote the guidelines about the mental procedures used by operators in real tasks. In their environment, this had not been studied before. The Tayloristic assumptions that the manager, planner, or designer can say how work should be done (and design both the system and the procedures in that image), had simply not allowed it, or even raised the need. In one of the first explicit studies of work-as-done, Rasmussen spent time with electronics repairmen of his own organization, a nuclear research establishment west of Copenhagen, Denmark. Rasmussen asked the technicians to verbalize their mental procedures during a repair and documented the results.

These 'verbal protocols' revealed significant differences between the work-as-imagined (in the design, information availability, and written guidance) and work-as-done. Technicians, in their hunt for faulty component(s) inside the piece of equipment, were not interested to explain why the system displayed a faulty response. Rather, they saw their task as locating a fault in a system that was known to work properly before, that is, to find the source of the discrepancy between the normal and actual system state. Rasmussen identified various search routines, none of them documented. One was topographic (focusing on a particular physical area), one was functional (tracing the supposed function through the device), and the third was guided by fault evaluation. He found that in all of these search routines, experience with electronics—and particularly with similar faults—played a strong role.

Rasmussen's studies foreshadowed later American work on what became known as recognition-primed decision-making (Klein, 1993). The technicians relied on redundant observations (i.e., confirming the same thing from multiple perspectives or angles) and impulsive decisions to follow a particular lead—neither of which were "rational" when compared to what the designer might have expected them to do. Yet these routines were driven by experience and in the end much faster than the "official" guidance documentation.

A basis for the distinction between work-as-imagined and work-as-done can be found in Rasmussen's work. In one image, he showed, there is a set of rational choices to be made by the worker in order to reach a goal. These decisions are supported or directed by written procedures that the worker should follow. In the other image of work, there is not a one-to-one mapping from written guidance to task, but rather:

> a many-to-many mapping between means and ends present in the work situation as perceived by the individual ... there are many degrees of freedom which have to be resolved at the worker's discretion. ... Know-how will develop..., 'normal ways' of doing things will emerge. ... Effective work performance includes continuous exploration of the available degrees of freedom, together with effective strategies for making choices, in addition to the task of controlling the chosen path to the goal.
>
> **Rasmussen (1990, pp. 454–455)**

This constituted a local rationality in the way the individual does and develops the most sensible way to work—even if that is not the rationality of the designer, the manager, or the procedure:

> Humans tend to resolve their degrees of freedom to get rid of choice and decision during normal work and errors are a necessary part of this adaptation, the trick in the design of reliable systems is to make sure that human actors maintain sufficient flexibility to cope with system aberrations, i.e., not to constrain them by an inadequate rule system.
>
> *(p. 458)*

In their study of electronics, technicians Rasmussen and Jensen found support for this in how these technicians actually worked (1974). They saw rapid sequences of simple decisions on informationally redundant observations. These seemed inefficient in terms of information economy (i.e., why check the same thing three times from different angles or approaches?) and thus not at all rational. But seen from the inside of the job, with goals such as minimizing time on task or mental workload, these informal procedures were locally rational. And indeed, they were more characteristic for human problem-solving and decision-making in general, something that would later be reproduced through the research of, among others, Klein (1998) and Orasanu, Martin, and Davison (1996).

2.5 MODEL 2 AND SAFETY

2.5.1 THE LIMITS OF PRESPECIFIED GUIDANCE

Model 2 suggests that there is always a distance between the logics dictated in written guidance and real actions to be taken in the world. Adapting procedures to fit unusual circumstances is a substantive cognitive activity. This takes us to the outlines of a different model:

- Safety is not the result of rote rule following; it is the result of people's insight into the features of situations that demand certain actions, and people being skillful at finding and using a variety of resources (including written guidance) to accomplish their goals. This suggests a second model on procedures and safety.
- Procedures are resources for action. Procedures do not specify all circumstances to which they apply. Procedures cannot dictate their own application.
- Applying procedures successfully across situations can be a substantive and skillful cognitive activity.
- Procedures can, in themselves, not guarantee safety. Safety results from people being skillful at judging when and how (and when not) to adapt procedures to local circumstances.
- For progress on safety, organizations must monitor and understand the reasons behind the gap between procedures and practice. Additionally, organizations must develop ways that support people's skill at judging when and how to adapt.

The limits of prespecified guidance are especially visible in the face of novelty and uncertainty. Take, for instance, the crash of a large passenger aircraft near Halifax, Nova Scotia in 1998 (TSB, 2003). After an uneventful departure, a burning smell was detected and, not much later, smoke was reported inside the cockpit. Newspaper accounts characterized (perhaps unfairly or stereotypically) the two pilots as respective embodiments of the models of procedures and safety: the copilot preferred a rapid descent and suggested dumping fuel early so that the aircraft would not be too heavy to land. But the captain told the copilot, who was flying the plane, not to descend too fast, and insisted they cover applicable procedures (checklists) for dealing with smoke and fire. The captain delayed a decision on dumping fuel. With the fire developing, the aircraft became uncontrollable and crashed into the sea, taking all 229 lives onboard with it. There were many good reasons for not immediately diverting to Halifax: neither pilot was familiar with the airport; they would have to fly an approach procedure that they were not very proficient at; applicable charts and information on the airport was not easily available, and an extensive meal service had just been started in the cabin.

2.5.2 Failing to Adapt or Adaptations That Fail

Part of the example illustrates a fundamental double bind for those who encounter surprise and have to apply procedures in practice (Woods & Shattuck, 2000):

- If rote rule following persists in the face of cues that suggests procedures should be adapted, this may lead to unsafe outcomes. People can get blamed for their inflexibility; their application of rules without sensitivity to context.
- If adaptations to unanticipated conditions are attempted without complete knowledge of circumstance or certainty of outcome, unsafe results may occur too. In this case, people get blamed for their deviations; their nonadherence.

In other words, people can fail to adapt, or attempt adaptations that may fail. Rule following can become a desynchronized and increasingly irrelevant activity decoupled from how events and breakdowns are really unfolding and multiplying throughout a system. In the Halifax crash, as is often the case, there was uncertainty about the very need for adaptations (how badly ailing was the aircraft, really?) as well as uncertainty about the effect and safety of adapting: How much time would the crew have to change their plans? Could they skip fuel dumping and still attempt a landing? Potential adaptations, and the ability to project their potential for success, were not necessarily supported by specific training or overall professional indoctrination. Civil aviation, after all, tends to emphasize Model 1: stick with procedures and you will most likely be safe (e.g., Lautman & Gallimore, 1987). Tightening procedural adherence, through threats of punishment or other supervisory interventions, does not remove the double bind. In fact, it may tighten the double bind—making it more difficult for people to develop judgment at how and when to adapt. Increasing the pressure to comply increases the probability of failures to adapt—compelling people to adopt a more conservative response criterion. People will require more evidence for the need to adapt, which takes time, and time may be scarce in cases that call for adaptation (as in the crash above). Merely stressing the importance of following procedures can increase the number of cases in which people fail to adapt in the face of surprise.

Letting people adapt without adequate skill or preparation, on the other hand, can increase the number of failed adaptations. One way out of the double bind is to develop people's skill at adapting. This means giving them the ability to balance the risks between the two possible types of failure: failing to adapt or attempting adaptations that may fail. It requires the development of judgment about local conditions and the opportunities and risks they present, as well as an awareness of larger goals and constraints that operate on the situation. Development of this skill could be construed, to paraphrase Rochlin, as planning for surprise. Indeed, as Rochlin (1999,

p. 1549) has observed: the culture of safety in high reliability organizations anticipate and plan for possible failures in "the continuing expectation of future surprise".

2.5.3 CLOSING THE GAP OR UNDERSTANDING IT?

Postmortems can quickly reveal a gap between procedures and local practice, and
hindsight inflates the causal role played by unofficial action (McDonald et al., 2002).
Accident stories are developed in which procedural deviations play a major role. Of
course, this does not mean that there actually *is* a causal link between not following
procedures and having a bad outcome. Think of all the times that a good outcome
followed (even when procedures were not followed in much the same way). The
response, though, is often to try to forcibly close the gap between procedures and
practice, by issuing more procedures or policing practice more closely. The 'official'
reading of how the system works or is supposed to work is reaffirmed: Rules mean
safety, and people should follow them. Actual practice, as done in those vast informal work systems discovered by McDonald and colleagues, can be driven and kept
underground.

Model 2, in contrast, would suggest that we need to invest in trying to monitor
and understand the gap between procedures and practice. The common reflex is not
to try to close the gap, but to understand why it exists. The role of informal patterns
of behavior, and what they represent (e.g., resource constraints, organizational deficiencies or managerial ignorance, countervailing goals, peer pressure, professionalism, perhaps even better ways of working) are all interesting to learn about. Such
understanding provides insight into the grounds for informal patterns of activity and
opens ways to improve safety by being sensitive to people's local operational context.

2.6 SCIENTIFIC MANAGEMENT IN SAFETY TODAY

2.6.1 WORKERS ARE DUMB, MANAGERS ARE SMART

Scientific management was a genuine attempt to bring the tools of experimentation
and empiricism to the improvement of productivity, if not necessarily work itself.
The study programs of Taylor and the Gilbreths held a commitment to grass roots,
real-world engagement that would appeal to many modern safety researchers. Taylor
was a prototypical consultant-researcher, offering the possibility of short-term efficiency gains to his industrial partners in return for the opportunity to build a body
of knowledge that he later offered back to the whole community. Frank Gilbreth
had an ethnographer's eye for detail, capturing the intricate details of work and then
theorizing models that could be used to improve every type of work.

The Tayloristic worldview, though, has left deep traces in our thinking about the
safety of work that has started to work against us. The first is, coarsely put, the idea
that workers are dumb, and that managers and planners are smart. Workers need to
be told what to do. Their work has to be specified in great detail otherwise they will
do the wrong thing. Autonomy and initiative are bad. Workers need to be supervised
and monitored closely in the execution of their work (for which we now have increasingly sophisticated electronic means). Taylor, in other words, has helped cement the

idea that workers are a problem to control, and that detailed specification of what they do and close monitoring of their doing it are the solution.

"PEOPLE DON'T ALWAYS DO WHAT THEY ARE SUPPOSED TO DO"

Have a look at the following quote from 2000, in an article about the creation of a safety culture in nuclear power plants. You will find all the assumptions and premises of the Tayloristic model. People and their frailties, errors, and violations represent the problem that needs to be controlled. Such control is offered by procedures, prescriptions, management treatises, and rules. But still, people do not always do what they are supposed to do: they depart from the one best method. It is this noncompliance that undermines the structures and processes that an organization has in place to prevent accidents:

> It is now generally acknowledged that individual human frailties … lie behind the majority of the remaining accidents. Although many of these have been anticipated in safety rules, prescriptive procedures and management treatises, people don't always do what they are supposed to do. Some employees have negative attitudes to safety which adversely affect their behaviours. This undermines the system of multiple defences that an organisation constructs and maintains to guard against injury to its workers and damage to its property.

> **Lee and Harrison (2000), pp. 61–62**

You can recognize the notion in this quote that managers and planners are "smart" and workers are "dumb." Managers carry the responsibility to set up the work processes, the treatises, the rules, and the prescriptive procedures. Worker compliance with those is necessary because that is the way to guarantee uniformity, reliability, and ultimately safety. Workers themselves are not supposed to figure out better ways of working:

> Managers assume, for instance, the burden of gathering together all of the traditional knowledge which in the past has been possessed by the workmen and then of classifying, tabulating and reducing this knowledge to rules, laws and formulae which are immensely helpful to the workmen in doing their daily work.

> **Taylor (1911), p. 36**

Related to this is the conclusion that to understand work in all its facets and improve it, it simply needs to be decomposed into the most elemental bits. Deconstruct, reorganize, and put the bits back together again in a better order: these are the keys to improved efficiency, quality, and supervisability. And perhaps most obviously, in a Tayloristic world, there is perfect overlap between work-as-imagined and work-as-done. There is no distinction between the two. If there is, then improved work design, better specified procedures, and more stringent supervision should solve it. It is the latter idea that continues to undermine our ability to engage with the lived experience of work in a complex world. As Rasmussen's studies and everyday experience show, better procedures cannot solve everything, nor can more compliance with how work has been prespecified.

2.6.2 Taylor and Linear, Closed, Predictable Work

When does Model 1 work, if at all? The work studied by Taylor had specific characteristics that made his approach 'fit.' That could still be the case for work that has those same characteristics. Demanding compliance with guidance developed beforehand (and likely by someone else who does not do the work) might make sense when a task is:

- Linear: one step necessarily precedes another, in one fixed order;
- Closed: no influences from outside can change the demands on how the work is to be accomplished;
- Predictable and uniform: the task will not vary from day-to-day or from worker to worker.

There may be pockets of closed, linear, and predictable work in a shop, site, factory, or other workplace, where compliance demands can lead to safe and predictable outcomes. It might also be possible to change certain aspects of work and workplace so as to *create* pockets of linearity, closeness, and predictability. Consider the often surprising, messy, and unpredictable nature of emergency care, and the enormous importance attached to the expertise of physicians and other clinical staff. After a number of unnecessary patient deaths, the so-called ABC protocol was introduced (or later ABCDE or an even longer acronym). In the swirl of activity and possible diagnostic uncertainty that may surround a new patient arrival, it focuses clinicians on the crucial parameters that are necessary to guarantee any human being's survival: airway, breathing, and circulation. Disabilities and Environment (e.g., cold, heat) were added later, as well as further letters.

The presurgical checklist introduced in the 2000s could be another example. It was meant to reduce the risks of complications and wrong-site surgeries, by asking operating room staff to go through a relatively straightforward list of questions and checks about the patient, site, procedure, instruments, and staff required. If this checklist is to be applied just before surgery, then a hospital could, in principle, make arrangements to have that situation be as closed, linear, and predictable as possible. Make sure everybody is scrubbed in and present and that all conceivably necessary instruments and materials are at hand, have the patient be prepped and ready, and close the doors to the operating room. This can push the work environment more closely to the Tayloristic idea. The checklist has been credited with halving surgical deaths in one study (Haynes et al., 2009).

But reality is not so easily subjugated. Surgeons, for example, have complained of "checklist fatigue" and a lack of buy-in. They have remarked that if any more checklists would be instituted, they "would need a checklist for all the checklists" (Stock & Sundt, 2015, p. 841). Checklists might serve a purpose in preventing a memory lapse or managing interruptions, even though that can be disputed (Degani & Wiener, 1990; Raman et al., 2016). When surveyed, 20% of the staff said the list was not easy to use and that it did not improve safety. Yet 93% wanted to have the checklist used when they were undergoing an operation (Gawande, 2010). In another example:

> Some health care organizations have successfully tested highly specifying processes. The Shock, Trauma, and Respiratory intensive care unit at LDS Hospital in Salt Lake City, Utah, developed protocols to better control glucose levels, decrease nosocomial infection rates, and reduce costs. These protocols are noteworthy because once developed, they were often changed as users encountered problems applying them.

Spear and Schmidhofer (2005, p. 628)

Even if you are able to carve out little pockets of Tayloristic stability where you can demand worker compliance with a fixed rule or procedure, that still does not automatically mean that this is the *one* best way. Take the pre-takeoff checklist on airliners as an example. On big jets some decades ago, such a list could contain over ten items. These had to be read aloud and cross-checked when it might be safer to direct a part of pilots' attention outside the cockpit: the runway environment where other aircraft may be taking off or landing. Over the decades, the typical pre-takeoff checklist has shrunk. On some airliners, it now only contains two items. Of course, this has been accompanied by increasing automation. But insight into the role of rules, lists, procedures, and compliance in such operational phases has also changed. Even the 'one best method' for doing the work has to evolve along with developing insight and experience.

2.6.3 METHODOLOGICAL INDIVIDUALISM

Taylor's research and its implications for how we organize work today have also been critiqued for methodological individualism (Newlan, 1990). Methodological individualism is the belief that social processes can be explained by reducing them to the level of an individual. Take for instance safety culture. According to this belief, it can be made intelligible by referring to the attitudes, behaviors, perceptions, and relational styles of individual workers. This does not take into account macro-level processes such as institutionalized role expectations, nor any other emergent characteristics of the social order:

> As a result of this fixation with surface, individual-level analysis, it is common practice for industrial psychologists to blame the individual worker for behaviors incongruent with the goals of the firm. Worker performance deemed inappropriate or unsatisfactory by a given set of work-related criteria is perceived to result from a 'lack of fit' between individual psychological characteristics and the expectations of the firm which are to reflect 'normal' standards".

Newlan (1990, p. 39)

By the 1920s, the world of work had drifted significantly toward the view that work can be planned to perfection by smart managers and planners, and that individual workers just need to be supervised closely so as to comply with the one best method for how to do the work. If things went wrong, it was because of worker deviation, noncompliance, violations, and unsafe acts. Chaney summed up the prevailing view. There is something almost desperate about his argument, trying as he is to reassert the importance of what he called engineering mitigation.

> The conviction that there must always remain an 'irreducible minimum' of accident rests largely on the idea that the main cause of accident is human recklessness. Since a perfected humanity can scarcely be looked for in the lifetime of the present generation, the hope of an industry measurably free from accidental death has seemed an 'iridescent dream'.
>
> **Chaney (1922, p. 203)**

In other words, there is only so much we can do by trying to influence the worker. If things go wrong, it is not necessarily because the worker did not comply, it is because we have left untapped the potential to mitigate accident and injury risk through design and engineering.

Chaney's plea was not going to get much of a hearing in the two decades following. The next two chapters will show an even more decided turn to methodological individualism. First, in Chapter 3, you will learn about accident proneness—the supposed tendency of *some* individuals to err more, to make more mistakes, to be involved in more accidents than others. The proneness to accidents has almost exclusively been seen as an individual characteristic. The best way to deal with it is to screen workers so as to keep them away from safety-critical tasks, or, better still, to keep them out of a workforce altogether. Chapter 4 deals with the work of Heinrich and the basis it created for behavior-based safety. You will find that this foundation of Safety Science is less focused on the personal characteristics of certain workers, but it still tends to reduce safety problems to individual worker behavior. Individuals are the target for safety interventions, and behavior change is the ultimate solution of an organization's safety problems. It was not until World War II that this commitment to methodological individualism was broken.

STUDY QUESTIONS

1. When you consider how safety is managed in your own organization, or an organization you are familiar with, which aspects of Taylorism can you recognize in that?
2. Why is it probably not very informative for an incident investigation to conclude that deviation from procedures was a factor in causing the incident?
3. This chapter has introduced you to different models of procedures. Can you identify an operational area in your own organization or elsewhere that matches the requirements for Model 1 to be successful?
4. Why does Model 2 not speak of procedural 'violations?' And why does it believe that applying procedures successfully can actually take substantive cognitive work?
5. How do you think a regulator can best relate to Model 1 and Model 2, given that both are probably going on (and various combinations of the two) in all organizations at any one time?

6. What alternatives would you develop for an organization that wants to stop writing a new rule or procedure in response to each incident they learn about?
7. What is the 'job perception gap,' or the difference between work-as-imagined, and work-as-done?
8. Is routine nonconformity a risk or an opportunity for an organization (and a risk or opportunity for *what*)? Does that also depend on whom you ask?
9. How can a reliance on, and deference to, expertise lead to the congealing of new, inflexible norms for how to get work done? What risks does this present for organizations and workers?

REFERENCES AND FURTHER READING

Amalberti, R. (2001). The paradoxes of almost totally safe transportation systems. *Safety Science, 37*(2–3), 109–126.

Amalberti, R. (2013). *Navigating safety: Necessary compromises and trade-offs — theory and practice.* Heidelberg, Germany: Springer.

Baritz, L. (1960). *Servants of power: A history of the use of social science in American industry.* Middletown, CT: Wesleyan University Press.

Barnes, R. M., & Mundel, M. E. (1939). *A study of hand motions used in small assembly work.* Iowa City, IA: University of Iowa.

Becker, M. C. (2005). The concept of routines: Some clarifications. *Cambridge Journal of Economics, 29*, 249–262.

Bieder, C., & Bourrier, M. (Eds.). (2013). *Trapping safety into rules: How desirable or avoidable is proceduralization?* Farnham, UK: Ashgate Publishing Co.

Billings, C. E. (1997). *Aviation automation: The search for a human-centered approach.* Mahwah, NJ: Lawrence Erlbaum Associates.

Chaney, L. W. (1922). *Causes and prevention of accidents in the iron and steel industry, 1910–1919.* Washington, DC: Government Printing Office.

Cwiek, S. (2014). The middle class took off 100 years ago... Thanks to Henry Ford? National Public Radio. Retrieved from https://www.npr.org/2014/01/27/267145552/the-middle-class-took-off-100-years-ago-thanks-to-henry-ford

De Keyser, V., Decortis, F., & Van Daele, A. (1988). The approach of Francophone ergonomy: Studying new technologies. In V. De Keyser, T. Qvale, B. Wilpert, & S. A. Ruiz-Quintallina (Eds.), *The meaning of work and technological options* (pp. 148–163). New York, NY: Wiley.

Degani, A., & Wiener, E. L. (1990). *Human factors of flight-deck checklists: The normal checklist.* Moffett Field, CA: Ames Research Center.

Dekker, S. W. A. (2001). Follow the procedure or survive. *Human Factors and Aerospace Safety, 1*(4), 381–385.

Dekker, S. W. A. (2003). Failure to adapt or adaptations that fail: Contrasting models on procedures and safety. *Applied Ergonomics, 34*(3), 233–238.

Dekker, S. W. A. (2005). *Ten questions about human error: A new view of human factors and system safety.* Mahwah, NJ: Lawrence Erlbaum Associates.

Dekker, S. W. A. (2011). *Drift into failure: From hunting broken components to understanding complex systems.* Farnham, UK: Ashgate Publishing Co.

Dekker, S. W. A. (2013). *Second victim: Error, guilt, trauma and resilience.* Boca Raton, FL: CRC Press/Taylor & Francis Group.

Dekker, S. W. A. (2014). The bureaucratization of safety. *Safety Science, 70*(12), 348–357.

Dekker, S. W. A., & Woods, D. D. (2002). MABA-MABA or abracadabra? Progress on human-automation co-ordination. *Cognition, Technology & Work, 4*(4), 240–244.

Derksen, M. (2014). Turning men into machines? Scientific management, industrial psychology, and the "human factor". *Journal of the History of the Behavioral Sciences, 50*(2), 148–165. doi:10.1002/jhbs.21650

Evans, C., & Holmes, L. (Eds.). (2013). *Re-Tayloring management: Scientific management a century on.* Farnham, UK: Gower.

Gawande, A. (2010). *The checklist manifesto: How to get things right* (1st ed.). New York, NY: Metropolitan Books.

Geller, E. S. (2001). *Working safe: How to help people actively care for health and safety.* Boca Raton, FL: CRC Press.

Gilbreth, F. B. (1911). *Motion study: A method for increasing the efficiency of the workman.* Whitefish, MT: Kessinger Publishing.

Gilbreth, F. B., & Gilbreth, L. M. (1919). *Fatigue study: The elimination of humanity's greatest waste, a first step in motion study.* Easton, PA: Hive.

Hale, A. R. (1990). Safety rules O.K.? Possibilities and limitations in behavioral safety strategies. *Journal of Occupational Accidents, 12*, 3–20.

Hale, A. R., & Borys, D. (2013). Working to rule, or working safely? Part 1: A state of the art review. *Safety Science, 55*, 207–221.

Hale, A. R., & Swuste, P. (1998). Safety rules: Procedural freedom or action constraint? *Safety Science, 29*, 163–177.

Hales, C. (2013). Stem cell, pathogen or fatal remedy: The relationship of Taylor's principles of management to the wider management movement. In C. Evans & L. Holmes (Eds.), *Re-Tayloring management: Scientific management a century on* (pp. 15–39). Farnham, UK: Gower.

Haynes, A. B., Weiser, T. G., Berry, W. R., Lipsitz, S. R., Breizat, A. H., Dellinger, E. P., … Gawande, A. (2009). A surgical safety checklist to reduce morbidity and mortality in a global population. *New England Journal of Medicine, 360*(5), 491–499. doi:NEJMsa0810119 [pii] 10.1056/NEJMsa0810119

Hollnagel, E. (2014). *Safety I and safety II: The past and future of safety management.* Farnham, UK: Ashgate Publishing Co.

Hopkins, A. (2011). Risk-management and rule-compliance: Decision-making in hazardous industries. *Safety Science, 49*, 110–120.

Hutchins, E. L. (1995). How a cockpit remembers its speeds. *Cognitive Science, 19*(3), 265–288.

JAA. (2001). *Human factors in maintenance working group report.* Hoofddorp, The Netherlands: Joint Aviation Authorities.

Kanigel, R. (2005). *The one best way: Frederick Winslow Taylor and the enigma of efficiency* (Vol. 1). Cambridge, MA: MIT Press, p. 676.

Katz, D. (1964). The motivational basis of organizational behavior. *Systems Research and Behavioral Science, 9*(2), 131–146. doi:10.1002/bs.3830090206

Klein, G. A. (1993). A recognition-primed decision (RPD) model of rapid decision making. In G. A. Klein, J. Orasanu, R. Calderwood, & C. E. Zsambok (Eds.), *Decision making in action: Models and methods* (pp. 138–147). Norwood, NJ: Ablex.

Klein, G. A. (1998). *Sources of power: How people make decisions.* Cambridge, MA: MIT Press.

Lancaster, J. (2004). *Making time: Lillian Moller Gilbreth ~ a life beyond "Cheaper by the dozen".* Lebanon, NH: Northeastern University Press.

Lautman, L., & Gallimore, P. L. (1987). *Control of the crew caused accident: Results of a 12-operator survey.* Seattle, WA: Boeing Airliner, pp. 1–6.

Lee, T., & Harrison, K. (2000). Assessing safety culture in nuclear power stations. *Safety Science, 34*, 61–97.

Leplat, J. (1998). About implementation of safety rules. *Safety Science, 29*, 189–204.

Loukopoulos, L. D., Dismukes, K., & Barshi, I. (2009). *The multitasking myth: Handling complexity in real-world operations*. Farnham, UK; Burlington, VT: Ashgate Publishing Co.

Martin, L. (2001). Bending the rules or fudging the paperwork? Documenting learning in SME's. *Journal of Workplace Learning, 13*(5), 189–197.

McDonald, N., Corrigan, S., & Ward, M. (2002). *Well-intentioned people in dysfunctional systems*. Paper Presented at the 5th Workshop on Human Error, Safety and Systems Development, Newcastle, Australia.

McGinty, S. (2008). *Fire in the night: The piper alpha disaster*. London, UK: Pan Books.

Morgan, G. (1998). *Images of organization*. Thousand Oaks, CA: Sage Publishing.

Newlan, C. J. (1990). *Late capitalism and industrial psychology: A Marxian critique*. (Master of Arts), San Jose State University, San Jose, CA. p. 1340534.

NTSB. (2002). *Loss of control and impact with Pacific Ocean, Alaska airlines flight 261 McDonnell Douglas MD-83, N963AS, about 2.7 miles North of Anacapa Island, California, January 31, 2000* (AAR-02/01). Washington, DC: National Transportation Safety Board.

Orasanu, J. M., Martin, L., & Davison, J. (1996). Cognitive and contextual factors in aviation accidents: Decision errors. In E. Salas & G. A. Klein (Eds.), *Applications of naturalistic decision making*. Mahwah, NJ: Lawrence Erlbaum Associates.

Raman, J., Leveson, N. G., Samost, A. L., Dobrilovic, N., Oldham, M., Dekker, S. W. A., & Finkelstein, S. (2016). When a checklist is not enough: How to improve them and what else is needed. *The Journal of Thoracic and Cardiovascular Surgery, 152*(2), 585–592.

Rasmussen, J. (1990). Human error and the problem of causality in analysis of accidents. *Philosophical Transactions of the Royal Society of London, B, 327*(1241), 449–462.

Rasmussen, J., & Jensen, A. (1974). Mental procedures in real-life tasks: A case study of electronic trouble shooting. *Ergonomics, 17*(3), 293–307.

Reason, J. T. (1990). *Human error*. New York, NY: Cambridge University Press.

Ricci, T. (2012). Frank Bunker Gilbreth. Retrieved from https://www.asme.org/engineering-topics/articles/construction-and-building/frank-bunker-gilbreth

Rochlin, G. I. (1999). Safe operation as a social construct. *Ergonomics, 42*(11), 1549–1560.

Rochlin, G. I., LaPorte, T. R., & Roberts, K. H. (1987). The self-designing high reliability organization: Aircraft carrier flight operations at sea. *Naval War College Review, 40*, 76–90.

Snook, S. A. (2000). *Friendly fire: The accidental shootdown of US black hawks over Northern Iraq*. Princeton, NJ: Princeton University Press.

Spear, S. J., & Schmidhofer, M. (2005). Ambiguity and workarounds as contributors to medical error. *Annals of Internal Medicine, 142*, 627–630.

Starbuck, W. H., & Milliken, F. J. (1988). Challenger: Fine-tuning the odds until something breaks. *The Journal of Management Studies, 25*(4), 319–341.

Stock, C. T., & Sundt, T. (2015). Timeout for checklists? *Annals of Surgery, 261*(5), 841–842.

Suchman, L. A. (1987). *Plans and situated actions: The problem of human-machine communication*. New York, NY: Cambridge University Press.

Taylor, F. W. (1903). Shop management. *Transactions of the American Society of Mechanical Engineers, 24*, 1337–1408.

Taylor, F. W. (1906). *On the art of cutting metals*. New York, NY: The American Society of Mechanical Engineers.

Taylor, F. W. (1911). *The principles of scientific management*. New York, NY: Harper & Brothers.

Taylor, F. W. (1912, January 25). *Hearing before special committee of the house of representatives to investigate the Taylor and other systems of shop management under authority of house resolution 90* (Vol. III, pp. 1377–1508). Washington, DC: US Government Printing Office.

TSB. (2003). *Aviation investigation report: In-flight fire leading to collision with water, Swissair transport limited, McDonnell Douglas MD-11 HB-IWF, Peggy's Cove, Nova Scotia 5 nm SW, 2 September 1998* (A98H0003). Gatineau, QC: Transportation Safety Board of Canada.

Vaughan, D. (1996). *The challenger launch decision: Risky technology, culture, and deviance at NASA*. Chicago, IL: University of Chicago Press.

Weick, K. E. (1993). The collapse of sensemaking in organizations: The Mann Gulch disaster. *Administrative Science Quarterly, 38*(4), 628–652.

Woods, D. D., & Shattuck, L. G. (2000). Distant supervision-local action given the potential for surprise. *Cognition, Technology & Work, 2*(4), 242–245.

Wrege, C. D., & Perroni, A. G. (1974). Taylor's pig-tale: A historical analysis of Frederick W. Taylor's pig-iron experiments. *Academy of Management Journal, 17*(1), 6–27. doi:10.2307/254767

Wright, P. C., & McCarthy, J. (2003). Analysis of procedure following as concerned work. In E. Hollnagel (Ed.), *Handbook of cognitive task design* (pp. 679–700). Mahwah, NJ: Lawrence Erlbaum Associates.

3 The 1920s and Onward
Accident Prone

KEY POINTS

- Accident-proneness was one of the earliest attempts to scientifically investigate the 'human element' in safety. It took the emerging sciences of psychology and eugenics, and applied them to explain patterns in industrial data.
- In doing so, it provided an object lesson in the limitations and moral perils of treating human capacities in a reductionist, 'scientific' way.
- Accident-proneness builds on two patterns that were hard to dispute: (1) Human performance is variable, in ways that can be relevant for accidents and (2) Some people are involved in more accidents than others, in ways that are unlikely to arise purely by chance.
- Accident proneness is ultimately difficult to prove. Safety outcomes are never purely due to specific individual attributes. If the same activities take place in different contexts (busy times versus low workload, or night versus day), even the same individual may well have different safety outcomes. It can never be determined with certainty how much of the outcome is attributable to the individual, and how much to context.

3.1 INTRODUCTION

By the early 20th century, industrial accidents had gone mainstream (Green, 2003). No longer were accidents seen as freak occurrences, or as unfortunate but unavoidable coincidences of time and space. The more we learned about them, the more we started to see patterns. One of those patterns was that some people seemed to suffer more incidents and accidents than others. If this was indeed true, was that because they were more accident prone? As early as 1913, Tolman and Kendall, two pioneers of the American safety movement, strongly recommended managers to be on the lookout for men who always hurt themselves and to take the hard decision to get rid of them. In the long run that would be cheaper and less arduous than keeping them around the workplace.

Behavior-based safety, which we will discuss in Chapter 4, recommends sanction or retraining—perhaps termination—as a form of negative feedback to improve individual safety. At least it holds out the possibility that an individual can be rehabilitated or improved. Accident-proneness takes the more pessimistic view: some people are innately and incorrigibly more dangerous than others. The precise explanation for this has varied according to the fashionable psychology of the day. Accident-proneness has, at various times, been considered as:

- an inherited condition;
- a personality trait;
- a moral deficiency;
- an incorrectly calibrated perception of risk;
- an undesirable tolerance for risk;
- part of a normal distribution of human performance.

Under all of these guises, accident-proneness provides a compellingly simple explanation for accidents, couched in the sophisticated language of statistics and psychology. Accident-proneness is in fact a concept that illustrates the contributions of interdisciplinary understanding in Safety Science:

- Epidemiological work discovered an apparent pattern in the causation of accidents. It appeared that if an individual was involved previously in an accident, they were statistically more likely to be involved in future accidents.
- Psychology provided a number of candidate explanations for this pattern. Maybe the individuals had an inherited predisposition for clumsiness. Maybe they were morally deficient, incorrigible risk takers. Maybe they had deviant personalities that caused them to break rules.

Social Science, however, warned that accident-proneness was not so much a 'discovered fact' as it was a 'constructed category.' It suggested that accident causes were assigned (or attributed), rather than determined. If this is the case, then the statistics revealed patterns, certainly, but these were patterns in how people thought and talked about accidents, rather than in the definitive causes of accidents.

3.2 THE DISCOVERY (OR CONSTRUCTION) OF ACCIDENT-PRONENESS

3.2.1 ACCIDENT-PRONE WORKERS

Around World War I (WWI), researchers in Germany and the United Kingdom began to believe that there were particularly "accident-prone" workers. It dawned on psychologists in these two countries independently. In history, this is known as a case of simultaneous discovery. Neither originator in these countries had heard about the work of the other. In fact, even by 1935 the English researchers were still not referring to the German research literature on the very same topic and findings. Simultaneous, independent construction of the same thing in different places typically occurs when there are shared, widespread cultural and social circumstances that create the conditions of possibility. More will be said about those circumstances later in this chapter.

Concern with industrial accidents, in part because of their sheer numbers, would receive boost during the Great Depression. Large-scale government-sponsored employment projects (the Hoover dam, for example) generated ample data, and the idea of accident-proneness would last successfully through the 1930s and into the 1940s. The disparity between the supply of, and demand for, labor in depressed western economies might have helped. Oversupply of labor offered the possibility of selecting the very best (i.e., the least accident prone), if the trait could effectively be tested for. Apart from the many accidents in construction, mining, and manufacturing, transportation became a major and growing producer of accidents too—first by rail and then road. As early as in 1906, 374 people died in automobile crashes in the United States alone. In 1919, the number for the year had climbed to 10,000 (Burnham, 2008). The idea that workers themselves are the cause of accident and injury took hold in the first half of the 20th century.

In 1922, a U.S. management consultant named Boyd Fisher published *Mental Causes of Accidents*, a book that explained how various kinds of individual actions and habits were behind errors and accidents. This included, he found, ignorance, predispositions, inattention, and preoccupation. Some of these were transient mental states, while some were more enduring character traits. Boyd also commented on the connection between age and the tendency to accident, pointing out that the young were more liable to accident (as were very old men). The latter reflects concern with an entire cohort, rather than individuals who are accident prone. But it sustains the same idea: some people are more likely to get things wrong than others.

3.2.2 GERMAN ORIGINS OF ACCIDENT-PRONENESS

Karl Marbe, one of the prominent accident-proneness researchers in Germany, traced his belief in the idea back to his school days. Writing in 1916, he said:

> it occurred to me that several of my fellow students again and again would suffer small accidents, while other students remained wholly free of them... And in life one meets a number of people who have broken every possible limb, while, again, others never have such a misfortune.

Burnham (2008, p. 103)

Marbe came from the distinguished Würzburg School in psychology, which was committed to rigorous experimental research, and not to the applied problems of safety. In 1913, however, Marbe became involved in the legal aftermath of a train crash near the Swiss border. It presented him with a train engineer who had a history of unreliable behavior, and a train guard who did not stop the train because he was said not to possess the mental capacity to make the appropriate judgment. Marbe's investigation of the case triggered his interest in the psychological causes of accidents. Soon, the idea that someone who had suffered an accident would be more likely to suffer subsequent accidents started appearing in his writings. Marbe determined that such consistently hapless behavior, with its predictable effects, resulted from a combination of hereditary factors and acquired habit. Together, he thought, these formed someone's personality. By 1925, Marbe's use of the word *Unfallneigung*, or 'accident tendency,' became 'accident-proneness.' A book on the psychology of accidents published in 1926 summarized his thinking (Marbe, 1926). It introduced his innovations to the German-speaking world and helped popularize the notion of accident-prone workers:

> There was one element that was new in 1926: Marbe explicitly extended the idea of accidents from personal injury to production errors, picking up on previous work that he had done on psychomotor errors. Errors made by children sorting peas showed the same incidence pattern found among workers who suffered injuries. Finally, Marbe emphasized accident proneness as an endogenous factor, and, despite earlier words about training, he wrote pessimistically. These were not ordinary people. The statistics for criminal convictions, Marbe pointed out, showed the same pattern as for accidents: the more convictions, the more chance of recidivism. And accident statistics, showing the law of repetition of behavior, therefore suggested action, Marbe believed. Insurance premiums for accident prone people ought to increase. Personnel policies should likewise take accident records into account. An accident prone chauffeur, Marbe contended, needed to find another kind of occupation.
>
> **Burnham (2008, p. 105)**

Some industries had already incorporated the idea by then. The German Railways, for example, had been conducting psychological testing since 1917. These concerned alertness and reaction times at first, but the railways added tests for estimating distance and speed necessary for quick decision-making. Physical strength, intelligence, color blindness, night blindness, and optical and auditory acuity were added later. Marbe turned these practical adoptions of his ideas to his advantage, using them as further evidence for the appropriateness of his ideas and methods. The presence or absence of a number of personality traits was believed to ensure the safe operation of transportation systems and industrial machinery. Marbe added the statistical operations that showed how accident-proneness depended not only on a particular mix of personality traits, but also on whether a person had suffered an accident before. The inclination to 'accident' (used as a verb back then) was proportional to the number of accidents previously suffered, Marbe pointed out. The opposite seemed even more convincing: those who consistently had no accidents were significantly less likely to have them in the future as well.

If specific workers were the problem, then the target for intervention was clear. And it should interest not only workers or their employers, but insurance companies and government regulators as well. Marbe proposed that production damage and costs could be minimized by finding the right worker for the job, or—perhaps more accurately—keep the dangerous ones out. This should be done by testing workers and carefully vetting and selecting them, excluding the accident-prone ones from the jobs where they could do most harm. Moede, a colleague of Karl Marbe

> ...enthusiastically claimed that standard psychotechnology could solve the problem of excesses of accidents by addressing the components of accident proneness. Psychotechnologists, he explained, started by doing a detailed analysis of accidents

and errors to find out exactly what was happening. Workers with reaction and attention deficiencies could be detected by psychotechnological testing, either before or during employment. Testing, especially aptitude testing, could also identify people with other personality traits that made them predisposed to have accidents. And at that point, wrote Moede optimistically, psychotechnologists could provide either ways to screen out unsafe job candidates, or they could suggest safety training that would counter undesirable traits. Thus, the problem and the cure were already being handled by psychotechnologists.

Burnham (2009, p. 48)

Another idea suggested by Marbe was that workers should have a personal card on which accidents and errors would be noted—a record for employers and others to base hiring and firing decisions on. This proposal was not followed concretely at the time, but today similar employment records are used across many industries to guide managerial decision-making about human resources.

3.2.3 English Origins of Accident-Proneness

As in Germany, the original idea of accident-proneness in England was data driven. On the back of the WWI, which had been accompanied by a vast increase in the production of munitions and other equipment for the war, great amounts of data about incidents became available:

The Industrial Fatigue Research Board was set up during World War I to facilitate war production in England. In 1919... the Board reported individual differences as one of the findings of British investigators [the university title of one of them was 'medical statistician']... their work was based on extensive records held by the Ministry of Munitions... They summarized their findings: "the bulk of the accidents occur to a limited number of individuals who have a special susceptibility to accidents, and suggests that the explanation of this susceptibility is to be found in the personality of the individual.

Burnham (2008, p. 107)

These findings showed that accidents did not just occur by chance. Statistics revealed a "varying personal liability" as important factor. The English-speaking world, too, realized that personal, individual factors contributed to accidents. The parallel to Marbe's categories of *Unfäller* and *Nichtunfäller* was obvious. As was the way to improve safety: weed out those who were susceptible to accidents, the *Unfäller* or accident repeaters. Greenwood, one of the researchers, involved in the original Ministry of Munitions work, put his ideas forward in a 1921 general book called *The Health of the Industrial Worker*. In it, he

repeated and expanded his conclusion that "in industrial life we may remove from risk those who show a tendency to accidents." He reported on the ways in which varying conditions could change "the human factor" in accidents, but he then took up the findings he reported jointly with Woods, that multiple accidents happened to the same individuals. To the earlier findings, Greenwood added, from further research by the group, confirmation that some individual workers have a "personal predisposition to accidents".

Burnham (2008, p. 108)

What was needed was a test that would systematically weed out the accident prone. Eric Farmer, industrial psychologist at Cambridge, left his work on time-and-motion studies and began devising tests to identify people who were likely to have accidents. The human factor at the time was thought of as exactly that: not generalizable psychological or environmental conditions (like fatigue or error traps in equipment design), but characteristics specific to the individual, or, in the vocabulary of deficit of the time, somebody's "physical, mental, or moral defects" (Burnham, 2009, p. 61). These negative characteristics could be identified by carefully testing and screening employees—something for which psychological institutes across Europe, the United Kingdom, and the United States were developing ever cleverer tests, simulations, and contraptions. Some of these could test a prospective tram driver's application of accelerator and brake forces, for example, to see whether movements would be too brusque to qualify for the job. Even a battery of relatively simple psychological tests was thought to predict accident-proneness. As measured at the time, a worker with a muscular reaction above their level of perception, for instance, was more likely to have frequent and severe accidents than someone whose muscular reactions were below their perceptual level.

Some of the researchers in accident-proneness referred back to the putative successes of pilot selection during WWI: the scientific selection of pilot candidates helped eliminate those unsuitable to pilot airplanes and reduced human factor-related accidents from 80% to 20% (Miles, 1925).The idea appealed to common sense and practical experience as well. The medical director of the Selby shoe company in England said in 1933 that:

> Anyone connected with an industrial clinic or hospital soon becomes acquainted with workers who make frequent visits because of accidents or illness. During the past seventeen years I have repeatedly remarked, 'Here comes the fellow who is always getting hurt.' At short intervals he reports with a cut, bruise, sprain or burn. Interspersed with his numerous minor accidents will be an occasional serious accident. Often you hear him remark, 'I don't believe anybody has more bad luck than I have.' He is the type of fellow who attracts the attention of safety engineers and safety committees. He and other such workers have been classed as 'accident prone' ... The accident prone workers are interesting people because they present so many problems and because the best authorities today admit they know so little about them.

Burnham (2009, p. 35)

Keeping people out of positions where they were more likely to suffer accidents was the way to improve safety and control costs. The public transport company in Boston, for example, found in the mid-1920s that 27% of its drivers on subways, street cars, or buses were causing 55% of all accidents. Psychologically testing new employees and eliminating those who tested badly on efficiency and power use, helped reduce their overall accident rate—something that caught the attention of transportation operators around the world. Farmer, based on his analyses of statistical records, divided workers up into the following categories:

- controlled workers
- good workers
- thinkers
- inept workers
- miscellaneous

Some workers, he concluded, simply did not adapt well to the industrial age. They reacted with accident-proneness and sickness. Perhaps the humane thing to do was to not have them participate in industry at all.

3.2.4 FRENCH ORIGINS OF ACCIDENT-PRONENESS

This conclusion was shared by French researchers, particularly Jean-Maurice Lahy, professor at the Institute for Psychology at the University of Paris and a major contributor to the safety literature of the time. Lahy was a founding figure of psychotechnology, who also concluded that accident-prone people did not adapt well to the new rhythms of work demanded by industrialization. Using a controlled design in 1936, he divided workers into those with frequent accidents (*sujets fréquement blessés*) and a control condition that contained 'normal' workers. He subjected them to a series of experimental tests to see what differentiated the two groups. This is the kind of research which, today, we might critique for selecting people on the dependent variable (number of accidents suffered) and subsequently going on a 'fishing expedition' to see if anything underlying can be found that separates them.

Interestingly, Lahy did not believe that fixed personality factors were responsible for someone's accident-proneness. This belief sets him apart from his German and English colleagues, who had committed to accident-proneness as an inborn and lifelong characteristic. This had been a point of contention since the invention of the term, and continues to be so to this day (see Section 3.4 of this chapter):

> One of the questions that challenged investigators before and during World War II was whether accident proneness was a temporary or a persistent tendency. As one expert noted, 'Many accidents are due to temporary emotional reactions that even a psychiatrist might have trouble in locating. There probably is a much closer relation between such common matters as unpaid grocery bills and industrial injuries than has ever been suggested in the accident records of any plant.' Another expert offered as an example from his own cases the worker who 'rushes to work without breakfast and on a sultry August morning faints while working in a manhole.' Other commentators remarked on temporary personal upsets that preceded accidents, with the assumption, often justified with happy-ending case histories in which managers intervened, that the tendency to have an accident was temporary. The injured workman who had argued with his wife made up with her and did not have any further industrial accidents.

> **Burnham (2009, p. 111)**

Accident-proneness was, according to Lahy, something that could be modified by amending the conditions of the person's work, and by educating and intervening in the behavior of the worker. No consensus ever emerged on whether accident-proneness was a single trait or factor that could be captured in one quotient (like IQ), or that it was a combination of traits and factors. For much of the era, however, and for many of the producers and consumers of the research, accident-proneness had become firmly established as an individual and undesirable psychological trait. It kept a growing machinery of *psychotechnik*, of testing and selection, busy well into the WWII. The tone of the English psychologists had become bullish, claiming that:

> Accident proneness is no longer a theory but an established fact, and must be recognized as an important element in determining accident incidence. This does not mean that knowledge of the subject is complete, or that the liability of any particular individual to accident can with certainty be predicted. What has been shown, so far, is that it is to some extent possible to detect those most liable to sustain accidents.

> **Farmer (1945, p. 224)**

The development and refinement of tests that could identify people who were either temporarily or permanently accident prone continued for decades after. The results, however, remained unsatisfying. And, as you will see in Section 3.4 below as well, the different operationalizations of accident-proneness by different researchers led to results that were often inconsistent with each other. The accident-proneness thesis had reached its high point when Farmer wrote his strident words. A slow decline in scientific publications on accident pronenenss had already started. It accelerated through the 1950s and 1960s, slowing to just a trickle today (Burnham, 2009).

3.3 THE SOCIAL CONDITIONS OF POSSIBILITY

3.3.1 Modernization, Measurement, and Statistics

What were the social conditions and ways of thinking in Europe and the United States during the time that gave rise to the notion of accident-prone people? The sense that some people were a bad 'fit' for society was due in part to industrialization and mechanization. These had been modernizing the West well since the previous century, and imposed new kinds of rigor, production demands, and rhythms of life that were unfamiliar to many. Think of, for example:

- repetitive indoor factory work;
- faceless bureaucratic interactions;
- nonnegotiable speed of a production line;
- clocking in and out of the workday;
- technical and mechanical hazards that did not exist before.

Another critical influence was the possibility and necessity for measurement by modern bureaucratic institutions. At the time the notion of accident-proneness was born, it was actually possible and deemed necessary to record workplace injuries and accidents and do statistical analyses on them. In an economy increasingly run on a division between labor and capital, where productivity was key, it mattered to keep track of injuries and damaged goods or tools. The goods or tools, after all, likely did not belong to the worker (in the same way that they might still have done when that worker was at a farm or in a small workshop), and time lost through injury potentially hurt someone else's bottom line, not just the worker's. General efforts by managers to reduce accidents were not only (if at all) concerned with preventing harm and suffering, but also driven by economic motives. Their intuitive observations that some employees were more likely to get into trouble gradually made room for the systematic measurement and record-keeping that bureaucracies inside and outside corporations do to this day.

Measurement, particularly on large scales, was a by-product of the scientific and industrial revolutions. Roughly, the former made it possible; the latter made it necessary. The ability to measure and make sense of data resulted from mathematical and statistical advances, some of which were quite recent at the time. German natural scientists in particular contributed in the later years of the Enlightenment.

LEARNING TO MEASURE: SCIENCE AND ENLIGHTENMENT

Alexander von Humboldt (1769–1859) was a natural scientist and adventurer who established geography as an empirical science, for example. This involved a lot of discovery but, more importantly, a lot of measurement (which he often did himself, on many journeys to Europe, South America, and Russia where nothing quantifiable escaped measurement by him). His field research extended into disciplines such as physics, chemistry, geology, climatology, astronomy, and botany. He is said to be one of the first who really set out to 'measure the world' (Kehlmann, 2005).

Humboldt's slightly later English counterpart was Francis Galton (1822–1911), a first cousin of Charles Darwin. Able to read at the mere age of two, he went to medical school in Oxford at 16. Like Humboldt, Galton had a penchant for measuring everything. He even measured the behinds of women he encountered in his travels in Africa. Abiding by the Victorian norms of the time, he of course had to do this from a distance, by means of triangulation. His interest in measurement led to his invention of the weather map, which included highs, lows, and fronts—all terms he introduced. He also suggested the use of fingerprints to Scotland Yard. His obsession with measurement eventually led to his efforts at measuring intelligence. In 1869, he published *Hereditary Genius: An Inquiry into its Laws and Consequences*, in which he demonstrated that the children of geniuses tend to be geniuses themselves.

In France, Alfred Binet (1851–1911) was born in Nice. After a start in medicine, he moved into psychology because it interested him more. In 1891, he joined a physiological-psychology laboratory at the Sorbonne, where he developed research into individual differences. In 1899, he and his graduate student Theodore Simon (1873–1961) were commissioned by the French government to study retardation in the French schools, and to create a test that separated normal from retarded children. After marriage, he began studying his own two daughters and testing them with Piaget-like tasks and other tests. It led to the publication of *The Experimental Study of Intelligence* in 1903. In 1905, Binet and Simon came out with the *Binet-Simon Scale of Intelligence*, the first test that allowed the graduated, direct testing of intelligence. They expanded the test to normal children in 1908 and to adults in 1911.

Another towering German figure was Carl Friedrich Gauss (1777–1855), who, at the age of 18, figured out the basis for modern mathematical statistics (think of the 'normal' or Gaussian distribution, for example) and regression analysis. The English Francis Galton and Karl Pearson (1857–1936) would further develop and invent other formulations of correlation.

Without these capabilities for measurement and analysis, the statistical basis for psychological notions ranging from IQ to accident-proneness would not have existed at the time.

3.3.2 INDIVIDUAL DIFFERENCES AND EUGENICS

The ability and desire to measure and do statistical operations on the data represented only one condition of possibility. The first half of the 20th century gave rise to a more sinister social one, but one that was necessary too for the idea of accident-proneness to be commonsensical. Proponents of accident-proneness might certainly have believed that they were helping industry alleviate a terrible social problem. This problem, however, was one posed by *certain* people, who in many cases were thought to have a hereditary basis for their accident-proneness:

> In 1920, a member of the State Safety Board in the state of Washington who had experience in investigating industrial accidents had concluded that there were individual differences in the realm of accidents. He observed "many unfortunate human beings engaged in very hazardous work who, through intemperance, hereditary mental disturbances, domestic troubles and generally speaking 'all around hard luck victims,' have grown disgruntled, careless and indifferent, thus jeopardizing their own lives and the lives of those surrounding them."
>
> **Burnham (2008, p. 113)**

These 'individual differences' made intuitive and psychological sense. But the notion both matched and helped promote the idea that some people were fit for the modern world, and some were not. Slocombe, a U.S. industrial psychologist, explicitly linked accident-proneness to social desirability and conformity. Accident-prone people were those who did not conform to social standards, he believed. Only a few years later, these same people would be labeled 'social deviants.' These deviants were beyond the reach of any safety campaign. Such campaigns, after all, targeted worker behavior and worker motivation to do the right thing. This, Slocombe argued, would never reach the deviants, because they did not possess the ability to conform. Their willingness or motivation to do so was beside the point. Writing in *Personnel Journal*, he proposed that these people required special, individual treatment instead (Slocombe, 1941).

By 1941, such reference to 'special, individual treatment' had sinister connotations indeed. To understand how the early 20th century West could gradually drift from the measurement to the identification and separation of those deemed unfit for its modern industrial age, let's go back to the intelligence test developed by Binet and Simon. The test, and what it stood for, soon became deployed in ways that its originators had warned against. Binet cautioned that intelligence tests should be used with restraint (Gottfredson, 2007). He believed that intelligence was very complex, influenced by many factors that were not inherent in the person, or not under their control. Intelligence should never be collapsed into one entity, he said, and only part of it could be explained on the basis of hereditary factors. For sure, Binet said, genetics might impose some upper limits on intelligence. But there was usually plenty of room for improvement and development with the right kind of education and exposure. Binet had observed that even children who were years behind their age level could catch up and prove smarter than most later on in life. He feared that any test and their numeric output could bias the environment (particularly teachers) into adapting the approach and prematurely giving up on certain children or people, making the judgment of low achievement self-fulfilling.

3.3.3 IDIOTS, IMBECILES, AND MORONS

Many ignored Binet's warnings. In 1911 in Germany, William Stern developed the intelligence quotient (IQ), which would later be used by U.S. psychologists for the development of IQ tests. (Stern himself had to flee Nazi persecution and ended up teaching at Duke in the United States until his death in 1938.) Charles Spearman (1863–1945), an English psychologist who was a pioneer of factor analysis and the creator of the Spearman correlation coefficient, believed that general intelligence was real, unitary, and inherited—quite to the contrary of Binet. In the United States, Henry Goddard (1866–1957) went much further. A prominent psychologist and eugenicist, Goddard translated the Binet and Simon test into English in 1908 and proposed a system for classifying people with mental retardation. He introduced terms that remained common in the field for many years afterward:

- Idiot — IQ between 0 and 25
- Imbecile — IQ between 26 and 50
- Moron — IQ between 51 and 70

Even the relatively gifted morons, Goddard decided, were unfit for society and should be removed either through institutionalization, sterilization, or both. One of his famous studies into 'feeble-mindedness' concerned a New Jersey family, the Kallikaks, which had descended from a single Revolutionary War soldier (Goddard, 1914). A dalliance with a feeble-minded barmaid on the way back from battle, Goddard found out, had led to an entire family branch of poor, insane, delinquent, and mentally retarded offspring. Goddard's 1912 book about it was a tremendous success and went through multiple print runs. In 1913, Goddard established an intelligence testing program on Ellis Island, with the purpose of screening immigrants for feeble-mindedness. Goddard's tests were applied only to those who were traveling steerage, and not those who came by first or second class. His testers found up to 40% or 50% of immigrants to be feeble-minded. They were often immediately deported. Some countries or groups were thought to bring more feeble-minded applicants to U.S. shores, including Jews, Hungarians, Italians, and Russians (Goddard, 1914). Of course, the tests were in English. Those traveling steerage had just endured a grueling journey, followed by the miserable bureaucracy of processing at Ellis Island.

Goddard, together with U.S. scientists Francis Galton and Lewis Terman, thought there was a close connection between feeble-mindedness and criminality. Intervening in the presence of feeble-minded people in society, or their ability to procreate, was a logical next step. Galton coined the term *eugenics*, the policy of intentionally breeding human beings according to some standard(s). States were recommended to institute programs of sterilization to manage the problem. Many did. In 1927, the Supreme Court weighed in, deciding that compulsory sterilization of the unfit was consistent with the U.S. Constitution. Writing for the majority, Oliver Wendell Holmes endorsed eugenics:

> It is better for all the world, if instead of waiting to execute degenerate offspring for crime, or to let them starve for their imbecility, society can prevent those who are manifestly unfit from continuing their kind. The principle that sustains compulsory vaccination is broad enough to cover cutting the Fallopian tubes. Three generations of imbeciles are enough.

Supreme Court Buck *v.* Bell, 274 US 200 (1927)

Thirty-three states adopted statutes for compulsory sterilization, and some 65,000 American citizens were affected. Such was the mindset in the first decades of the 20th century, which helped shape the conditions for the possibility of 'accident-proneness.' Some people were deemed 'fit' or deserving to be in society (or in a workplace). Others were not. Whether persons were 'fit' was increasingly measurable through a variety of tests that determined their suitability, and through statistical data that determined their history of performance in a workplace or elsewhere. The characteristics that determined whether someone was 'fit' or not, were generally believed to be hereditary and fixed (though recall that Binet, among others, had voiced a different opinion). Interestingly, to this day, the U.S. Supreme Court has not yet reversed its 1927 decision.

3.4 ACCIDENT-PRONENESS TODAY

3.4.1 The Growth of Dissent

Toward the middle of the 20th century dissent grew. Scientists started to question the measurements and statistics that had been the trigger and the foundation of the accident-proneness thesis. Recall the percentages, such as those used by the Boston public transport system: 27% of its drivers were causing 55% of all accidents. There was also the 1939 study by Cobb who pointed out that 4% of his automobile driver sample were responsible for 36% of the accidents (McKenna, 1983). These figures now came under fire. For these figures to have any validity, after, everyone needed to expect the same number of accidents. Otherwise deviations from that norm could never be found. This was an untenable assumption, as the mix of processes and factors (including chance) that went into the causation of accidents was too complex. In other words, different drivers on different shifts in different sorts of traffic or street architectures could never be exposed to the same accident potential: the environment alone made that their accident exposure varied wildly. Thus, concluding on the basis of statistics that one person is more accident prone than the other is untenable.

This critique had been foreseen in 1920, when Greenwood and his colleagues in the United Kingdom concluded that accident-proneness would not work for Air Force personnel, because they were subject to 'unequal exposure.' Data on individuals would, therefore, not be comparable. They believed that the thesis, however, might hold up for workers doing the same work every day. But even there, others found that the "complexities of the uncontrolled world surrounding operational accidents" ruled out that we could at all predict future accidents on the basis of having had past ones (Webb, 1956, p. 144). It was concluded that:

> the evidence so far available does not enable one to make categorical statements in regard to accident-proneness, either one way or the other; and as long as we choose to deceive ourselves that they do, just so long will we stagnate in our abysmal ignorance of the real factors involved in the personal liability to accidents.

Arbous and Kerrich (1951, p. 373)

Some findings were outright contradictory. The *Journal of Aviation Medicine* reported a study in the mid-1950s, for example, that showed that pilots who had a greater number of aircraft accidents than expected tended to be younger, better adjusted to their jobs and status, and more skillful than pilots who have fewer accidents (Webb, 1956). Was it something about the situations they got themselves into, or the airplanes or missions they were asked to fly precisely because they were the better-adjusted and more skillful pilots? The link between accident-proneness and fixed characteristics of someone's personality no longer seemed to hold. Personal proneness as an explanatory factor for accident causation paled in comparison to other contributors. A worker's proneness to having an accident proved to be much more a function of the tools and tasks he or she was given, then the result of any personal characteristics. In 1975, road safety researcher Colin Cameron wrote that:

> For the past 20 years or so, reviewers have concluded, without exception, that individual susceptibility to accidents varies to some degree, but that attempts to reduce accident frequency by eliminating from risk those who have a high susceptibility are unlikely to be effective. The reduction which can be achieved in this way represents only a small fraction of the total. Attempts to design safer man-machine systems are likely to be of considerably more value.

Cameron (1975, p. 49)

As you will learn in Chapter 5, WWII cemented the idea of intervening in the system surrounding people, rather than intervening in the people themselves. Technological development in the years of the war would be so rapid, pervasive, and complex that no amount of intervention toward the human worker alone could solve its emerging safety problems. Instead of trying to change the human so that accidents became less likely, engineers and others realized that they could, and should, change the technologies and tasks so as to make error and accident less likely. In addition, it became increasingly obvious that incidents and accidents were the result of complex interactions between people, organizations, systems, and machines (see Chapter 12) and that trying to trace their cause to one factor was an oversimplification. Roger Green, one of the early aviation psychologists, commented:

> Closer examination of individual accidents reveals why this is so, for it is very rare for an accident to be caused entirely because of deficiencies or other discrete human factors problems; culpability almost invariably lies with an interaction between these and other factors. The possible number of such interactions are virtually limitless; consequently, the determination of the most likely aetiology of a given accident is likely to remain a non-precise 'science'.

> *(1977, p. 923)*

Humans would no longer be seen as just the cause of trouble—they were the recipients of trouble: trouble that could be engineered and organized away. Organizational resources spent on measuring and tabulating statistics to weed out accident-prone individuals were increasingly seen as waste, hampering safety research and getting in the way of more effective considerations. Safety conferences and professional meetings in the decades after the war started to call attention more to managerial responsibilities and engineering solutions for safe work, leaving the focus on individual workers to human resource professionals or supervisors. The moral imperative shone through clearly as well. As one review said, this focus on the individual worker as cause of injury and incident

> …may allow managements to escape their responsibilities in machine design, in selection and training of personnel in safe operating procedures, and in meticulous attention to the environment where high energy transfer can impinge on the human.

> **Connolly (1981, p. 474)**

3.4.2   RECENT STUDIES OF ACCIDENT-PRONENESS

The original view of accident-proneness, born from statistical data, was that the distribution of accidents among a population of workers is not random. This suggested that:

- There were people with certain attributes that made it more likely for them to be involved in accidents than others. If these attributes could be identified and measured, it should be possible to reduce the overall accident rate by keeping accident-prone people out of the population.

- A small number of 'accident repeaters' were involved in a large number of accidents. The identification and elimination of this 'hard core' would see the accident problem greatly reduced.

Subsequent statistical data of interventions based on this in the 1960s (e.g., by revoking driving licenses from accident repeaters in the United States) was not so encouraging: most accidents turned out to be the first ones for the people involved (Cameron, 1975). Nonetheless, the idea "that some people are more liable to accidents than others due to innate personal characteristics" or that "it would be most surprising if individuals did not vary innately in their tendency to accident when exposed to the same risk" has been resilient, if controversial (McKenna, 1983, p. 68).

A review and meta-analysis from 2007 provides an overview of more recent research on accident-proneness (Visser, Pijl, Stolk, Neeleman, & Rosmalen, 2007). Seventy-nine studies with empirical data on accident rates were identified in this review. As could be expected:

- Definitions of accidents varied highly, but most studies focused on accidents resulting in injuries requiring medical attention;
- Operationalizations of accident-proneness varied greatly. Studies categorized individuals into groups with ascending accident rates or made non-accident, accident, and repetitive accident groups;
- The studies included in the meta-analysis examined accidents in specific contexts (traffic, work, sports) or populations (children, students, workers, patients).

The prevalence of accident-proneness was hard to establish because of the large variety in operationalizations (Visser et al., 2007). Nonetheless, and interestingly, a meta-analysis of the distribution of accidents in the general population showed that the observed number of individuals with repeated accidents was higher than the number expected by chance. Accident-proneness exists, the authors concluded, but its study is severely hampered by the variation in operationalizations of the concept.

The consideration of IQ as a predictor of accident-proneness has also made a modest comeback through cognitive psychology and its links to human factors and ergonomics. This line of thinking reminds us that human performance ultimately comes down to the assessments and actions of individual people (Campbell & Bagshaw, 1991). The cognitive abilities that these people bring to the task matter. In an increasingly complex, dynamic, technologically sophisticated and hyperspecialized world, such cognitive abilities are going to matter ever more. To the extent that IQ represents someone's ability to recognize, plan and solve problems, and think abstractly, it should be a strong predictor of overall job performance. The lack of intelligence, accordingly, would make workers more accident prone. As Gottfredson concluded:

Avoiding accidental death … requires the same information processing skills as do complex jobs: continually monitoring large arrays of information, discerning patterns and anomalies, understanding causal relations, assessing probabilities, and forecasting future events.

(2007, p. 407)

The argument is made by stretch and analogy (complex work requires the same skills as avoiding accidental death: thus, lack of those skills makes someone more accident prone in the execution of such work). The ultimate appeal made in this literature is that of evolution. Accidental injury has been an important part of natural selection, researchers say, and they believe that it plays a similar role today, a belief also advertised by the infamous Darwin Awards which 'commemorate those who improve our gene pool by eliminating themselves from the human race' (The Darwin Awards, n.d.). Fields such as cognitive epidemiology, which use statistical methods to seek associations between various datasets, but do not demonstrate causal links, provide the foundation for such beliefs. Intelligence is invariably seen as unitary and hereditary (despite Binet's cautions on the matter) and higher intelligence is associated with longevity (Deary, 2008; Gottfredson, 2007).

3.4.3 Accident-Proneness Versus Systems Thinking

Another appeal of today's accident-proneness research is, if anything, a moral one. There is, for instance, "increasing disquiet at how the importance of individual conduct, performance and responsibility was written out of the ... safety story" (Shojania & Dixon-Woods, 2013, p. 528). As you will learn in Chapter 5 and onward, the focus of much of Safety Science since WWII has been on the systems in which people work, and not so much the people as individuals. This has some concerned, particularly with respect to 'accountability' (which remains a rather elusive term (Dekker, 2012; Sharpe, 2004)). Conceptually aligned with the original 1920s work on accident-proneness, the *British Medical Journal* in 2013 reported how a

> study of formal patient complaints filed with health service ombudsmen in Australia found that a small number of doctors account for a very large number of complaints from patients: 3% of doctors generated 49% of complaints, and 1% of doctors accounted for 25% of all complaints. Moreover, clinician characteristics and past complaints predicted future complaints, with the authors' model identifying those doctors relatively unlikely to generate future complaints within 2 years as well as those highly likely to attract further complaints. These findings are consistent with other recent research, including work showing that some doctors are repeat offenders in surgical never-events.

> **Shojania and Dixon-Woods (2013, p. 528)**

Explanations of failure based on individual accident-proneness remain seductive. But as with the original statistical analyses of accident-proneness, no mention is made of the problem of 'unequal exposure.' What if the surgeons who are more likely to be 'repeat offenders in surgical never-events' are precisely those who take on the riskiest cases which their colleagues do not dare touch? Also, patient complaints are generated for a variety of reasons, many of which are not related to clinicians' competence, intelligence, or even the clinical outcome, but rather linked with their perceived interest in the patient and attentiveness to their problem, or the time they could spend with the patient. It could well be that those who are likely to attract further complaints are those working in a highly pressurized, underfunded system, with a clinically demanding population; focusing on the individuals who work there is not likely to resolve any of that. Thus, skepticism of 'accident-proneness' remains strong. The difficulties in operationalizing accident-proneness, the statistical flaws in demonstrating it, and in linking it to fixed, stable individual characteristics, are perhaps only a footnote to the ethical implications of the idea. Its use has continued to generate calls for caution. For example:

> All the evidence we have so far tends to the conclusion that proper design of the man-machine system, with proper attention to the sources of human error, and proper protective devices for those who are unavoidably exposed to some risk, is more effective and gives better value for money.
>
> **Cameron (1975, p. 52)**

The safety literature has long accepted that systemic interventions have better or larger or more sustainable safety effects than the excision of individuals from a particular practice. For example:

> The addition of anti-hypoxic devices to anesthetic machines and the widespread adoption of pulse oximetry have been much more effective in reducing accidents in relation to the administration of adequate concentrations of oxygen to anesthetized patients than has the conviction for manslaughter of an anesthetist who omitted to give oxygen to a child in 1982.
>
> **Merry and Peck (1995, p. 84)**

3.5   EXPERTISE AND ACCIDENT-PRONENESS

3.5.1   ARE EXPERTS MORE ACCIDENT PRONE?

Many people who are active in safety can probably accept the conclusion that intervening in the general design of a tool, system, or organization is the best way to mitigate accident risk. After all, it covers an entire population of workers. It is a more cost-effective way than chasing after putatively accident-prone people and weeding them out. But is the potential of having an injury, incident, or accident not at all related to someone's personal characteristics? What about the risk perception and risk appetite of a certain surgeon, for instance? And something as simple as age, for example, is known to play a role. It is commonly reflected in insurance premiums for under-25 drivers. The latter contains an important difference with original accident-prone research, however. The characteristics identified are population characteristics, not individual ones. And interventions target the entire population, not some individual. But age, even as a population characteristic, is complicated. After all, it is linked to both experience and health. At a lower age, people typically have less experience, but also less risk of health issues contributing to accidents. The higher the age, the more this association is reversed: experience tends to mitigate risk, whereas health problems increase risk (Bazargan & Guzhva, 2011).

That said, the relationship between experience and accident-proneness itself is contested. Experience is an important ingredient for expertise: the development and possession of skills and knowledge in a particular field. The Webb study of 1956 was among the first to systematically point to a link between expertise and accident-proneness. Even though the more accident-prone pilots tended to be younger, they were also more skillful than the ones who did not have accidents. This might suggest that at least some accidents happen to those who possess more expertise. As Webb pointed out, this could well be because supervisors and others expected more of these skillful pilots, and exposed them to situations of greater risk.

A recent study of mortality in patients hospitalized with acute cardiovascular problems seemed to suggest as much too. The study showed that during dates of national cardiology meetings, which are typically attended by the most senior cardiologists who then leave their patients in the hands of less experienced residents, mortality is actually lower. So, if the experts are gone, fewer patients die. This could be related to treatment rates, of course: when senior cardiologists are not there, patients could be less likely to be subjected to risky treatment of their conditions. The authors of the study found that to be true only for a limited number of treatments: other treatment rates were similar, whether senior doctors were off to a national meeting or not (Jena, Prasad, Goldman, & Romley, 2015).

Another hypothesis is that accidents happen to experts precisely *because* they are more expert. One counterintuitive but intriguing Canadian study showed that injury professionals are 70% more likely to be injured than the overall population. One hypothesis put forward was that of 'risk compensation.' Feeling that their knowledge and precautions will protect them, injury prevention professionals may actually take greater risks (Ezzat, Brussoni, Schneeberg, & Jones, 2013). It could also be that these professionals are more physically active than the rest, increasing their exposure to risk (not unlike Webb's young but very skillful pilots, or the senior cardiologists).

Even more nuanced and less counterintuitive explanations for why experts get it wrong are possible, particularly in the diagnosis of risky problems (Johnson, Grazioli, Jamal, & Zualkernan, 1992; Montgomery, 2006). The experts' strength, according to this line of thinking, is also their weakness. Take, for example, the pattern recognition that imposes ready-formed templates onto complex situations, which can cause garden-pathing toward a plausible but wrong solution (De Keyser & Woods, 1990).

Contemporary safety thinking suggests that expert practitioners have learned to cope with the complexity of multiple simultaneous goals and imperfect information. They know how to adapt and create safety in a world fraught with hazards, limited data, trade-offs, and goal conflicts (Woods, Dekker, Cook, Johannesen, & Sarter, 2010). They also know where and how routines, written guidance, or rules apply and where they do not:

> The expert is someone who has done the job for a considerable period of time, knows the rules and procedures and the range of routines required very well, but also knows how and when to sidestep these and to intervene and apply strategies of their own.

Farrington-Darby and Wilson (2006, p. 18)

Experts typically see more nuances in a situation than novices do, and also recognize patterns more quickly. This makes experts more sensitive to the weaker cues that can tell them what to do, and also allows them to experience fewer degrees of freedom in a decision situation. They have likely been in similar situations before and are able to swiftly separate signal from noise, and know what to do (Jagacinski & Flach, 2003; Orasanu & Connolly, 1993). There is a price to pay for this success, however. This is a notion at the heart of psychological links between experts and accident-proneness. As Ernst Mach (after whom the 'Mach number' is named) famously put it in 1905:

> Knowledge and error flow from the same mental sources, only success can tell one from the other.
>
> *(1976, p. 29)*

The idea is that experts are more likely to form routines. In fact, they build a large stock of routines that represent both knowledge and rules at a more abstract level than novices have access to. In certain situations, these routines can be "strong but wrong" (Reason, 1990, p. 57). In complex settings, with large problem spaces, it is possible that the more skilled and experienced practitioners carry out work in ways that is mostly successful but occasionally flawed. In other words, experts sometimes fail precisely because they mostly succeed.

Vaughan showed how expertise can get absorbed by cultures of production; how subtle normalization of signs of deviance in technical and operational details affects experts' perception of risk as much as everybody else's (Vaughan, 1996). Other researchers have also shown how the supposed technical neutrality that comes with expertise is often powerless in the face of larger organizational goals and production pressures (Weingart, 1991; Wynne, 1988). Before the fact, the accidents are largely inconceivable to the experts, to the engineering and technological communities. As a result, Wagenaar and Groeneweg termed these accidents 'impossible' (1987): they fell outside the scope of experts' technical language and prediction (see also Chapter 9).

3.5.2 Expertise and Organizational Vulnerability to Accidents

Recent literature on accidents and disasters, however, prefers to speak of *expertise* rather than experts (Weick & Sutcliffe, 2007). This takes into account the socially interactive nature of what experts know and do, or rather what others *believe* them to know or expect them to do. Expertise, according to this kind of thinking, is not a personal possession. It is relational; it is something that gets constructed between people. This changes the way we think about expertise and accident-proneness. It is not the experts who are accident-prone because of personal risk perception and appetite. Rather, expertise makes an organization accident-prone because others sometimes falsely legitimatize, and count on, the knowledge and decisions that are localized with certain individuals or groups of people.

This is one explanation that lies behind the much larger proportion of accidents (over 80%) that happen when the captain is flying the airplane, as opposed to when the copilot is flying the airplane (flying duties are typically shared 50-50). As Barton and Sutcliffe (2009, p. 1339) found, "a key difference between incidents that ended badly and those that did not was the extent to which individuals voiced their concerns about the early warning signs." But who voices their concern to whom matters. Captains are more willing to stop a copilot from doing something unsafe than the other way around: a result of hierarchy and presumed expertise (Orasanu & Martin, 1998). It is also related to the limits of expertise (Dismukes, Berman, & Loukopoulos, 2007): the impossibility for an expert to know everything in complex and dynamic situations:

> First, when organizations create a culture of deference to expertise, low status individuals may become overly reliant on 'the experts' and abdicate their own responsibility for monitoring and contributing to the safety of the situation. This becomes particularly dangerous if the perceived expert is not, in fact, terribly knowledgeable. The second danger of deference to expertise arises when individuals and groups mistake general expertise for situational knowledge. That is, especially in very dynamic and complex situations, it is unlikely that one person holds all the necessary knowledge to managing that situation, regardless of their years of experience or training. There are simply too many moving parts and changing realities. When individuals equate general expertise with situational knowledge they create and rely on unrealistic expectations of those 'experts'.

> **Barton and Sutcliffe (2009, p. 1341)**

What is more, the very nature of expertise implies specialization of knowledge. Specialization of knowledge, in turn, means that knowledge about a problem is distributed across people. In some cases, knowledge is distributed across many people:

> This involves both an epistemic problem and a social problem of organizing the decision-making involved... The 'social' problem is also epistemic: a problem of the aggregation of knowledge. The knowledge relevant ... was distributed among many people, with specialist knowledge and perhaps tacit knowledge that was not possessed by other members of the team.

> **Turner (2010, p. 247)**

Sometimes there is very little that organizations can do about this. Their formal mechanisms of intelligence gathering, safety regulation and auditing will always somehow, somewhere fall short in foreseeing and meeting the shifting or hitherto hidden risks created by a world of limited resources, uncertainty and multiple conflicting goals. That said, often the problem is not that an industry actually lacks the data. After all, the electronic footprint left by many of today's operations is huge. And pockets of expertise that may have predicted what could go wrong often existed in some corner of the industry long before any accident. The problem is an accumulation of noise as well as signals, which can muddle both the perception and conception of "risk." This means that experts may not have privileged insight into the hazards ahead. The complexity of large-scale organizations can render risk invisible to experts too. Jensen (1996) described it as such:

We should not expect the experts to intervene, nor should we believe that they always know what they are doing. Often, they have no idea, having been blinded to the situation in which they are involved. These days, it is not unusual for engineers and scientists working within systems to be so specialized that they have long given up trying to understand the system as a whole, with all its technical, political, financial and social aspects.

(p. 368)

In the next chapter, we will discuss behavior-based safety, which begins a long march away from focus on individual workers toward a focus on the conditions and systems that guide work, including the increasing complexity of their work environments. The first stage of this march is a small shift, moving away from treating behavior not as an innate tendency arising from personality and moral defects, but as a choice that can be influenced.

STUDY QUESTIONS

1. What were some of the social conditions of possibility that allowed the idea of accident-proneness to flourish in the 1920s and 1930s? And today, what are some of the conditions of possibility that give current credence to the idea?
2. What is one of the fatal statistical flaws in the accident-proneness thesis that basically makes proof of its existence impossible?
3. Why might expert practitioners in a safety-critical domain actually be more accident-prone than novices?
4. Why are there different relationships between (1) experts and accident-proneness and (2) expertise and accident-proneness?
5. Can you recognize ideas about accident-proneness in how people in your own organization talk about errors and failures?

REFERENCES AND FURTHER READING

Arbous, A. G., & Kerrich, J. E. (1951). Accident statistics and the concept of accident-proneness. *Biometrics, 7,* 340–432.

Barton, M. A., & Sutcliffe, K. M. (2009). Overcoming dysfunctional momentum: Organizational safety as a social achievement. *Human Relations, 62*(9), 1327–1356.

Bazargan, M., & Guzhva, V. S. (2011). Impact of gender, age and experience of pilots on general aviation accidents. *Accident Analysis and Prevention, 43*(3), 962–970.

Burnham, J. C. (2008). Accident proneness (unfallneigung): A classic case of simultaneous discovery/construction in psychology. *Science in context, 21*(1), 99–118.

Burnham, J. C. (2009). *Accident prone: A history of technology, psychology and misfits of the machine age.* Chicago, IL: University of Chicago Press.

Cameron, C. (1975). Accident proneness. *Accident Analysis and Prevention, 7,* 49–53.

Campbell, R. D., & Bagshaw, M. (1991). *Human performance and limitations in aviation.* Oxford, UK: Blackwell Science.

Connolly, J. (1981). Accident proneness. *British Journal of Hospital Medicine, 26,* 474.

De Keyser, V., & Woods, D. D. (1990). Fixation errors: Failures to revise situation assessment in dynamic and risky systems. In A. G. Colombo & A. Saiz de Bustamante

(Eds.), *System reliability assessment* (pp. 231–251). Amsterdam, The Netherlands: Kluwer Academic.

Deary, I. J. (2008). Why do intelligent people live longer. *Nature, 456*, 175–176.

Dekker, S. W. A. (2012). *Just culture: Balancing safety and accountability* (2nd ed.). Farnham, UK: Ashgate Publishing Co.

Dismukes, K., Berman, B. A., & Loukopoulos, L. D. (2007). The limits of expertise: Rethinking pilot error and the causes of airline accidents. Aldershot, Hampshire, UK; Burlington, VT: Ashgate Publishing Co.

Ezzat, A., Brussoni, M., Schneeberg, A., & Jones, S. J. (2013, September). 'Do as we say, not as we do': A cross-sectional survey of injuries in injury prevention professionals. *Injury Prevention, 20*, 172–176. doi:10.1136/injuryprev-2013-040913

Farmer, E. (1945). Accident-proneness on the road. *Practitioner, 154*, 221–226.

Farrington-Darby, T., & Wilson, J. R. (2006). The nature of expertise: A review. *Applied Ergonomics, 37*, 17–32.

Goddard, H. H. (1914). *Feeble-mindedness: Its causes and consequences.* New York, NY: Macmillan.

Gottfredson, L. S. (2007). Innovation, fatal accidents and the evolution of general intelligence. In M. J. Roberts (Ed.), *Integrating the mind: Domain-general versus domain-specific processes in higher cognition* (pp. 387–425). Hove, UK: Psychology Press.

Green, J. (2003). The ultimate challenge for risk technologies: Controlling the accidental. In J. Summerton & B. Berner (Eds.), *Constructing risk and safety in technological practice* (pp. 29–42). London, UK: Routledge.

Green, R. G. (1977). The psychologist and flying accidents. *Aviation, Space and Environmental Medicine, 48*, 923.

Jagacinski, R. J., & Flach, J. M. (2003). *Control theory for humans: Quantitative approaches to modeling performance.* Mahwah, NJ: Lawrence Erlbaum Associates.

Jena, A. B., Prasad, V., Goldman, D. P., & Romley, J. (2015). Mortality and treatment patterns among patients hospitalized with acute cardiovascular conditions during dates of national cardiology meetings. *Journal of the American Medical Association Internal Medicine, 175*(2), 237–244.

Jensen, C. (1996). *No downlink: A dramatic narrative about the challenger accident and our time* (1st ed.). New York, NY: Farrar, Straus, Giroux.

Johnson, P. E., Grazioli, S., Jamal, K., & Zualkernan, I. A. (1992). Success and failure in expert reasoning. *Organizational Behavior and Human Decision Processes, 53*(2), 173–204.

Kehlmann, D. (2005). *Die Vermessung der welt (Measuring the world).* Reinbek bei Hamburg: Rowohlt Verlag GmbH.

Mach, E. (1976). *Knowledge and error* (English ed.). Dordrecht, The Netherlands: Reidel.

Marbe, K. (1926). *Praktische Psychologie der Unfälle und Betriebsschaden.* Munchen, Germany: R. Oldenbourg.

McKenna, F. P. (1983). Accident proneness: A conceptual analysis. *Accident Analysis and Prevention, 15*(1), 65–71.

Merry, A. F., & Peck, D. J. (1995). Anaesthetists, errors in drug administration and the law. *New Zealand Medical Journal, 108*, 185–187.

Miles, G. H. (1925). Economy and safety in transport. *Journal of the National Institute of Industrial Psychology, 2*, 192–193.

Montgomery, K. (2006). *How doctors think: Clinical judgment and the practice of medicine.* Oxford, UK: Oxford University Press.

Orasanu, J. M., & Connolly, T. (1993). The reinvention of decision making. In G. A. Klein, J. M. Orasanu, R. Calderwood, & C. E. Zsambok (Eds.), *Decision making in action: Models and methods* (pp. 3–20). Norwood, NJ: Ablex.

Orasanu, J. M., & Martin, L. (1998). *Errors in aviation decision making: A factor in accidents and incidents.* Human Error, Safety and Systems Development Workshop (HESSD) 1998. Retrieved from http://www.dcs.gla.ac.uk/~johnson/papers/seattle_hessd/judithlynnep

Reason, J. T. (1990). *Human error.* New York, NY: Cambridge University Press.

Sharpe, V. A. (2004). *Accountability: Patient safety and policy reform.* Washington, DC: Georgetown University Press.

Shojania, K. G., & Dixon-Woods, M. (2013). 'Bad apples': Time to redefine as a type of systems problem? *BMJ Quality and Safety, 22*(7), 528–531.

Slocombe, C. S. (1941). The psychology of safety. *Personnel Journal, 20,* 42–50.

The Darwin Awards. (n.d.). History & rules. Retrieved from http://www.darwinawards.com/rules/

Turner, S. (2010). Normal accidents of expertise. *Minerva, 48,* 239–258.

Vaughan, D. (1996). *The challenger launch decision: Risky technology, culture, and deviance at NASA.* Chicago, IL: University of Chicago Press.

Visser, E., Pijl, Y. J., Stolk, R. P., Neeleman, J., & Rosmalen, J. G. M. (2007). Accident proneness, does it exist? A review and meta-analysis. *Accident Analysis and Prevention, 39,* 556–564.

Wagenaar, W. A., & Groeneweg, J. (1987). Accidents at sea: Multiple causes and impossible consequences. *International Journal of Man-Machine Studies, 27*(5–6), 587–598.

Webb, W. B. (1956). The prediction of aircraft accidents from pilot-centered measures. *Journal of Aviation Medicine, 27,* 141–147.

Weick, K. E., & Sutcliffe, K. M. (2007). *Managing the unexpected: Resilient performance in an age of uncertainty* (2nd ed.). San Francisco, CA: Jossey-Bass.

Weingart, P. (1991). Large technical systems, real life experiments, and the legitimation trap of technology assessment: The contribution of science and technology to constituting risk perception. In T. R. LaPorte (Ed.), *Social responses to large technical systems: Control or anticipation* (pp. 8–9). Amsterdam, The Netherlands: Kluwer Academic.

Woods, D. D., Dekker, S. W. A., Cook, R. I., Johannesen, L. J., & Sarter, N. B. (2010). *Behind human error.* Aldershot, UK: Ashgate Publishing Co.

Wynne, B. (1988). Unruly technology: Practical rules, impractical discourses and public understanding. *Social Studies of Science, 18*(1), 147–167.

4 The 1930s and Onward
Heinrich and Behavior-Based Safety

KEY POINTS

- Heinrich used the analogy of a row of dominoes to explain how distant causes lead to injuries. Ancestry and the social environment give rise to character flaws in a worker, such as bad temper, ignorance, or carelessness. Character flaws give rise to unsafe conditions, mechanical hazards, and unsafe acts. These factors in turn lead to accidents, which lead to injuries and fatalities.
- Like a row of falling dominoes, the sequence could be interrupted by removing a factor from the sequence. Heinrich reestablished the idea that many accidents and injuries are preventable.
- Heinrich's opinion on the best point for intervention shifted throughout his career. Early on, he placed a strong emphasis on improving the physical conditions and physical safeguards of work. Later he placed increasing emphasis on eliminating unsafe acts by workers. He advocated creating an environment where even small undesirable acts are not tolerated.
- It was mostly through these later, human-error-focused ideas that Heinrich influenced the theory and practice of safety. Behavior-based safety (BBS) is one of the most visible expressions of it, with us to this day.
- Three key ideas of Heinrich's have influenced safety practices (and even some theories) for decades: (1) Injuries are the result of linear, single causation; (2) there is a fixed ratio between accidents (or simply "occurrences"), minor injuries and major injuries; and (3) worker unsafe acts are responsible for 88% of industrial accidents. All three have been proven false.
- As the prominence of behaviorism and BBS wore off, the focus of safety thinkers shifted dramatically to the environment from the 1940s onward. Human errors were understood as locally logical acts, as systematically connected to features of people's tools, tasks, and working conditions. This in turn gave rise to the idea that problems *and* solutions needed to be sought in the human-machine system: the relationship, or ensemble of human and environment.

4.1 INTRODUCTION

Even if accidents were to occur because some individuals are fundamentally and irredeemably unsuited for safe work, that idea offers little to a practically minded safety practitioner. All that accident proneness theory suggests is that organizations should identify and weed out accident-prone individuals.

Herbert William Heinrich (1886–1962) believed in the scientific validity of accident proneness research, but he rejected its utility for safety practice. Whereas much of Heinrich's work is now considered outdated and controversial on a number of grounds, he reestablished the idea—somewhat lost during the age of accident proneness—that many accidents and injuries are preventable.

Heinrich, working in a corporation that insured industrial plants and factories against various risks, was probably a practical man. He needed to find things that could work, that the insurers' clients could use in their daily practice, and that could ultimately save his company money. These would have to be things that could be changed in practice so that risks were reduced. In his own words from 1931:

> Selection of remedies is based on practical cause-analysis that stops at the selection of the first proximate and most easily prevented cause.

> *(p. 128)*

As you will learn in more detail in this chapter, Heinrich used the metaphor of a row of dominoes to explain how distant causes lead to injuries. According to his model, ancestry and the social environment give rise to character flaws in a worker, such as bad temper, ignorance, or carelessness. Character flaws give rise to unsafe conditions, mechanical hazards, and unsafe acts. These factors in turn lead to accidents, which lead to injuries and fatalities. Like a row of falling dominoes, the sequence could be interrupted by removing the right factor in the sequence.

Heinrich suggested that focusing on innate traits of the workforce was futile. Whereas accident proneness might cause accidents, it could not be effectively managed. Better, Heinrich argued, to focus on causes closer to the accident. Heinrich's opinion on the best point for intervention shifted throughout his long career. His early writing placed a strong emphasis on improving the physical conditions of work, and he was influential in improving standards for mechanical safeguards. Later versions of Heinrich's work placed increasing emphasis on eliminating unsafe acts by workers. He advocated creating an environment where even small undesirable acts are not tolerated. It was mostly through these later, human-error-focused ideas, that Heinrich influenced the theory and practice of safety.

Three aspects of Heinrich's work have had significant impact on the application of Safety Science. These aspects show up explicitly in much safety work today. They also, though more subtly, pervade popular beliefs about the links between behavior and accidents. And they even reappear, sometimes under the very same labels, in more modern models of accident causation (see e.g., Chapter 9). Here they are:

1. Injuries are the result of linear, single causation. In the words of Heinrich (1931, p. 21): "The occurrence of an injury invariably results from a completed sequence of factors – the last one of these being the accident itself.

The accident in turn is invariably caused or permitted by the unsafe act of a person and/or a mechanical or physical hazard."

2. There is a fixed ratio between what Heinrich called accidents (or simply "occurrences"), minor injuries and major injuries. For all 300 occurrences, there are 29 minor injuries, and 1 major injury. Focusing on compliance – reducing occurrences – will therefore automatically reduce the number of major injuries.

3. Worker unsafe acts are responsible for 88% of industrial accidents. 2% are deemed to be unpreventable. A remaining 10% are the result of unsafe mechanical or physical conditions.

4.2 A 'SCIENTIFIC' EXAMINATION OF ACCIDENT CAUSATION

4.2.1 HEINRICH'S STUDY

In 1931, Heinrich was working as Assistant Superintendent of the Engineering and Inspection Division of the Travelers Insurance Company. Travelers Insurance had been founded in 1864 in Hartford, Connecticut, originally to provide travel insurance for railroad passengers. Travel by rail was obviously far more risky and dangerous than it is today. Known for innovation, Travelers went on to be the first insurer to provide an automobile policy, a commercial airline policy and, recently, a policy for space travel. It exists to this day, employing approximately 30,000 people and generating annual revenue of some $27 billion with assets of over $100 billion.

Even in Heinrich's time, Travelers' insurance umbrella already covered much more than just travel. Heinrich's idea was based on an analysis of industrial insurance claims he gathered for the company in the late 1920s. Even though Heinrich claimed his approach to be 'scientific,' a description of his method would not stand up to scientific peer-review scrutiny today (or back then). In 1931, Heinrich told his readers that:

Twelve thousand cases were taken at random from closed-claim-file insurance records. They covered a wide spread of territory and a great variety of industrial classifications. Sixty-three thousand other cases were taken from the records of plant owners.

(p. 44)

One could argue that this lack of methodological specification might not matter, as Heinrich was a corporate employee. And the point of his study was surely to help his insurance company save money in the long run. But that argument only goes so far. After all, the subtitle of his book was "A Scientific Approach." That would supposedly raise the expectation for him to divulge the basis of his selections, or the statistical power calculations behind his sample sizes. You might even think that his company would be interested to know, if they were going to base actuarial and policy decisions on his study. But he did not provide these things, at least not in his published writings. All we know is that as his source material, Heinrich used closed insurance claim files and records of industrial plant and factory owners and operators.

The raw data has been lost to history. It is not offered in Heinrich's book. There is no evidence of other analysts pouring over the same data and coming up with either similar or contrasting conclusions. This lack of raw data echoes through the subsequent editions. Even the coauthors of the 1980 edition of Heinrich's book never saw the files or records (Manuele, 2011). What we do know, however, is that these files and records did not provide for the insertion of causal data, and thus rarely contained them. In other words, there was no space in these claim forms to specifically note down the personal cause of incidents or injuries. In addition, the reports were completed by supervisors, not workers. They may have had certain incentives to blame the worker—not themselves or their bosses—for things going wrong in the workplace.

An interesting detail, seldom mentioned, is how Heinrich obtained knowledge of the occurrences that did not have any consequences. One can wonder how Heinrich learned about the occurrences that did not result in an injury. These occurrences did not lead to an insurance claim, as there would be nothing to claim, so they would not have shown up in his samples of reports and claim files. So how did he find this out? How did he determine that number? We do not know. He might have asked supervisors. He might have used his intuition, imagination, or experience. It took until the 1959 edition of his book, 3 years before his death, for this to be somewhat clarified by a reference to "over 5000 cases" (p. 31):

> The determination of this no-injury accident frequency followed a study of over 5,000 cases. The difficulties can be readily imagined. There were few existing data on minor injuries—to say nothing of no-injury accidents.

Indeed, "the difficulties can be readily imagined." How could he have had any confidence in the number or rate of incidents with no notable outcome (no damage, no injuries, no insurance claims)? In his much-replicated or uncritically copied triangular figure (e.g., Krause, 1997, p. 32), Heinrich uses the number 300 (see Figure 4.1). But he could not have known. It must have been a guess, or just a number inspired by the beauty of the approximate order of magnitude (300:29:1).

This creates another problem. Without knowing this base rate (or without us knowing how Heinrich could know it), it becomes very difficult to make a case for the "broken windows" idea of safety. This, of course, is something that bedevils Heinrich and the vast swath of behavioral safety interventions based on his ideas. The principle affects almost everyone in Safety Science. As long as we take the number of negative events as our measure of safety, we have the problem of not knowing the base rate: similar or identical events that did not produce a noteworthy or measurable negative outcome. The alternative that is being developed circumvents the central problem (e.g., Hollnagel, 2014) by understanding safety as the *presence* of something—not the absence. Earlier moves in that direction can be found as well, and are covered elsewhere in this book.

4.2.2 Bird and 'Damage Control'

Heinrich's work was replicated, extended, and further popularized by Frank E. Bird. Bird was safety manager for a steel plant before working as Director of Engineering

FIGURE 4.1 Heinrich's triangular figure was based on insurance claims of 1920s. Heinrich could not have known the number for 300 no-injury accidents, since these did not produce a claim. The triangle has since been proven wrong. (Drawing by author.)

Services for the Insurance Company of North America, and based his books on two major studies that sought to examine not just the ratio between incidents and accidents, but the overall rate of occurrence of incidents over a population of workers.

In his second and larger study he analyzed 1,753,498 accidents reported by 297 participating companies. They represented 21 different kinds of industries, employing a total of 1,750,000 people who worked over 3 billion hours during the period he studied. Bird also tried to be more secure in determining the base rate. He oversaw some 4,000 hours of confidential interviews by trained supervisors on the occurrence of incidents that—under slightly different circumstances—could have resulted in injury or property damage. What he found from these was that there were approximately 600 incidents for every reported major injury.

Bird used the metaphor of an iceberg both to report these results, and to convey his conclusion that focusing on property damage was just as important as focusing on actual injuries when it came to preventing major accidents. The Bird triangle, as some call it, was born. Bird's sample size was impressive, and the methodological trace left by him was more detailed than Heinrich's. Bird suggested that removing enough from the base of the triangle could ensure that nothing would rise to the level of severe incidents, injuries, or worse. By starting at the bottom, and slicing off something from the side of the triangle, all levels of injury and incident risk could get reduced.

Bird added two important aspects missing from Heinrich's domino train:

- The influence of management and managerial error;
- Unwanted outcomes in addition to injuries, such as the loss of production, property damage, or wastage of other assets.

With these additions and modifications, the domino model became known as the International Loss Control Institute (ILCI) model in 1985. Like Heinrich's domino model, the ILCI model was based on the idea of a linear, single sequence of events (Bird & Germain, 1985). But it represented an important update—indeed reflecting the *Zeitgeist* of the postwar period. No longer was it deemed so legitimate, or useful, to begin the consideration of an accident sequence with the inherited personality shortcomings of the worker. Instead, Bird put a "lack of control" at the start of the line of dominoes. It was the first domino in his sequence. His dominoes were:

- **Lack of control** (because of superficial or inadequate programs, organization or compliance standards);
- This allows **basic causes** to exist and grow, which include personal factors and job factors;
- These in turn enable **immediate causes** which themselves stem from substandard practices and substandard conditions;
- These trigger the **incident or accident**;
- Which can then result in a **loss**

What did lack of control mean? For Bird, this referred to a lack of standards. Or it pointed to problems with the measurement and evaluation of those standards. It meant that standards might not be followed by members of the organization. This in turn could be due to inadequacies in the continuous updating of standards to make sure they reflected evolving technology and working conditions. Bird considered *control* as one of the four essential functions of management. The other three were planning, organizing, and leading. What were the most common reasons for such a lack of control? For Bird, these were (Storbakken, 2002):

- An inadequate or superficial safety or loss prevention program. This happens when there are too few program activities to address the needs of the organization. Programs are needed to provide management and employee training, selection, and use of PPE (Personal Protective Equipment), engineering controls, planned inspections, task analysis, emergency preparedness, and incident investigations.
- Inadequate safety or loss prevention program organization and standards. This is needed to let people know what is expected of them, as well as to provide them with a tool to measure their performance against the standard.
- Inadequate compliance with those standards. Consistent with the behavioral thinking from which the model owes its existence, inadequate compliance was seen as the single greatest reason for loss. Poor compliance with effective program standards is due to ineffective communication of standards to employees or a failure to enforce standards.

If there is no adequate control, then basic causes related to the person or the job are allowed to germinate and grow (along with lack of control, such basic causes would be renamed 'latent factors, errors or conditions,' or 'resident pathogens' in the Swiss cheese model in the 1990s: see Chapter 9). Personal factors include lack of knowledge, skill, or an inability to handle pressures of the job. Job factors include inadequate training, inappropriate equipment and tools, worn equipment and tools, or inadequate equipment and tools. Such basic causes can enable immediate causes, which show up in a substandard condition or the performance of a substandard practice. These are of course renaming of Heinrich's unsafe conditions and unsafe acts, respectively, and contain the same essential things.

Bird's use of the term 'substandard,' however, acknowledged that organizations have responsibilities for ensuring that there are (or should be) standards of performance that are to be followed by all employees. The point for Bird was not to sanction individuals for committing substandard acts or for allowing substandard conditions to exist, but rather to encourage managers to evaluate their own programs and processes that influenced human behavior. Evidence of substandard practices and conditions should motivate managers to ask why these occurred or existed, and what failure in their own supervisory or management system permitted this in the first place (Storbakken, 2002). Bird offered the following examples:

Substandard practices
- Operating equipment without authority
- Improper loading
- Horseplay
- Being under the influence of alcohol and/or other drugs

Substandard conditions
- Inadequate guards or barriers
- Defective tools, equipment, or materials
- Poor housekeeping; disorderly workplace
- Inadequate ventilation

There is the potential for overlap between these two: where does improper loading become poor housekeeping, for instance? And why is a disorderly workplace a substandard condition, rather than a substandard practice? Perhaps the practice is its cause, and the condition its effect, but such subtleties were beyond the model. Like Heinrich, Bird did not split the unsafe acts from the unsafe conditions. They were part of the same domino, perhaps in part because of the difficulty of determining where one ended and the other began. This is a continuing problem introduced by models that have 'unsafe acts' as one of the final enablers of failure, or that see human error as primary contributor: where do those human errors begin and their mechanical or organizational precursors end? You will find more reflections on that question toward the end of this chapter.

Bird proposed that when substandard conditions or practices are allowed to exist, there is always the potential for an energy transfer that is beyond the person's or object's ability to absorb without damage. This not only applied to kinetic energy but also electrical energy, acoustic energy, thermal energy, radiation energy, and chemical energy. Some of the more common types of energy transfers included:

- Struck against (running or bumping into)
- Struck by (hit by moving object)

- Fall to lower level (either the body falls or the object falls and hits the body)
- Fall on same level (slip and fall, tip over)
- Caught in (pinch and nip points)
- Caught on (snagged, hung)
- Caught between (crushed or amputated)
- Contact with (electricity, heat cold, radiation, caustics, toxics, noise)
- Overstress/overexertion/overload

For Bird, immediate causes led to the incident itself, and concluded with the loss. The loss might be to people, property, product, the environment, or the organization's ability to provide its services. In an important addition to Heinrich, Bird also invoked the indirect costs to business as a loss, which included increased training to replace injured employees, legal expenses, investigation time, and loss of business due to an unfavorable reputation. The ILCI model estimated that for every dollar of direct loss, the indirect costs may be 6–53 times as much.

4.3 THREE PILLARS OF HEINRICH'S THEORY

Let's look in more detail now at three pillars of Heinrich's theory. They are, you will recall:

1. Injuries are the result of linear, single causation.
2. There is a fixed ratio between what Heinrich called accidents (or simply 'occurrences'), minor injuries and major injuries.
3. Worker unsafe acts are responsible for 88% of industrial accidents.

4.3.1 Injuries Are the Result of Linear, Single Causation

Accidents, according to Heinrich, are the result of linear sequence of events. He called this a "completed sequence of factors—the last one being the accident itself." It was presented as not just an idea, but as an axiom. An axiom is a statement or proposition that is regarded as being established, or accepted as self-evidently true. At the time, such an axiom would have been quite innovative. It suggested that incidents or accidents actually came from somewhere. Their causes are traceable. We could walk back from an accident and reconstruct what had led up to it. Only a few decades before the appearance of Heinrich's book, popular belief had it that accidents come out of the blue as unfortunate coincidences of space and time, or that they were acts of God (quite literally so: remember people's beliefs about the Tay train accident from Chapter 1).

To suggest, then, that there was a sequence of identifiable and potentially preventable factors and fixable problems that preceded a bad outcome, was rather empowering. The idea was consistent with the rise of modernism itself (which you might also recall from Chapter 1). To a modernist mindset, the world was makeable, controllable. And through the Industrial Revolution and growing scientific insight, it had indeed gradually become more so. Humans could go in and change and adjust things on increasingly consequential scales, so as to affect outcomes in the real world.

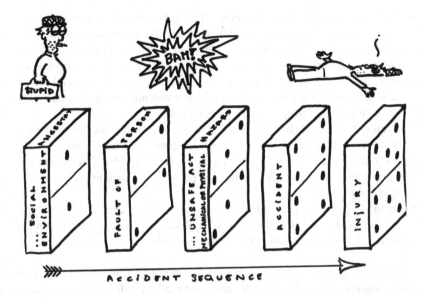

FIGURE 4.2 Linear accident sequences after Heinrich. The social environment and ancestry sets a person up for faults, which lead to unsafe acts. In combination with physical or mechanical hazards these can become an accident, which in turn can lead to injury. (Drawing by author.)

They could exercise influence over what happened there: promote good outcomes, avoid bad ones. This idea either drove, or was driven by, the very title of Heinrich's book: Industrial Accident *Prevention* (Figure 4.2).

So how should accidents be prevented? Heinrich got his readers to imagine a row of five dominoes. If set up correctly, then dominoes topple each other, in a fixed sequence. The last of Heinrich's dominoes was the injury to the worker. Each preceding domino represented a 'factor' that could (or was in fact needed to) contribute to that injury. Look at how Heinrich set up his dominoes, and how indeed each preceding domino needed to fall to take out the next one:

- **Social Environment and Ancestry**: The first domino in the sequence dealt with worker personality. In keeping with the *Zeitgeist* (recall the previous decade's thinking about accident-proneness from Chapter 3), Heinrich explained that undesirable personality traits, such as stubbornness, greed, and recklessness can be 'passed along through inheritance' or develop from a person's social environment. Both inheritance and environment (today we refer to these as 'nature' and 'nurture') contributed to the next domino: Faults of Person.
- **Fault of Person**: The second domino also dealt with worker personality traits. Heinrich explained that inborn or obtained character flaws such as bad temper, inconsiderateness, ignorance, and recklessness contributed to accident causation. According to Heinrich, natural or environmental flaws in the worker's family or life caused these secondary personal defects, which were themselves contributors to Unsafe Acts.

- **Unsafe Acts or Unsafe Conditions**: The third domino dealt with Heinrich's direct causes of incidents. He did not limit this to unsafe acts, but also warned about unsafe conditions. Heinrich identified as unsafe acts and unsafe conditions things like starting machinery without warning and the absence of rail guards. In Heinrich's experience, unsafe acts and unsafe conditions were a central factor in causing accidents, though they depended on the toppling of the previous two dominoes.
- **Accident**: In Heinrich's own words, the occurrence of a preventable injury is the natural culmination of a series of events or circumstances which invariably occur in a fixed and logical order. He defined accidents that cause injury as events such as falls of persons and striking of persons by flying objects.
- **Injury**: Heinrich saw an injury as the outcome of an accident. Some types of injuries Heinrich specified in his 'Explanation of Factors' were cuts and broken bones, though ultimately fatality would be part of this too.

Dominoes had been around for some eight centuries. And though it is hard to establish when a row of domino tiles was first made to fall over in a sequence, it would not take much for players to come up with that variation on normal games (in which the domino tiles are kept flat and the number of eyes has an important role). By presenting industrial accidents through a popular and commonly known metaphor, Heinrich would have fed the imagination of his readers. Preventing an injury, he suggested, was as simple as lifting one of the dominoes out of the line.

But which tile should be lifted? Another important idea was introduced: the closer you got to the accident, the easier it became to lift the tile:

- Fixed traits such as stubbornness are difficult to excise from a person (even though we try to do that to this day in, for example, crew resource management training in aviation).
- Faults of a person might be slightly easier, because they are more situational. People can be made less ignorant by training. People can learn to control a bad temper by being told to step back from a problem and not try to remove or solve something in anger.
- Even easier, Heinrich would have suggested, is it to prevent unsafe acts. This is done by intervening in people's behavior, and by giving them procedures and rules to follow, and more training.

Even though Heinrich had come to the conclusion that 88% of all workplace accidents are due to 'human error,' he did not argue that human behavior should be the only *target* for intervention. This might have been one reason for him to combine unsafe acts and unsafe conditions onto a single tile. Lifting out a domino tile with unsafe acts alone would not have been so promising: No matter how strongly the statistical records emphasize personal faults or how imperatively the need for educational activity is shown, he wrote in 1931, no safety procedure is complete or satisfactory that does not provide for the correction or elimination of physical hazards. For a factory owner or manager, the removal of unsafe conditions was perhaps the most promising accident prevention strategy. Putting his effort where his mouth was, he devoted some 100 pages of his writing to the topic of machine guarding. Telling the worker to not stick their hands or other body parts in certain places was not very useful if it was not combined with making such actions impossible or difficult with engineering controls.

4.3.2 The Ratio between Occurrences, Minor Injuries and Major Injuries

Both Heinrich and Bird, as well as many subsequent authors, placed considerable attention on the precise ratios between different types of events. This is what Heinrich derived from his analysis, even though we cannot trace precisely how:

- 0.03% of all accidents produce major injuries;
- 8.8% of all accidents produce minor injuries;
- 90.9% of all accidents produce no injuries.

In the words of Heinrich, the ratios—1-29-300—show that in a unit group of 330 similar accidents, 300 will produce no injury whatsoever, 29 will result only in minor injuries, and 1 will result in a serious one. The major injury may result from the very first accident or from any other accident in the group.

These were Bird's conclusions:

- For every reported major injury (resulting in fatality, disability, lost time, or medical treatment), there were 9.8 reported minor injuries (ones only requiring first aid).
- For each major injury, 30.2 property damage accidents were reported.
  The resulting ratio was 600-30-10-1, or 600 incidents with no consequences, 30 property damage incidents, 10 minor injuries, and 1 major injury or fatality.

Even the possibility of a fixed ratio relies on an inevitable collapse of dominoes once an unsafe act or condition occurs. Beyond that point, not only is an adverse outcome inevitable, but the type of outcome is a chance event with fixed odds. The exact same unsafe act may give rise to a fatality, a major injury, a minor injury, property damage, or no consequence at all.

Some followers of Heinrich and Bird have challenged the generic use of the ratios. They suggest that every type of accident has its own ratios, but they accept the underlying logic of the original model. This logic has been described as the "common cause" hypothesis. Heinrich's reasoning assumes that major and minor accidents have the same causes. This assumption is what gives the ratios their value. If the causes are the same and minor accidents are far more frequent than major accidents, then minor accidents are a very useful metric or "indicator" of the state of safety. It is harder to measure major accidents because they are very rare—but it is also unnecessary to do so if minor accidents reveal the same problems that cause major accidents.

You can take the logic one step further. If minor and major accidents are in turn caused by unsafe acts and unsafe conditions, then safety can be measured and managed using acts and conditions, rather than accidents at all. It is not necessary to agree with common cause reasoning in order to accept Heinrich's ideas. The same conclusion can be reached by a slightly more circuitous path, through what has been called the 'broken window' approach to safety.

THE BROKEN WINDOWS THEORY

'Broken windows' originally refers to a criminological idea about the norm-setting and signaling effect of small signs of disorder. The broken windows theory in criminology was first tested by Phil Zimbardo in 1969, and further developed by Wilson and Kelling in the 1980s. The theory proposes that the more petty crime and low-level antisocial behavior are deterred and eliminated, the more major crime will be prevented as a result. The thinking behind it is this. Disorder, such as graffiti, abandoned cars, and broken windows, leads to increased fear and withdrawal from residents and others in the community. This then permits more serious crime to move in or develop. Informal social control erodes and allows the growth of a culture where increasingly criminal behavior is tolerated or least not stopped. Particularly in areas that are large, anonymous, and with few other people around, we look for signals in the environment to tell us about the social norms and the risks of getting caught in following or violating those norms. The area's general appearance and the behavior of those in it are important sources of such signals. Whether interventions based on the broken windows theory actually work has been controversial.

The popular interpretation of Heinrich's and Bird's ideas follows the same logic; eliminate the small stuff, the undesirable little acts by people at the sharp end, and the big safety problems (fatalities, containment losses, big crashes) will be reduced proportionally. Focus on risk-taking behaviors, on housekeeping, the appearance of a workplace, the use of personal protective equipment and compliance with all kinds of small rules and procedures, and you can create the kind of culture that will intercept the growth of bigger risks: fatalities, process accidents.

4.3.3   WORKER UNSAFE ACTS

According to the results of Heinrich's study, 88% of the easily preventable occurrences were caused by 'human error.' The remaining proportion was attributable to unsafe conditions 10% of the time, and 2% of incidents were considered to be unpreventable. Heinrich actually did not call it 'human error,' but rather 'man-failure.' He did not define what he meant by 'man-failure.' But on the basis of his efforts to control personal behaviors such as unsafe acts, the use of unsafe tools, and the willful disregard of instructions, we can see that his 'man-failure' comprised both errors and violations (as they became popularized in the 1990s (see Chapter 9)). His observation was that:

In the occurrence of accidental injury, it is apparent that man failure is the heart of the problem; equally apparent is the conclusion that methods of control must be directed toward man failure.

(1980, p. 4)

For Heinrich, particularly toward the later part of his career, human error was the heart of the problem. The reports and claim files that formed his data told him so. But we should remember who wrote those reports and claims. They were typically written by supervisors, with a particular goal in mind: an insurance claim. It is probably not too far-fetched to suggest that this might have 'biased' those writing the reports or claims toward blaming the worker. This, after all, would relieve a company and its managers and supervisors from the kind of responsibility that might have nullified the claim.

Despite these origins, the figure has had deep and wide consequences for Safety Science. The popularity, or almost mythical status, of a 70%–90% human error proportion is visible in almost every field of practice. It is, simply, "that 70 to 80 percent of human error referred to in the literature" (Shappell & Wiegmann, 2001, p. 60). The problem is that the literature refers mostly to itself. Practical and scientific publications alike seem to uncritically copy the figure, without referring to original data. "Approximately 80% of airplane accidents are due to human error (pilots, air traffic controllers, mechanics, etc.)" (Rankin, 2007, p. 16). Or 79% of fatal general aviation accidents are caused by pilot error (Salas & Maurino, 2010). Whether as researchers or practitioners, we seem to take the figure as given—literally. "*Given* that human error plays such a central role in aviation accidents, it is crucial to understand the causes of error and the ways errors contribute to accidents" (Salas & Maurino, 2010, p. 337) (italics added). This could have been said by Heinrich in 1931, and was in fact said by him in so many ways. But it was written in 2010. Heinrich's ideas are alive and well.

Before we reflect more on these pillars and their impact on Safety Science, do consider this. Heinrich's most important innovation is sometimes overlooked, and that is that we can do things to make certain incidents and injuries actually preventable. This is not surprising coming from an insurer—they would rather their clients prevent the occurrences that lead to claims in the first place. Insurers still do a lot of work in prevention today. But recall from earlier in the book that in the 1920s and 1930s, this was a somewhat novel idea. Accidents were turning into objects of scientific study and targets for possible prevention during the very time that Heinrich was writing. From his perspective, the approach he took made good sense. In order to engage managers and factory owners in preventive activities, he would have to know more about the causes of incidents and injuries. This is where Heinrich's work came in and was innovative at the time.

4.4 BEHAVIORISM AND BBS

The combination of Heinrich's ideas—that human error is responsible for most accidents, that errors or at-risk behaviors represent the base of an iceberg or triangle that eventually produces failure on a grand scale, and that worker errors come at the end of a sequence of other events—has created the foundation for a popular idea in

Safety Science. It is the idea, as was concluded above, that safety can be improved by targeting people's behaviors. Today, the approaches based on this are known by many labels, but most have something of "behavior-based safety" in them. Here is one justification for this approach:

> Whilst good safety management systems, policies, procedures and other tools have pre-vented hundreds and thousands of injuries at work, and brought accident and injury rates down for companies around the globe, many organizations find themselves on a performance plateau. Close, but not quite close enough to their target of zero injuries. It's usually at this point that these organizations shift their attention beyond engineering controls, such as machinery guarding, and administrative controls, like procedures and training courses, and turn to what they perceive to be the panacea: influencing behavior.
>
> **Sharman (2014, p. 20)**

BBS has once again become quite popular—with regulators and employers and consultants alike, though not often with workers themselves:

> Behaviour modification programs are now widely advocated as a means of enhancing safety at work. A variety of proprietary programs are on the market, for example, DuPont's STOP, and Chevron Texaco's POWER, all aimed at encouraging workers to behave more safely. These programs are highly controversial, with unions arguing that they amount to a return to the strategy of blaming workers for the accidents that befall them, especially when they are associated with programs that punish workers who have accidents. On the other hand, companies are hoping such programs will prove the key to driving accident rates lower, and they criticize the union viewpoint as being merely obstructionist.
>
> **Hopkins (2006, p. 584)**

LET'S PUT UP A POSTER

A common way to try to include behavior change is to put up posters in work-places that remind them of the safety commitments their company has made and to appeal to people to behave consistent with them. The so-called "dirty dozen" can serve as one example (out of many). It is a list of mostly behavioral problems identified particularly in maintenance work. It is used as a reminder during training, as a checklist for task assessment, as a tool for incident inves-tigation, and as a call for vigilance on part of the workers.

The Dirty dozen
stress
insecurity
no cooperation
complacency
lack of communication
distraction
inattention
fatigue

pressure
inadequate means
insufficient knowledge
norms

Interestingly, the target of such posters is not often the manager whose job it is to institute a safety program or have it audited. In fact, putting up posters in the workplaces of reports may well constitute such a program. It is the kind of managerial control that was once advocated by Bird: creating standards and instituting a program to manage any deviation or slippage from those standards. Many of those dirty dozen are to be found with the frontline worker. They are Heinrich's unsafe conditions or unsafe acts—from complacency to lack of communication, from distraction to insufficient knowledge. As Bird and Heinrich would remind us, though, some of these can be traced back to the organization. Fatigue, pressure, and insufficient knowledge, for example, come from somewhere: they are consequences of organizational trade-offs and decisions, rather than causes of trouble brought to the workplace by frontline operators.

Yet, you would think that if they still have to print it on a poster that is put up in a workplace, then earlier control (Bird's precontact control) through policies, safety-by-design or managerial interventions has not been very effective. And some others clearly start and stop at the sharp end in their entirety: complacency, distraction, and inattention. The poster is a behavioral intervention meant to help fix those dirty behaviors. On last examination, the safety-scientific literature still offered no empirical study or data to prove the efficacy of such posters or of the moral-behavioral appeals on them.

Heinrich's ideas formed an original foundation for BBS, but they themselves were based in, and coproduced by, the larger cultural, social and scientific setting of the time. This setting created the conditions of possibility for Heinrich's thinking to make sense, to be appealing and applicable. To examine this, we have to travel back to the 1930s. Behaviorism was the dominant school in psychology and management science at the time. It informed and served a technologically advancing and increasingly industrialized world. Behaviorism is fundamental to BBS approaches. It suggests that behavior can be influenced and changed by carefully studying and adjusting the (dis) incentives—regardless of the engineering and organizational features of the job.

PSYCHOLOGICAL BEHAVIORISM

Psychological behaviorism holds that behaviors are learned and maintained through positive and negative reinforcements. In order to change people's behavior, that is exactly what you target: their behavior. You do not worry about, nor should you try to directly influence, what goes on in people's minds.

This might sound intuitive or common-sensical today, but there were both practical and doctrinal reasons for the emergence and growth of psychological behaviorism in the first half of the 20th century. Let's deal with these now.

WATSON AND BEHAVIORISM

"Psychology as the behaviorist views it is a purely objective experimental branch of natural science. Its theoretical goal is the prediction and control of behavior. Introspection forms no essential part of its methods, nor is the scientific value of its data dependent upon the readiness with which they lend themselves to interpretation in terms of consciousness" (Watson, 1978, p. 435)

This is how John Watson opened his 1913 broadside against psychology as he saw it: preoccupied with internal, invisible things, using methods that were anything but scientific, and losing credibility and standing as a result.

To understand Watson, and the appeal of his ideas, we have to understand what went before. By the time Watson published his strident views, psychology as we know it had been around for about half a century, with most developments in Europe. Early names in experimental psychology include Hermann Ebbinghaus. Ebbinghaus studied memory by having his subjects memorize nonsense syllables, counting the trials it took and also investigating the effects of the meaningfulness of syllables on their memorability. As Ebbinghaus was going about his work, philosophers kept pointing out that psychology could never be a science, because the activities and contents of the mind could not be measured. It could, therefore, never achieve the objectivity found in, for example, the sciences of physics and chemistry. Early psychologists such as Ernst Weber and Gustav Fechner (both at Leipzig) made forays to prove the skeptical philosophers wrong. Weber, for instance, studied psychophysics and demonstrated the two-point threshold: the smallest distance noticeable to touch at various parts of the body. This led him to the just-noticeable difference, known even today as the JND: the smallest difference in stimulus a person was capable of perceiving (first demonstrated by having subjects hold two weights). He went on to prove that the JND is a constant fraction of the size of each individual stimulus. Fechner (a professor of physics) picked up on Weber's work, deriving what he called Weber's law. This law said that there is a constant proportion between the minimum change detectable in a stimulus and the strength of the stimulus. By the 1860s, such lawfulness gave lie to philosophers' skepticism about the scientific stature of psychology.

Wilhelm Wundt, who was born in Germany in 1832, studied medicine and served as an assistant to the famous physiologist Helmholtz, investigating the neurological and chemical stimulation of muscles. In 1864, inspired by Weber and Fechner, he launched a course on physiological psychology, which examined the border between physiology and psychology, or, more specifically, the links between stimuli, sensations, and reactions. Wundt subscribed to Spinoza's idea of psychophysical parallelism, which said that every physical event has a mental counterpart. This allowed him to think of consciousness as an activity of the brain, and not as something material. Wundt spent a great deal of effort refuting reductionism. The mental activities that produced consciousness, he said, could not be reduced to elementary or physically visible phenomena. Rather, Wundt thought, this was a matter of creative synthesis. In this, you can recognize early ideas about emergence, which is so important to understanding complexity (see Chapter 11). Emergence says that a phenomenon arises from the interaction and activities of many underlying components, but that it cannot be reduced to those components. When the components are studied in isolation, the phenomenon will not be there. This is, of course, true for consciousness: it cannot be seen (or does not even exist) at the level of an individual neuron. Today, the study of blood flows to areas of the brain (a correlate to neuronal activity) is, in a sense, an attempt to reduce consciousness to the visible workings of groups of individual components after all.

The research technique developed by Wundt in his Leipzig laboratory (1870) has been called introspection in English. This is perhaps a bit unfortunate as it might connote armchair speculation about one's thoughts and feelings. But Wundt's method, also known as experimental self-observation, in which highly trained observers were presented with carefully controlled sensory stimuli. The observers, in a state of high attention, then described their mental experiences of the repeated trials. Edward Titchener, an English doctoral student of Wundt's, proceeded to develop introspection, but in a departure from his advisor, focused on breaking down the mental experiences into individual elements.

Even Wundt noted the significant limitations of introspection as an experimental method. Its unreliability was the biggest problem. Different subjects gave very different introspective reports about the same stimulus, and even subjects trained in introspection offered different responses to the same stimulus from trial to trial. The method also did not work on children or animals, and complex topics that were starting to be the focus of other areas of psychology (mental disorders, development, personality, and learning) could not be studied through introspection.

Wundt was aware of these limitations. He was not very troubled. He found that psychology was not so different from philosophy (he actually published four books on the latter after founding his lab) and that it had so much more to offer.

For him, introspection merely scratched the surface. Of course, there were going to be significant differences between and within subjects, as these were often linked to the cultural practices, mythologies, rituals, literature, and art that people brought with them into the laboratory. In the last two decades of his life, Wundt wrote a ten-volume Völkerpsychologie (roughly translatable as Cultural Psychology), which included ideas about the stages of human cultural development. Ironically, this kind of 'cultural baggage,' so richly captured by Wundt, would be discounted from the experimental psychological laboratory for many decades to come (in the United States as elsewhere) and dismissed as a confound that introduced unreliable individual differences to any experiment. As you will see later on in this book, it is only after the 1990s that we have seen a resurgence of interest in culture as a hugely important determinant of safety in many worlds of practice. We will also learn more about Wundt in the next chapter, as we see some of his basic ideas return in how human factors approach psychological research questions.

In the meantime, though, Watson asked—for a decidedly more pragmatic U.S. audience—how to deal with these problems and shortcomings, and the lack of credibility it spelled for the field of psychology as a whole. In the United States, William James had established a psychological laboratory (at Harvard) around the same time as Wundt did in Leipzig, but the school of psychological thinking that grew out of that was much more pragmatic too. It is now known as functionalism. To find the meaning or purpose of an idea, one had to look at its consequences or impact. James was interested in cause and effect, prediction and control, and observations of environment and behavior instead of internal experiences. Watson was more radical than that. His behaviorism was almost explicitly a psychology of protest, coined in sharp contrast against the Wundtian experimental introspection that had preceded it. Before Wundt, Watson claimed, there was no psychology. And after Wundt, there was only confusion and anarchy. Behaviorism could change that: by refusing to study consciousness and instead dedicating itself to scientifically manipulating and influencing visible behavior, psychology would attain new rigor and a new future.

The solution, he declared, was indeed to study only visible things, using objective methods. Mental states should no longer be the object of observation or study: it only produced subjective theorizing. A departure from studying consciousness, or mental states, elevated the status of animal models in psychology. Consciousness would be difficult to study in animals, but behavior could be observed, and even more easily influenced than that of humans. Rats, pigeons, kittens, chickens, dogs—they could all generate valuable insights into how organisms learn and respond to incentives, constraints, and sanctions in their environment. Watson was not alone in this, and nor the first. Pavlov (who had actually studied under Wundt!) and Thorndike had gone before him with their studies of animal behavior. The term "trial and error" learning, for example, came from Thorndike, stemming from his animal work in the late 1890s. The basic psychological unit of analysis in behaviorism was that of stimulus-response (S-R). The brain was considered a "mystery box" in which the stimulus was connected to the response, but it did not matter how.

Behaviorism became an early 20th-century embodiment of the Baconian ideal of universal control. This time, the ideal was reflected in a late Industrial Revolution preoccupation with technology and maximum control over productive resources (workers). It appealed to an optimistic, pragmatic, rapidly developing, and result-oriented North America. Behavioral laws were extracted from simple experimental settings. And they were thought to carry over to more complex settings and to more experiential phenomena as well. Behaviorists thought that their field could meaningfully study things such as imagery, thinking, and emotions. It was fundamentally nomothetic. It aimed to derive general laws which were thought to be applicable across people and settings. All human expressions, even including art and religion, were reduced to no more than conditioned responses. Behaviorism turned psychology into something wonderfully Newtonian: a schedule of stimuli and responses, of mechanistic, predictable, and changeable couplings between inputs and outputs; or, in Newton terms, between causes and effects. The only legitimate characterization of psychology and mental life was one that conformed to the Newtonian framework of classical physics, and abided by its laws of action and reaction.

4.4.1 Behaviorism, Industrialization, and Progress

It was not just Watson's youthful optimism, his tough-mindedness or his uncompromising, self-confident expression in spoken and written word that propelled behaviorism to the forefront of psychology. Even more important was the era of industrialization and capitalism in which his ideas were able to gain traction. Behaviorism turned psychology into a highly practical, highly useful science. Published only 2 years after Frederick Taylor's *Scientific Management* and during a period of growing popularity of the production line, Watson's ideas about influencing human behavior were a welcome source of inspiration for managers who sought to maximize efficiencies of human labor. The tenets of behaviorism were as follows:

- Psychology is the science of behavior. It is not the science of mind, certainly not of mind as something different from behavior. This is an explicit rejection of Spinoza's notion of psychophysical parallelism, in which every physical event (like observable behavior) has a mental counterpart.
- Behavior can be described and explained without making any reference to internal or mental phenomena. Behavior is driven externally, by the environment, not internally by the mind. This gave behaviorism its evidential and pragmatic status.
- If mental concepts slip into the psychological lexicon in the process of studying behavior, then these should be eliminated and replaced by behavioral terms. Mental concepts like "belief" and "desire" were thus converted to S-R relationships, or as reactions to (dis)incentives. Learning is not seen as happening in the head either, but as represented in what people (or animals) *do* in response to stimuli and consequences.

The world, for a behaviorist, was essentially fixed—its contours and content lay beyond the reach of ordinary working humans. The world produced progress in ways that were difficult or impossible for them to influence. Recall from Taylorism how engineers and managers and planners were assumed to be the smart ones: they developed and designed the technology, and then specified the tasks for those who would operate it. Work processes could be standardized, and should be, because workers could be assumed to be 'dumb.' All they could be expected to do was go with the flow, follow the scripts written by others, and adapt to the demands of technology and production. This was a belief, a conviction, a guiding principle for progress. Consider, for a moment, the motto of the world fair in Chicago in 1933:

Science finds
Industry applies
Man adapts

Such was the vision of technological progress at that time, and the optimism that had been fed by successful industrialization during the century before. Man adapts to technology—and behaviorism was going to help industry do exactly that. It would make sure that man has got adapted to the demands and expectations of an industrial era.

4.4.2 BEHAVIORISM AND INDUSTRIAL PSYCHOLOGY

Behaviorism can be seen as one of the first instances of industrial psychology—a field that became an important addition to managing and meeting the interests of corporations and other organizations (Warr, 1987). Ironically, one of the first industrial psychologists, whose ideas aligned closely with Watson's, was a German from the prestigious Leipzig laboratory whose work Watson so reviled and disdained. In 1913, the same year as Watson's seminal publication, Hugo Munsterberg (actually a student of Wundt's and a highly qualified researcher) outlined the goals that industrial psychology should work toward and the interests it should serve. These were, he believed, to help select workers most beneficial to capitalist enterprises, to endeavor to increase per-worker productive output, and to adjust workers' behavior "in the interest of commerce and industry" (Newlan, 1990, p. 10).

The industrial psychologist was not to have any moral qualms about this—after all, modernism and industrialization were believed to usher in a new, better, more controllable and manageable world for all—but rather to be a scientifically disinterested party dedicated solely to the efficiency of the production process. Munsterberg's ideas got a lift from the huge ramp-up of industrial work generated by the U.S. entry into the World War I, where they were put to good work. Walter Scott, a psychologist and professor at Northwestern University, further popularized industrial psychology, explaining in his 1911 *Influencing Men in Business* that psychological principles be deployed in the service of increasing the quantity and quality of labor output, oriented to meet the needs of what became known as capital interests. Newlan (1990) quotes a 1960s analyst who observed that in the early part of the 20th century:

> American management came to believe in the importance of understanding human behavior because it became convinced that this was one sure way of improving its main weapon in the struggle for power: the profit margin.

(p. 14)

The scope and purpose of industrial psychology, and a good part of the behaviorist research enterprise with it, congealed around the interests of managers and factory owners. Often missing from the purpose was an awareness of, and commitment to, the well-being of the worker. In fact, worker well-being was not included or defined in the industrial-psychological lexicon of the time. And when it was considered, this well-being was mostly equated with labor efficiency. An economist at the time might have seen industrial psychology as a natural, morally neutral response to demands from a market that was seeking greater efficiencies.

4.4.3 Productivity Measures as Safety Measures

To understand this better, remember the shift that happened during the Industrial Revolution in the 19th century (e.g., as discussed in Chapter 2). In preindustrial labor relations, people generally purchased a worker's actual labor or directly the product produced by it. In an industrialized society, it no longer worked that way. Capital purchases a worker's labor time, or potential labor, as opposed to products or actual work. It becomes natural, in such an arrangement, to pursue strategies that regulate the productive processes in order to derive as much work (and ultimately value) as possible from a given amount of purchased labor time (Newlan, 1990).

This arrangement is still in place in most industries, from manufacturing to flying, to construction and beyond. The measurement and management of worker safety has gradually become an important part of it. Two of the most important (supposed) safety measures used across many industries—lost-time injuries (LTI's) and medical treatment injuries (MTI's) are a direct reflection of lost productive capacity and reduced surplus value, respectively. Even recent figures of the International Labor Organization show the size of the problem—if measured in this way. About 340 million work-related injuries (and a further 160 million illnesses) were sustained annually by workers around the world, with a subset of 2.3 million fatal injuries and illnesses each year; in other words, an average of 6,000 work-related deaths across the globe each and every day (O'Neill, McDonald, & Deegan, 2015). What is interesting about MTI or LTI figures is that they do not measure the lack of safety (if they do that at all) in terms of human hurt or suffering. To ponder that for a moment, consider that fatalities that happen on the worksite are not counted as LTI's. The worker loses his or her life, but there is no lost time, because a dead worker has no time to lose. Also, if the death is instantaneous and the worker required no medical treatment, the case would not be an MTI either. This way, a company can have a year with no LTI's, no MTI's, yet suffer a number of fatalities.

The LTI figure hides social conditions that reflect the distribution of hurt and suffering at work, and also hides the causes of such suffering and what might be done to reduce it. Fatalities in the workplace are actually not distributed equally at all:

> The branches with the highest level of fatalities are industries as mining, offshore activities and construction. More detailed structural aspects of deaths at work also appear. Fatal accidents at work happen most often to male workers at the lowest levels of hierarchies. Race seems to be an important explanatory factor, and it is not purely by chance that in many cases migrant workers are over-represented in fatal accident

statistics. Without stretching the point too much it can be argued that fatal accidents are the only one—and ultimate—indicator of general deprivation in working life.

Saloniemi and Oksanen (1998, p. 59)

None of this is really visible (or even much cared about in the more nuanced sense of social responsibility) in the ways in which safety is typically assessed and reported (see the example about Qatar).

DEATH TOLL HIGH AMONG QATAR'S 2022 WORLD CUP FOREIGN WORKERS (FROM *THE GUARDIAN*, DECEMBER 23, 2014)

Nepalese migrants building the infrastructure to host the 2022 World Cup have died at a rate of one every two days in 2014—despite Qatar's promises to improve their working conditions, *The Guardian* has learned. The figure excludes deaths of Indian, Sri Lankan, and Bangladeshi workers, raising fears that if fatalities among all migrants were taken into account the toll would almost certainly be more than one a day.

Qatar had vowed to reform the industry after *The Guardian* exposed the desperate plight of many of its migrant workers last year. The government commissioned an investigation by the international law firm DLA Piper and promised to implement recommendations listed in a report published in May.

But human rights organizations have accused Qatar of dragging its feet on the modest reforms, saying not enough is being done to investigate the effect of working long hours in temperatures that regularly top 50°C.

The Nepalese foreign employment promotion board said 157 of its workers in Qatar had died between January and mid-November this year—67 of sudden cardiac arrest and 8 of heart attacks. Thirty-four deaths were recorded as workplace accidents.

Figures sourced separately by *The Guardian* from Nepalese authorities suggest the total during that period could be as high as 188. In 2013, the figure from January to mid-November was 168.

"We know that people who work long hours in high temperatures are highly vulnerable to fatal heat strokes, so obviously these figures continue to cause alarm," said Nicholas McGeehan, the Middle East researcher at Human Rights Watch.

"It's Qatar's responsibility to determine if deaths are related to living and working conditions, but Qatar flatly rejected a DLA Piper recommendation to launch an immediate investigation into these deaths last year."

Some within Qatar suggest the cardiac arrest death rates could be comparable to those among Nepalese workers of a similar age at home. The Indian embassy argued this year that the number of deaths was in line with the average in their home country. But in the absence of robust research or any attempt to catalogue the cause of death, human rights organizations say it is impossible to properly compare figures. A series of stories in *The Guardian* has shown that migrant workers from Nepal, India, Sri Lanka, and elsewhere were dying in their hundreds.

Safety in large capital projects like the construction of soccer stadiums (see above) is measured through productivity or cost numbers, which is used to compare and contrast not only industries but also peers inside an industry. Reporting LTI and MTI figures is common practice, if not mandatory in many places, particularly if a company wants to win a government contract. As a direct descendant from practices and economic arrangements that congealed in the first half of the 20th century, safety is still measured in terms of productivity and surplus value. What matters is the impact that safety incidents have on productive capacity and efficiency.

IS IT A FIRST AID, AN LTI, AN MTI? IT DEPENDS ON CASE MANAGEMENT

Most industries consider it desirable to show low numbers of incidents with consequences for productivity. It "has led to increased attention to injury measures and a 'drift' in the way LTI performance measures, in particular, are conceptualised and employed"(O'Neill et al., 2015, p. 185). A whole substratum of practice and expertise, of job categories and consultancies, has grown up underneath the measurement of safety performance to help industries with that. It is often called case management, and industries recruit injury case managers with the explicit assignment to reduce LTIs and needless absences. They do this not by preventing the incident or injury from happening, but by calling it something else after it has happened. This renegotiation or reclassification is fundamental to case management, and crafty 'case management skills' are highly valued. A typical menu from which case managers get to pick—this one from my own university and based in national regulations—is shown below. First aid injuries do not carry the stigma of hurting an organization's productivity, and so it is the most desirable category for case managers to pursue. There is not an expectation among peer organizations to share first aid data the same way that they do with MTI's or LTI's. After all, a first aid intervention gets people back onto the job swiftly.

FIRST AID INJURY

A first aid injury is an injury that requires a single first aid treatment and a follow-up visit for subsequent observation involving only minor injuries, for example, minor scratches, burns, cuts and so forth, which do not ordinarily require medical care, and for which the person would typically return immediately to their normal activities. Such treatment and observation are considered first aid even if it is administered by a physician or registered medical professional. Typical treatments normally considered as first aid injuries include:

- Application of antiseptics during a first visit to medical personnel;
- Treatment of minor (first degree) burns;
- Application of bandages (including elastic bandages) during a first visit to medical personnel;

- Irrigation of eye injuries and removal of non-embedded objects;
- Removal of foreign bodies from a wound using tweezers or other simple first aid technique;
- Use of nonprescription medication (schedule 2 or 3 medications), and administration of a single dose of prescription medication on a first visit to medical personnel for minor injury or discomfort.
- Soaking, application of hot-cold compresses, and use of elastic bandage on sprains immediately after injury (initial treatment only);
- Application of ointments for abrasions to prevent drying or cracking;
- One-time administration of oxygen, for example, after exposure to toxic atmosphere; and
- Observations of injury during visits to medical personnel, including hospitalization (for less than 48 h) for observation only for a blow to the head or abdomen, or exposure to toxic substances.

The following are typical examples of diagnostic/preventive procedures that may also be classified as a first aid injury: X-ray examination with a negative diagnosis (it will be reclassified as an 'MTI' if positive);

- Physical examination, if no condition is identified or medical treatment is not administered; and
- One-time dose of prescribed medication, for example, a tetanus injection or pharmaceutical.

LTI
An LTI is a work-related injury or disease that resulted in:

- time lost from work of at least one day or shift;
- a permanent disability;
- or a fatality.

MEDICAL TREATMENT INJURY
An MTI is defined as an injury or disease that resulted in a certain level of treatment (not first aid treatment) given by a physician or other medical personnel under standing orders of a physician. Types of treatment that classify an injury under MTI are:

- Use of prescription medication (schedule 4 or 8 'prescription only' medication), except a single dose administered on a first visit for minor injury or discomfort;
- Therapeutic (physiotherapy or chiropractic) treatment, more than once;
- Stitches, sutures (including butterfly adhesive dressing in lieu of sutures);
- Removal of dead tissue or skin (surgical debridement);

- Treatment of infection;
- Application of antiseptic during a second or subsequent visit to medical personnel;
- Removal of foreign objects embedded in an eye;
- Removal of foreign objects embedded in a wound (not small splinters);
- Removal of embedded objects from an eye;
- Treatment of deep tissue (second or third degree) burns;
- Use of hot or cold soaking therapy or heat therapy during the second or subsequent visits to medical personnel;
- Positive x-ray diagnosis of fractures, broken bones, etc.; or
- Admission to hospital or equivalent medical facility for treatment.

Today, LTI and MTI, as essentially cost- and productivity figures, are used as stand-in for a lot of other things: workplace safety, injury frequency, injury severity, workplace culture, national safety culture, workplace health and safety cost, and individual worker performance (O'Neill et al., 2015):

> Rather than offering a measure of that subset of injuries indicative of lost workplace productivity, corporate reporters are increasingly presenting LTI numbers as measures of (total) injury performance, and even of occupational health and safety itself. Critics suggest injury data routinely forms the cornerstone of occupational health and safety performance reports with an almost exclusive status quo-reliance on recordable and lost time injury rates as safety performance measures.
>
> *(p. 185)*

There is a link between the role of behaviorist psychology and the measurement of LTI's and MTI's. Heinrich showed that occurrences that could potentially turn into LTIs or even MTIs (1) were mostly caused by human behavior and (2) needed to be avoided as much as possible, because they had a fixed, proportional relationship to real and worse productive loss. The pressure to achieve a speedy return to work and reduce the number of LTI's and MTI's is commonly felt by private and public organizations alike. In many countries, the burden to pay for lost time and to compensate work-related disabilities, illnesses, or injuries, is shared between employers, insurers, and governments/taxpayers. Think about the fact that "safe work" and "return to work" (after an incident or injury) are often mentioned in the same breath, and governed by the same company department or government regulator. You might deduce that even today, 'safety' measures foremost have to fit with the highly quantified treatment of the production process and the control of labor that stems from an earlier industrial age.

For Watson, the intricate relationship between behaviorist psychology, industrial applications, and business interests gradually became personal, even biographical. *Psychology from the standpoint of a behaviorist*, a major work by him, appeared in 1919, to be revised twice in the subsequent decade. Behaviorist animal psychology and the virtues of behavioral conditioning, received much attention. But then, in

1920, Watson's academic career abruptly ended. A divorce left his public image marred, and got him fired from Johns Hopkins University. He married his research assistant the same year, and entered the business world. His contact with psychology was now through the publication of popular writings in *Harper's* and *Collier's* magazines, and culminated a book on behaviorism meant to appeal to a popular (or business) audience. After publishing its revised version, Watson exclusively occupied himself with the business world until his retirement in 1946.

4.5   BBS

4.5.1   Impact across the Decades

A substantial spin-off from Heinrich's thinking, with impact on industries across the world, has been BBS. BBS programs target the worker, and his or her behavior, as the cause of the occurrences that may lead to incident or injury. Recall Heinrich's finding that 88% of occurrences are caused by 'man failure' or human error, or worker behaviors:

> The popularity of this approach stems in part from the widely held view that 'human factors' are the cause of the great majority of accidents … As the general manager of Dupont Australia once said, 'In our experience, 95 per cent of accidents occur because of the acts of people. They do something they're not supposed to do and are trained not to do, but they do it anyway'.
>
> **Hopkins (2006, p. 585)**

The U.S. Department of Energy, which manages safety-critical facilities such as the Los Alamos National Laboratory, the Strategic Petroleum Reserve, and the Lawrence Berkley National Laboratory, explained the motivation for its embrace of BBS as follows:

> Heinrich reported that about 90% of all accidents were caused by 'unsafe behavior' by workers. Subsequent studies by DuPont confirmed Heinrich's contention. Traditional engineering and management approaches to counter this, such as automation, procedure compliance, administrative controls and OSHA type standards and rules were successful in reducing the number of accidents significantly. There was however, a persistence of incidents and accidents that kept rates at a level that was still disturbing to customers, managers, and workers. Developed in the late 1970s, behavior-based safety has had an impressive record. Research has shown that as safe behaviors increase, safety incidents decrease.
>
> **DOE (2002, p. 7)**

BBS interventions typically center on observation of behaviors, and feedback to those performing them. The four basic steps of a typical behavior-based program are:

- Define the correct behaviors that eliminate unsafe acts and injuries;
- Train all personnel in these behaviors;
- Measure that personnel are indeed behaving correctly;
- Reward worker compliance with these correct behaviors.

There is a variety of ways to achieve these steps. In keeping with behaviorist principles, some involve penalties, demerits, or disincentives for undesirable behaviors. Many involve surveillance of worker behavior, either by people or by technology (e.g., cameras but also computerized monitoring systems installed on equipment, like vehicles). Others involve coaching and guidance. All are of course centered on Tayloristic ideas about codes and procedures to follow for a particular task or job, otherwise there would be no best method or behavior to point workers to (see Chapter 2). Today, many programs are also supportive and training-oriented. Some deliberately pursue worker commitment to injury- and incident-free behavior. In some cases, workers have described these sessions as akin to taking part in a religious conversion ritual.

BBS, however, typically has (or should have) a back-end program that gathers observational data for action planning to make system improvements. As did Taylor and the Gilbreths, there is little point to focusing on people's behaviors if the system in which they work simply does not contain the conditions for the sort of behavior you want to see. As Krause explained:

> This aspect of the behavior-based approach is the easiest to overlook and in many cases it is the most difficult to implement, but it makes the difference between a continuous process and a temporary program. The front end of behavior-based safety, namely the identification of critical behaviors and their systematic observation with feedback, has very positive effects for a number of years. After that period of time, however, observation and feedback *alone* are not enough to sustain an ongoing improvement process. The behavior-based initiatives that achieve that crucial success are the ones that bring about obvious improvements in their site's safety system (emphasis in original).
>
> **Krause (1997, p. 5)**

BBS programs can be driven by a variety of stakeholders. Some are initiated by the employer, and often brought in with the help of consultants. Others come from state regulatory bodies, or from social partner organizations. Some programs, though not many, recognize that the most worthwhile initiatives come from individual employees' own insights, experiences, and ideas. This diversity leads those who develop, implement, and manage BBS programs to argue that they are much more than just the four steps above, indeed far more than just a behavior-modification program. "But its emphasis is undeniably on behavior modification and that is how it is understood by many of its advocates as well as its critics" (Hopkins, 2006, p. 585). Indeed, a focus on behavior modification and the behaviorist legacy is obvious from the way BBS programs are promoted:

> Reinforcement occurs when a consequence that follows a behaviour makes it more likely that the behaviour will occur again in the future … For example, a toolbox talk addressing correct manual handling techniques might result in correct techniques on the day of the talk; however, over time employees will revert to old practices. This is because nothing has occurred after their correct behaviour to indicate that it is correct, or that it has benefitted the individual or the organisation to be so safety-conscious.
>
> **HSA (2013, p. 7)**

The principles on which behaviorism was developed are indeed easily recognizable in BBS. It is organized around behavior and how to influence it, without making reference to what goes on in the head. Motivation, for example, is measured not so much as an internal or mental phenomenon, but as observable property, visible through people's behavior, through what they do or can be triggered to do. The focus is on the stimulus (changing the motivation) and response (which is evidence for that change): the S-R concept made famous by behaviorism. The following is based on a figure from the Health and Safety Authority's guide on BBS (HSA, 2013), showing the S-R connection (though calling it antecedent and consequence). The connection between the two, visible behavior, treats whatever happens inside of the person as a black box. This is what behaviorism did too: only that behavior which we can see an individual do can be measured and treated as data.

BBS ACCORDING TO THE U.S. DEPARTMENT OF ENERGY AND A HEALTH AND SAFETY AUTHORITY

According to the Health and Safety Authority in a typical, industrialized (if not almost post-industrial) Western country (HSA, 2013), BBS is an approach that:

- Is based on solid principles about engaging, motivating, assisting, reinforcing, and sustaining safe behaviors.
- Takes a systematic approach, examining the motivation underlying behaviors, in order to increase safe behavior.
- Is an ongoing effort; not 'once-off' provisions, but a new way of working that the safety leader must continually promote for sustainable, positive results.
- Takes time to achieve; however, results can be observed immediately due to the nature of measurement involved.
- Emphasizes increasing safe behaviors rather than focusing on length of time without injury. BBS programs do not depend solely on 'lagging indicators' (after the fact), and instead shift the focus to 'leading indicators' (preventative).
- Is not a substitute for an already existing comprehensive health and safety program; it is a supplementary tool that will enhance the effect of already existing practices, and will allow for an objective measurement system.
- Aims to understand causes of incidents and near misses and correct them through the behavior of relevant people. For example, reducing hazards often requires behavior change of managers and front-line workers, and equipment redesign involves behavior change of engineers. The focus in BBS is behavior. The overarching theme in behavior analysis and BBS is that behavior is maintained by what occurs after it (consequences).

Many safety interventions in work settings focus on antecedents, or events that come before behavior that may evoke behavior. For example, many work settings rely heavily on training, safety signs, pep talks, toolbox talks, or pre-briefs. These can be effective in activating behaviors initially, but it is what occurs after our behavior that ensures the behavior will occur time and time again.

The U.S. Department of Energy (DOE, 2002) explains how sites developing and maintaining a BBS process must use several steps to define the scope of the work:

- Extract behaviors that were involved in past accidents/incidents
- Developed definitions that describe the safe behavior
- Compile datasheet using identified behaviors
- Determine observation boundaries
- Train observers
- Gather data
- Determine barrier removal process
- Form barrier removal teams
- Analyze the hazards

Analyzing hazards is built into the BBS process. Hazards are not only identified on the basis of things that went wrong in the past, but are being analyzed during each observation. The worker observed receives immediate feedback on how to minimize the risk. The assessment team and barrier removal team analyzes the data gathered through observations to determine workplace hazards. The teams then develop action plans to remove barriers to safe work.

DEVELOP AND IMPLEMENT HAZARD CONTROLS

Employees tasked with planning or designing work can also use the behavior assessment and data. By studying the definitions and data, barriers that could require a worker to perform at-risk behaviors can be designed-out up front. This forethought makes the workplace a much safer environment.

PERFORM WORK WITHIN CONTROLS

Although work has been designed and training conducted to help the employee know how to work safely, bad habits and shortcuts can introduce at-risk behaviors into the workplace. The ongoing observation process encourages the continued use of safe behaviors and reminds workers that one at-risk behavior could cause an accident, injury, or even a fatality.

PROVIDE FEEDBACK AND CONTINUOUS IMPROVEMENT

Feedback is provided each time an observation is performed. The feedback process reinforces the use of safe behaviors and helps determine why certain at-risk behaviors were performed. Collecting information about the at-risk behaviors helps the teams determine the root cause of a behavior and develop an action plan to remove the barrier causing the behavior.

4.5.2   Does BBS Work?

Evidence for the success of BBS approaches exists in the safety-scientific literature. One study, for example, reported a longitudinal evaluation of an employee-driven behavior-based accident prevention initiative implemented in industrial settings. Up to 5 years of injury data from 73 companies, drawn from a target population of 229 companies who implemented BBS, were examined. Comparisons of pre- to post-initiative incident levels across groups revealed a significant decrease in incidents following the BBS implementation. The average reduction from baseline amounted to 26% in the first year increasing to 69% by the fifth. These effect sizes were estimated from the average percentage reduction from baseline (Krause & Seymour, 1999).

In another, one-on-one interviews and focus-group meetings were held at 20 organizations that had implemented a BBS process in order to find reasons for program successes or failures. A total of 31 focus groups gave 629 answers to six different questions. A content analysis of these responses uncovered critical information for understanding what employees were looking for in a BBS program. A perception survey administered to individual employees ($n = 701$) at these organizations measured a variety of variables identified in prior research to influence success in BBS efforts. The survey data showed five variables to be significantly predictive of employee involvement in a BBS process: (1) perceptions that BBS training was effective; (2) trust in management abilities; (3) accountability for BBS through performance appraisals; (4) whether or not one had received education in BBS; and (5) tenure with the organization (DePasquale & Geller, 1999).

In the study above, it might be seen as tautological that—broadly speaking—employees' perception of the effectiveness a BBS program is driven primarily by their perception of its effectiveness (finding 1). Interventions through BBS can of course generate short-term returns, if anything because of the Hawthorne effect. Employees might feel valued if they and their work get an intensive burst of attention that helps them be more efficient or focused. The effects, however, will wear off quickly if there is no investment in understanding or addressing the environmental causes that drive the behavior in the first place.

In addition, studies like those above typically miss important scientific checks and balances. One of those is randomized control. This means that the intervention (BBS) is randomly assigned to certain organizations and randomly withheld from others (even if the organizations in the different conditions have to have important aspects in common so as to make the results comparable). Many social sciences employ randomized controlled trials to produce 'objective' evidence of the effectiveness of an intervention. It is a kind of assurance that the researcher did not just 'pick the winners' beforehand. Also, if there is only one condition, that is, the study only contains organizations that underwent the treatment or intervention (i.e., the introduction of BBS), there is no convincing way to attribute any positive results to that intervention. The reduction in incident figures over 5 years, for instance, may be due to factors external to the experiment—factors (economic, demographic, technological, environmental, etc.) that were not measured or even known. In a well-planned randomized controlled trial, all conditions are (in principle) exposed to such factors—organizations that had the intervention and organizations that did not.

This means that any difference in outcome between them can be more accurately attributed to the intervention rather than one or more of those (unknown or perhaps even unknowable) confounds. In addition, reductions in reported incident injuries can be achieved, as has been discussed above, by more assertive case management.

What matters even more is that authors of the success of behavior-based approaches might have a stake in publicizing such success. As is well-known from the experiences of the medical and pharmaceutical literature, such a stake can form a disincentive against publishing negative or neutral findings (Kerridge, Lowe, & McPhee, 2005). The possible conflicts of interest, however, are not typically disclosed in the publications of the safety studies above, or related ones (e.g., Geller, 2001). This is of course not unique to proponents of BBS. Conflicts of interest can occur in the promotion of and in the presentation of research results about, any safety-scientific approach as long as the ethics of publishing and publicizing them are not managed carefully.

THE GOVERNMENT ACCOUNTABILITY OFFICE REPORT ON BBS PROGRAMS

Another perspective on behavior-based approaches comes from a report of the U.S. Government Accountability Office (GAO) to congressional requesters (GAO, 2012). GAO is the audit, evaluation, and investigative arm of the U.S. Congress. It aims to improve the performance and accountability of the federal government, providing analyses, recommendations, and other assistance to help Congress make oversight, policy, and funding decisions. The federal Occupational Safety and Health Administration (OSHA) is one such agency falling under Congress oversight and funding. The immediate trigger for the report was the explosion at the BP Texas City refinery in March 2005. Fifteen workers died, and 180 others were injured. The refinery, it turned out, had a safety incentive program that tied workers' bonuses to achieving low rates of injuries and illnesses. GAO found how workers feared reprisals for reporting potentially risky conditions at the refinery, and that the behavioral programs in place created disincentives for workers to report injuries or incidents. Interestingly, in managing its limited resources, OSHA targeted those industries and employers for inspection that had a high number of reported workplace injuries and illnesses.

GAO estimated that 25% of U.S. manufacturers had safety incentive programs, and most had other workplace safety policies that, according to experts and industry officials, may affect injury and illness reporting. GAO estimated that 22% of manufacturers had rate-based safety incentive programs, and 14% had behavior-based programs. Almost 70% of manufacturers also had demerit systems, which discipline workers for unsafe behaviors, and 56% had post-incident drug and alcohol testing policies according to GAO's estimates. Most manufacturers had more than one safety incentive program or other workplace safety policy and more than 20% had several.

In its analysis, GAO used the term incentive programs to cover:

- behavior-based programs, which reward workers for certain behaviors or sanction them for those behaviors believed to cause work-related incidents and injuries, and
- rate-based programs, which reward workers for certain outcomes (low counts of incidents and injuries).

Even though a full 75% of manufacturing employers in the United States were using incentive programs at the time, and nearly three-quarters of them rewarded workers for having no reported injuries or illnesses. Despite this intuitive link between the existence of incentive programs and low or nonexistent injury/illness figures, GAO points out that there is little conclusive academic research on whether safety incentive programs of either kind affect workers' injury and illness reporting. The few studies that exist contain such methodological limitations, according to GAO, that their results are not generalizable to the general worker population. The remaining studies reached divergent conclusions—from not definitive, to no effect to a reduction in injuries. The latter got the attention in particular, as authors of these studies told GAO that workers may "intentionally fail to report injuries in an effort to preserve potential bonuses for their work groups" (p. 8).

GAO confirmed this 'unintended consequence' of behavior based or other incentive programs through 50 expert interviews. Many of them agreed that such programs can discourage workers from reporting injuries, incidents, and illnesses in an effort to preserve incentives and avoid disincentives. Disincentives that were found by GAO ranged from financial losses, demerit systems, fear of discipline, career jeopardy, peer pressure, mandatory drug or alcohol testing upon reporting, all the way to possible dismissal from the job or removal from site. What makes the results difficult to entangle, GAO admits, is that very few employers only have one kind of incentive program. Some have post-incident drug and alcohol testing programs, demerit systems and safety incentive systems in addition to behavioral safety programs in place. This makes it difficult to attribute the results of (and reasons for) non-reporting to just one of those programs or systems.

GAO did find that contractual relationships are an important contributor to putting incentive programs in place:

> Companies sometimes request information on manufacturers' injury and illness rates before signing a contract with them to manufacture goods. According to some workplace safety experts, such contractors may feel pressure to lower injury and illness rates to avoid the risk of losing bids for contracted work. Manufacturers whose injury and illness rates were requested by potential contracting companies were more than twice as likely to have rate-based safety incentive programs than manufacturers whose rates were not requested.

(p. 17)

Alluding to Texas City, and on the possible role of the regulator in providing meta-incentives for organizations to report low numbers of injuries and incidents, GAO told Congress that:

> Because OSHA relies heavily on accurate injury and illness reporting in tailoring its programs and allocating its finite enforcement resources, it is important for the agency to assess the impact of safety incentive programs and certain workplace safety policies on injury and illness reporting, particularly given their prevalence. Without accurate data, employers engaged in hazardous activities can avoid inspections and may be allowed to participate in voluntary programs that reward employers with exemplary safety and health management systems by exempting them from routine inspections.

On the basis of its report, GAO recommended that OSHA provides consistent guidance about safety incentive programs across the agency's cooperative programs. There was little such guidance at the time. It also suggested that OSHA add something about safety incentive programs and other workplace safety policies to the guidance provided to inspectors in its field operations manual. That way, they might be better equipped to look for the various disincentives on disclosure of injuries and incidents when they visit workplaces. OSHA agreed with the recommendations.

4.6 CRITIQUES OF HEINRICH, BEHAVIORISM AND BBS

Heinrich's work has had such an impact on Safety Science that it is difficult to get a good overview of all the consequences—either positive or negative. His ideas have seeped deeply into safety policies, interventions, procedures, practices, insurance, consultancies, regulations, safety folklore, and more. They have also helped firm up the paradigm from which subsequent safety theory and models have lent their legitimacy. Perhaps Heinrich was not wrong when he thought about his ideas as axiomatic—as true or self-evident without needing substantial empirical proof. His ideas, and the beliefs and practices that followed from them, have certainly been taken to be axiomatic in subsequent models and theories. Safety models, pictures, training, and interventions are often taken as self-evident because they are based on the models and pictures made by Heinrich; not because they are supported by empirical evidence of their own value or efficacy.

In this last section of the chapter, I once again (though in a different order, for sake of how the arguments are built up) divide these critiques across the three broad ideas that Heinrich brought into Safety Science: (1) the primacy of human error, (2) the pyramid, and (3) the sequence of events with causes and contributions, with a particular focus on the putative separation between unsafe conditions and unsafe acts.

4.6.1 THE PRIMACY OF 'HUMAN ERROR'

For Heinrich, the end of the chain contains a human error or a mechanical failure: "the accident in turn is invariably caused or permitted by the unsafe act of a person and/or a mechanical or physical hazard" (Heinrich, Petersen, & Roos, 1980, p. 21). As systems have become more reliable, the remaining portion has increasingly become attributed to 'human error' (Hollnagel, 2004). Human error is seen by many as the main remaining cause of incidents, accidents, and fatalities. It turns human error (or human behavior) into the target. Human error becomes something that needs to be 'eliminated':

> ...the elimination of human error is of particular importance in high-risk industries that demand reliability. For example, errors in commercial aviation, the military, and healthcare can be particularly costly in terms of loss of life.

Krokos and Baker (2007, p. 175)

The frailties of individuals are seen as the primary source of risk: they undermine the engineered and managed features of already safe systems. Much has been said about 'human error' in the years since Heinrich. This will be covered throughout the book. For now, let's summarize the most important arguments that have been raised:

1. 'Human error' is not a thing in itself. It is just an attribution, a label that we stick on other people's behavior. This is the argument that you will find in the next chapter. The field of human factors was built on it. As one of the recent publications suggests, we have to go *behind* the label human error in order to understand our own reasons for attributing bad outcomes to it, and to understand the normal cognitive processes that produce that behavior (Woods, Dekker, Cook, Johannesen, & Sarter, 2010).
2. 'Human error' is a political act. This is a special case of the first argument. It shows up in, for example, Perrow's book on Normal Accidents in the 1980s (see Chapter 8) (Perrow, 1984). There is a secondary literature on accidents, often written by practitioners or investigative journalists, who point out that 'human error' was blamed out of political expediency. The verdict of 'human error' served to hide a host of managerial, political, and organizational issues that could honestly be identified as the 'real' causes of the tragedy (Byrne, 2002; Jensen, 1996; Vette, 1984).
3. 'Human error' is a real thing that we can separate out and identify, but it is only the outcome; the end of a whole chain of events and factors and contributors. Those distal factors are more likely to be interesting targets for intervention. Humans at the end of this chain are more the inheritors than the instigators of trouble. This is the argument of the Swiss cheese model (see Chapter 9), and in a sense even of Heinrich himself.

The idea of 'unsafe acts' has survived safety-scientific thinking for many decades. Heinrich's chain-of-events model, which allows you to separate out the human contribution from other events, has contributed in the following ways (among others):

- Human error, or an 'unsafe act' by a worker at the end of the chain, is necessary for an accident together. This is also the assumption that needs to be true to hold together the Swiss cheese model and make it explain accidents at all.

- 'Human error' has become seen as a managerial problem, as a target of an organization's or regulator's intervention. The system can be well-engineered, but it still gets undermined by unreliable people at the end of the chain.
- It is this kind of thinking that has driven the custodian role of safety management: that we are custodians of already safe systems that just need to be protected against the vagaries of human behavior.
- A consulting industry of behavioral modification and error prevention has responded to the perceived need. Some have declared a 'war on error' (Kern, 1998).

Historically, this thinking gets its legitimacy from Heinrich's problematic numbers about 'human error' and its dominant role in the accident chain. A substantial part of the work of Safety Science in the decades since Heinrich has been to undo the 'myth' of a human error problem (Besnard & Hollnagel, 2014)—from human factors in the 1940s (see Chapter 5) to resilience engineering today (see Chapter 11).

4.6.2 THE TRIANGLE (OR PYRAMID)

Recall the idea of the triangle. It was first raised by Heinrich, and then putatively confirmed by Bird's figures. It says that for every x number of low level or inconsequential occurrences, there is a proportional, smaller number of higher-consequence events, all the way up to a fatality.

Critiques of—and data to disprove—this triangle idea are typically of two kinds:

- The proportions (e.g., 30-10-1) do not apply, or look very differently for the world or profession studied in a particular case. This critique was anticipated by Heinrich, and even more by Bird. They would readily admit that there are going to be local differences in these proportions. But the basic idea stood: take care of stuff at the bottom, and the stuff up the way to the top of the triangle will take care of itself.
- The second critique, however, makes that impossible. For the triangle idea to be sustainable, a second and much more important criterion needs to be met. And that is that unwanted events at any level of the triangle have the same causes. If you believe, after all, that you can prevent the ultimate high-consequence outcome, a fatality (at the top of the triangle), by preventing low-consequence, higher-frequency incidents or injuries lower down the triangle (which in turn means stopping certain behaviors at the bottom of the triangle), then bad outcomes—no matter their consequences—all have the same or similar causal pattern. This has become known as the common cause hypothesis.

Wright and van der Schaaf (2004) pointed out that these two get conflated in many critiques. If you do not know what you are trying to prove or disprove, then it is hard to produce evidence for the triangle being either right or wrong. More about their study will be said below. But for the moment, consider a case where the proportions not only fail to conform to Heinrich's or Bird's figures, but actually become inverted to them. What if there are fatalities, but no incidents lower down the triangle? This would simultaneously disprove the proportions and the common cause hypothesis.

DEEPWATER HORIZON

One such case seems to be the 2010 Macondo (or Deepwater Horizon) well blowout in the Gulf of Mexico, which killed 11 people and caused the biggest oil spill in the history of humanity. It was preceded by a celebrated 6 years of injury-free and incident-free performance on the platform (or boat, really) (BP, 2010; Graham et al., 2011). So 11 deaths at the top of the triangle and nothing noteworthy below it—nothing. If Heinrich or Bird were to have been right, then 6 years of not having anything reportable happen in the lower parts of the triangle should have assured thousands of years of fatality-free performance—enough to outlast known oil reserves in the Gulf many times over.

For each such example, however, triangle and behavioral-safety researchers and consultants come up with counterexamples. To affirm the belief in the common cause hypothesis, they borrow from a proposed connection between worker behavior and workplace culture. For example, Marsh (2013) writes about frequent and trivial behaviors on the one hand and fatalities on the other:

> The two things aren't separate, but interlink and overlap. A good example of an overlap would be the housekeeping on the Piper Alpha oil platform, which was notoriously poor. (Piper Alpha was the North Sea oil platform that exploded in 1988 with the loss of 167 lives). Any meaningful analysis of the poor housekeeping before the accident would have taken the auditor straight to the permit-to-work system ... perhaps considered *the* key cause of the explosion. The permits contained a 'housekeeping put right?' element which, in the light of the poor housekeeping, would have demonstrated clearly a tick-box mentality. ... Something as 'relatively trivial' as housekeeping or PPE compliance could lead to an underlying cause that could be instrumental causing something catastrophic.
>
> *(p. 17)*

The idea put forward is that lower-level, 'relatively trivial' behaviors with no immediate consequences are evidence of the existence of a workplace culture (as expressed in sloppy housekeeping, tick-box mentality, inadequate auditing) which both hides and enables more sinister things to brew and explode into disaster. But is that indeed true? The work permit system used on the platform had actually been awarded a prize not long before the deadly fire. Instead of being obviously inadequate and porous (which might have become clear only after the fact), it was relied on, honored with awards, and not the pre-accident source of precursor incident or injury reports. In other words, no 'unsafe acts' related to this were visible at the base of the triangle until the top suddenly blew into view—167 times over (in other words, you could argue that there *was* no triangle: only a very wide top with little or nothing underneath).

And so back to the counterexample. If $n = 1$ evidence for Heinrich can be (problematically and perhaps falsely) constructed out of such examples, then more $n = 1$ evidence to the contrary is not difficult to find either. In his comments on a 1998 gas explosion at an Esso plant in Victoria, which killed two people and injured eight, Hopkins (2001) wrote:

> Ironically Esso's safety performance at the time, as measured by its Lost Time injury Frequency Rate, was enviable. The previous year, 1997, had passed without a single lost time injury and Esso Australia had won an industry award for this performance. It had completed five million work hours without a lost time injury to either an employee or contractor. LTI data are thus a measure of how well a company is managing the minor hazards which result in routine injuries; they tell us nothing about how well major hazards are being managed. Moreover, firms normally attend to what is being measured, at the expense of what is not. Thus a focus on LTIs can lead companies to become complacent about their management of major hazards. This is exactly what seems to have happened at Esso.
>
> *(p. 4)*

Other petrochemical accidents elicited the same reflections. For example, the Chemical Safety Board found that the

> BP Texas City explosions was an example of a low-frequency, high-consequence catastrophic accident. Total recordable incident rates and lost time incident rates do not effectively predict a facility's risk for a catastrophic event.
>
> **CSB (2007, p. 202)**

On the basis of its investigation, the CSB advised (like the GAO would do a few years later) that inspection targeting should not rely on traditional injury data.

Single cases, however, do not convincingly prove or disprove a theory. Saloniemi and colleagues studied the common cause hypothesis more systematically, referring to it as the isomorphism of causal chains:

> The empirical testing of Heinrich's ideas can be formulated as a problem of causation: are the causal chains behind fatal and non-fatal accidents identical or different? A reasoning based on Heinrichian ideas requires this isomorphism. On the other hand, the validity of this logic has also been challenged. For example, in their empirical comparison of causation processes between fatal and non-fatal accidents, [others] found conclusive support for the hypothesis of different causation. This is the angle which provides the justification for the analysis of the relationship between fatal and non-fatal accidents. The situation where the two are not increasing and declining simultaneously can be interpreted as evidence against identical causation and vice versa.
>
> **Saloniemi and Oksanen (1998, p. 60)**

Recall how Wright and van der Schaaf (2004) argued that the hypothesis of similarity of causes for major and minor accidents has become confounded with the interdependence of the ratio relationship between severity and frequency. In other words, there is a difference between:

- Asking whether a low-frequency high-consequence event has the same causes as a low-consequence high-frequency event.
- Asking whether there is a fixed ratio between low-consequence high-frequency and low-frequency high-consequence events.

This confounded view of the hypothesis, they suggested, has led to invalid tests of the hypothesis and thus to erroneous conclusions (i.e., the triangle is true or the triangle is not true). They carried out what they called a 'proper' test, with data from U.K. railways. Incidents were analyzed using the confidential incident reporting and analysis system cause taxonomy, which contains 21 causes. The results provided qualified support for the common cause hypothesis for only some of these causes.

The problem is that in studies like these, as in Heinrich's work, the results hinge on an analyst's choices and interpretations. They depend on what certain people call certain things, which in turn means that the results depend on who those people are, and on what their goals are. For example:

- Heinrich got his data about the prevalence of unsafe acts from supervisors, managers, and factory owners—not from the workers whose supposedly unsafe acts were being tabulated.
- Wright and van der Schaaf relied on reports from a confidential incident reporting system. Not only does this leave the denominator a guess (i.e., how many reports could be sent in based on what happens on the sharp end (the denominator) versus how many are actually sent in (the numerator)?), it also relies entirely on the unknown and unrecorded judgment of the (potential) reporter about what constitutes an unsafe event worthy of reporting (and thus counting and tabulating).

Analyzing data that was put into a database by other people creates what Giddens would call a double hermeneutic. (Hermeneutics refers to the meaning and interpretation or understanding of an event or a text.) In these studies, it is actually a triple hermeneutic:

- Causes have been constructed, divided up into categories, and given names by those who have made the reporting system or the bowtie model (the first hermeneutic).
- An event then happens and is interpreted by a reporter as it is being put into a database where the reporter needs to match their experience with the preexisting categories (the second hermeneutic).
- Some of these reports are then selected and reinterpreted by an analyst who picks certain things from it and assigns them to one or more of the 21 causal categories or 36 bowties (the third hermeneutic).

The results of such studies, then, are only as stable as the choices or attributions that (1) the designer of the bowtie/reporting system, (2) the reporter who uses the system, and (3) the subsequent analyst, respectively, make about a particular event. We know that attributions like these are generally not very stable in their relationship to the sequence of events as seen through the eyes of various participants or observers—and particularly not in hindsight (see Woods et al., 2010). Because of the triple hermeneutic, it is impossible to talk about 'results' of studies like this in an objective sense:

- The analyst was already limited to the attributions made by the reporter;
- Yet the reporter was already limited by the categories that the reporting system offered; and
- We generally do not know what inspired or limited the choices of categories in the models or reporting systems.

A systematic study used an industry-level dataset that evaluated over 25,000 enterprises over a 13-year time period, with the specific aim to answer (Yorio & Moore, 2018):

1. Are an increased number of lower severity incidents in an industry significantly associated with the probability of a fatal event over time?
2. At the industry level, do the effects of occupational health and safety incidents on the probability of a fatality over time decrease as the degree of severity decreases—thereby taking the form of a triangle?
3. Do distinct methods for delineating incidents by severity affect the existence of the safety triangle form?

As above, the answers to all three questions depended on how severity was delineated. It also depended on other factors such as total hours worked, and on what, indeed, the organization or industry meant by a particular severity level. In addition, it depended on whether an event was seen as a reportable or recordable incident. Hermeneutics are once again at work here:

> For example, in the current sample of 22,140 reportable noninjury cases that occurred between 2000 and 2012, about 75% were due to unplanned roof and face falls. Indeed, if the circumstances were different, these 16,445 reportable, noninjury events could have easily resulted in a fatality.
>
> **Yorio and Moore (2018, p. 850)**

A growing number of studies now show, instead, that fewer incidents are linked to more fatalities. In certain cases, they depend on the multiple hermeneutic problems as well, though some of these studies have tried to control for its diluting effect.

FEWER INCIDENTS, MORE FATALITIES (1)

In 2000, Arnold Barnett and Alexander Wang from MIT published a study that showed that passenger mortality risk is the highest in airlines that report the fewest incidents. This, of course, contraindicates a commitment to a zero target, since achieving a lower (or zero) number of incidents increases the risk that passengers get killed in an air crash. Barnett and Wang used data from 1990 to 1996, a period in which U.S. carriers actually had higher passenger mortality risks than their counterparts elsewhere in the developed world, so it offered a good statistical basis for their analysis and conclusions. Aviation is a bit unique, of course, because crashes tend to kill large numbers of people in one event, and the exact number of fatalities is in part a function of seat occupancy (or the so-called load factor) at the time. The load factor on any particular flight fluctuates with all kinds of factors, none of them having anything to do with the risk of incidents or fatalities:

> When [an aircraft] hits a mountain, killing all passengers, the implications about safety are not three times as grave if there were 150 passengers on board rather than 50. And a crash that kills 18 passengers out of 18 should be distinguished from another that kills 18 out of 104. (In the latter case, the high survival rate might reflect excellence in the airline's emergency procedures.) Statistics that weight crashes solely by their numbers of deaths, in other words, are vulnerable to irrelevant fluctuations in the fraction of seats occupied, yet insensitive to salient variations in the fraction of travelers saved.
>
> *(p. 2)*

To circumvent this problem, and still have meaningful statistics on the relationship between incidents and fatalities, Barnett and Wang came up with what they called the 'Q-statistic.' This represents passenger mortality risk per randomly chosen flight. To find Q, the probability of selecting a flight that results in passenger fatalities was multiplied by the average proportion of passengers who are killed aboard such flights (Barnett & Wang, 2000). Using the Q-statistic, they were able to show that the correlation between nonfatal incidents and passenger mortality risk is negative. That means that airlines with fewer incidents pose higher passenger mortality risks. Given the proliferation and sophistication of data gathering in the airline industry, they were able to do more still. As nonfatal incidents became more severe, the correlation with passenger mortality risk became increasingly negative. In other words, the more severe the nonfatal incident suffered by the airline, the less likely it was to subsequently kill its passengers in another one.

FEWER INCIDENTS, MORE FATALITIES (2)

In a study that examined the relationship between workplace fatalities in the construction industry from 1977 to 1991, Saloniemi and Oksanen of Tampere University in Finland found a negative correlation between the number of fatalities suffered in a year, and the incident rate (Saloniemi & Oksanen, 1998). The negative correlation between fatalities and nonfatal incidents in construction was in fact statistically highly significant (a correlation of -0.82, $p < 0.001$). In other words, in a year when the industry produced more incidents, a construction worker was much less likely to get killed and vice versa (Saloniemi & Oksanen, 1998). This again strongly contraindicates a focus on zero. Fewer incidents, after all, increased the probability of death.

Whenever occupational accident statistics are used, it is necessary to address the question of how actual accidents and accident statistics are related to each other. The definition and outcome of a fatal accident was deemed to be unambiguous, for sure. As for incidents and lost work time, those used for Saloniemi's and Oksanen's analysis here were compiled on the basis of sick leave and compensations granted. We can assume that the practices of granting such leave and compensation inside of a small, centrally governed, culturally homogeneous country, did not change much from year to year or site to site, and thus would not be responsible for producing much, if any, variance in the results. Even if case management and other ways to negotiate the granting of leave or compensation might have played a role, then this was deemed to be sufficiently constant across the study period. For Finland, the authors were confident that statistical data describing deaths at work avoid at least some of these problems. To correct for any possible influence of the volume of work in construction on the fatality rate, the researchers also ran a correlation between fatalities and economic activity (including the cubic meters under construction) in the given year. The fatality rate increased when the volume of work went down. So fatalities were not simply a product of the amount of work being done in a year—in fact, there was a negative correlation.

Studies like those by Saloniemi and colleagues may have trouble unseating the almost mythical status of the ideas of Heinrich and Bird. But the evidence tells us that what causes unwanted events toward the bottom of the triangle is not what causes events at the top. They are not predictive of each other. We should of course have suspected as much, given that Heinrich's or Bird's ratios are impossible to replicate in most studies (Manuele, 2011). Indeed, actual data often reveals the exact opposite: an inverted triangle (if that image even still makes sense). Saloniemi and Oksanen (1998) in fact concluded that:

> The present material does not corroborate the hypothesis of the iceberg-like constitution of safety problems. The results are consistent with earlier findings which emphasize the specific nature of fatal accidents, their own distinctive logic and their own causes.

(p. 63)

As you will see in Chapter 9, even the highly popular and much later Swiss cheese model can only predict accidents on the basis of incidents if the common cause hypothesis is true. The difference between an incident and an accident in that model, after all, is the presence or absence or functioning of the last layer(s) of defense. If they held, then it was an incident. If they did not, then it was an accident. The trajectory according to the model, however, is the same: the causes driving something bad through the layers of defense are common to both incidents and accidents.

Another expression of the iceberg myth (Besnard & Hollnagel, 2014) is 'Zero Harm' or the 'Zero Accident Vision.' It promotes a vision of zero tolerance for even small incidents or injuries and gets people to assume that everything is preventable. The idea is that pursuing a zero of everything (particularly the small stuff at the bottom of the triangle) will result in no triangle at all: no harm of any kind or severity.

What happens in an organization committing to zero harm depends heavily on its cultural proclivities. There are historical examples of organizational cultures that had no tolerance whatsoever for any deviation, failure, incident, or even indication of any possibility of harm, and that actively incentivized assertive, immediate conversations about such indications and a collective responsibility to create meaningful interventions or corrections.

Such was the case for Japan's 'zero-accident total participation' campaign in the 1960s, which predated the West's Zero Accident Vision (JICOSH, 1964). Pointing and calling out hazards while on dedicated walks (in which workers were instructed exactly how to 'strike a pose with spirit, straighten themselves and then briskly point to the hazard'), with workers taking responsibility for the collective good, worked in that cultural setting. There is, however, no data from that time about whether this helped prevent major incidents or fatalities.

But now we have data. And the result of a successful implementation of zero tolerance, of course, should not be surprising: fatalities actually become *more* likely (see the study below). Zero harm can foster a climate of risk secrecy and structural learning disabilities, in which workers will be reluctant to share near misses or injuries and incidents that can easily be hidden (Dekker, Long, & Wybo, 2016). In addition, once the vision of zero has (almost) been achieved, the

> ... system becomes mute, and its safety can no longer be tuned. Investments stop being directed at safety and are earmarked towards improving performance; the control and the spontaneous limitation induced by the occurrence of incidents no longer play their role. The system can then brutally and itself in a situation of disastrous accident because its over-stretched performance has given rise to new risks. Beyond a certain incident-reduction quota, the absence of incidents (known or visible) ... does not prevent disastrous accidents from occurring; it may actually even increase the risk of accident (the system gets out of control because of over performance).

Amalberti (2001, p. 120)

FEWER INCIDENTS, MORE FATALITIES (3)

A study of 20 top construction companies in the United Kingdom found that 9 had an explicit zero policy or program in place (Sheratt & Dainty, 2017). Correcting for work volumes and hours worked, accident data for 4 years revealed:

- There were four fatal accidents for companies with zero safety;
- There were zero fatal accidents for companies without zero safety.

And for the same period:

- There were 214 major injuries for companies with zero safety.
- There were 135 major injuries for companies without zero safety.

The authors concluded that:

Taken together, this analysis suggests the possibility of a 'Zero Paradox' in play within construction site safety for large firms: you are marginally more likely to have major/specified accident while working on a large construction site operated by a contractor mobilizing any form of zero safety, than if you are working on a site without it. Zero, for construction on large U.K. sites, actually means a greater risk of injury (or death) in practice.

Based on their study, Sheratt and Dainty concluded that:

These findings also therefore lend weight to the various challenges levelled at Zero; the need for Safety II over Safety I, the need to acknowledge Model 2 alongside Model 1 [see Chapter 2], the need to de-bureaucratize safety as well as broader critiques of the Zero mantra. It suggests that it is perhaps time to abandon the flawed discourse of Zero and instead consider and mobilize more contingent, nuanced and authentic approaches to safety in practice.

(2017, p. 8)

Presenting empirical evidence for and against the efficacy of Zero Vision is problematic. The main issue is that of confounds. Results at a New Zealand aluminum smelter, for instance, showed a reduction in injuries since a zero accident vision was introduced in 1990 (Young, 2014). But, as Zwetsloot et al. (2017) found in this case:

Automation, thereby eliminating hazardous work, was probably the most important successful intervention strategy over the years. Hazards were mostly ameliorated by long-term persistence following the principle of hierarchy of controls for injury prevention. The second most important factor was transformational leadership...

(p. 262)

The dominance of these other factors makes it difficult to attribute any success to the launch of Zero Vision. In fact, Zero Vision may have had little to do with the results. No studies have yet been able to single out the presence or absence of Zero Vision as a separate variable on a comparative basis as to determine its effect on safety outcomes. What we do know from the work of Sheratt (2014), for example, is that the introduction of zero harm creates "disenchantment, incoherence and inconsistency" (p. 737), where many "remain derisive of its achievement within current working contexts" (p. 747).

So why do workers submit to these kinds of programs? Arendt (1967) described the potent mix of submission and cynicism that typifies behavior under totalizing regimes that tell you what to believe in, and who will watch you practice that belief. It leads to psychological detachment where people believe that they are better off showing that they believe in the program, even if they do not at all. This has three components:

- *Optic compliance*: workers do as if they believe the program and its value. When under surveillance or in public, they will make it look as if they care and share the belief;
- *Resignation*: workers have reconciled themselves with the fact that there is little or nothing they can do to change the program or the organization's belief behind it;
- *Cynicism*: workers no longer believe (if we ever did) that such programs help them at all, or that they actually do what they are advertised to do. To them, the program makes other people in the organization feel, or look, good.

An organization, meanwhile, might make efforts to get people to believe in the virtue of the program, by printing posters and embossing shirts with the slogan. Of course, things can go wrong, and sometimes probably do. So an organization's commitment to reduce harm can come from a defensible moral position. But then that commitment can get translated into hiding evidence of (possible) harm, particularly when zero has become a statistical, bonus-linked target (Hopkins, 2015).

This might explain in part why serious, life-changing injuries and fatalities have remained stubbornly constant in many industries for the last 20 years or more. That figure has refused to budge under interventions that were intended to reduce unsafe acts, unwanted worker behaviors, lower-consequence incidents, and minor injuries. As long as Safety Science designs interventions and underwrites advice for industries based on the common cause hypothesis, it might actually be contributing to the unnecessary deaths of workers. What to believe? Some workers might shake their heads and just get on with their business. As Arendt (1967) explained:

> ... cynicism had been an outstanding characteristic... [people] had reached the point where they would, at the same time, believe everything and nothing, think that everything was possible and that nothing was true... [The] audience was ready at all times to believe the worst, no matter how absurd, and did not particularly object to being deceived because it held every statement to be a lie anyhow: they would take refuge in cynicism.

(p. 382)

The consensus in Safety Science is now growing that belief in the common cause hypothesis, and its reification through popular models and cartoons (pyramids, cheese slices) is holding back progress in Safety Science and practice alike (Amalberti, 2001). Recent safety-scientific thinking offers significant alternatives. Current and emergent ideas about drift into failure, about normal work, about 'safety differently' and about resilience will be presented in subsequent chapters. These theories and models suggest that the drivers for serious injuries and fatalities should be sought in normal work—rather than in noncompliant behavior.

4.6.3 CHAIN-OF-EVENTS THINKING AND DECOMPOSITION ASSUMPTIONS

Thinking about accidents in terms of chains-of-events has become intuitive to many people. It may be such common sense that we rarely wonder about its origins anymore, or question the validity and limits of the model. A recent book published by NASA was called *Breaking the Mishap Chain*, and the authors explained that:

> ...accidents and incidents rarely resulted from a single cause, but were the outcome of a chain of events in which altering at least one element might have prevented disaster.

Merlin, Bendrick, and Holland (2012, p. xiv)

This is consistent with Heinrich, if not almost synonymous with his argument. There is a logical pathway that took someone or something to a bad outcome. And when we trace it back, we can find failures and decisions along that pathway that necessarily led to the outcome. As Heinrich suggested, and as is repeated in the NASA quote above, eliminating one of the contributors or containing just one of the events would have stopped the entire chain. The dominoes would not all topple. Chain-of-events thinking, however, is riddled with risks and problems. Let's look at some of the obvious and important problems that show up when we apply this kind of thinking to a safety investigation. You can see how one of the foundations of Safety Science—Heinrich's domino model—has considerable consequences. Here we consider, in turn:

- The consequences and pitfalls of trying to impose a linear chain onto tangled events, including hindsight and outcome bias, counterfactual and judgmental reasoning;
- The decomposition assumptions that go into a chain of events and into separating out the human contribution.

There is much more to say about the pervasive impact of linear thinking on Safety Science. Chapter 9, for example, talks about the Swiss cheese model, which represents a late 20th-century reinvention of Heinrich's ideas and retains the same linear, decompositional assumptions about the creation of failure as Heinrich had. As we get to Chapter 11, the linear model that underlies Heinrich's ideas is reintroduced as the contrast for complexity theory: perhaps more easily understood if we look at the differences.

Another effect of seeing an accident as the outcome of a linear series of events is the pervasive effect of hindsight. The hindsight bias is one of the most consistent biases found in psychology:

> People who know the outcome of a complex prior history of tangled, indeterminate events, remember that history as being much more determinant, leading 'inevitably' to the outcome they already knew.

Weick (1995, p. 28)

Hindsight allows us to change past indeterminacy and complexity into order, structure, and oversimplified causality. A standard response after mishaps points to the data that would have revealed the true nature of the situation. In hindsight, there is an overwhelming array of evidence that did point to the real nature of the situation, and if only people had paid attention to even some of it, the outcome would have been different. Confronted with a slew of indications that could have prevented the accident, we wonder how people at the time could not have seen it coming.

When we trace a sequence of dominoes or events back from the outcome—the accident we as outsiders already know about—we invariably come across joints where people had opportunities to revise their assessment of the situation but failed to do so; where people were given the option to recover from their route to trouble, but did not take it. Such counterfactual reasoning may be a fruitful exercise when trying to uncover potential countermeasures against such failures in the future. But saying what people could have done in order to prevent a particular outcome does not explain why they did what they did. The hindsight bias makes it much easier for us to blame workers for unsafe behaviors, or supervisors for shortcomings, or managers for creating unsafe conditions.

And what about accidents being 'caused' by a combination of factors? Heinrich offered the following paragraph in the second through fourth editions of his book on *Industrial Accident Prevention*:

> In this research, major responsibility for each accident was assigned either to the unsafe act of a person or to an unsafe mechanical condition, but in no case were both personal and mechanical causes charged.

Manuele (2011, p. 57)

The choice of language itself is interesting, of course, and makes it difficult not to think about blame ("charging" a cause). But even more important is the fundamental premise that human and mechanical causes can be separated. Since Heinrich, Safety Science can default to 'human error' whenever no mechanical failures are found. It is a forced, inevitable choice that fits nicely into an equation, where human error is the inverse of the amount of mechanical failure. Equation 4.1 shows how we determine the ratio of causal responsibility:

$$\text{human error} = f(1 - \text{mechanical failure}) \qquad (4.1)$$

If there is no mechanical failure, then we know what to start looking for instead. As Leveson (2012) has pointed out, analytic reduction assumes the following:

- The separation of a whole into constituent parts is feasible;
- Subsystems operate independently;
- Analysis results are not distorted by taking the whole apart.
- This in turn implies that the components are not subject to feedback loops and other nonlinear interactions, and that they are essentially the same when examined singly as when they are playing their part in the whole.
- Moreover, it assumes that the principles governing the assembly of the components into the whole are straightforward; the interactions among components are simple enough that they can be considered separate from the behavior of the whole.

Separating material and human contributions in a chain of events is like separating *res extensa* away from *res cogitans*, like Descartes did in the 17th century. As long as we keep doing that we may get results that do not help us a whole lot further. Tom Krause, an erstwhile proponent of behavioral safety, explained that

> we now recognize that this dichotomy of causes, while ingrained in our culture generally and in large parts of the safety community, is not useful, and in fact can be harmful.
>
> **Manuele (2011, p. 57)**

As Snook (2000) explained it, the two classical Western scientific steps of analytic reduction (the whole into parts) and inductive synthesis (the parts back into a whole again) may *seem* to work, but simply putting the parts back together does not capture the rich complexity hiding inside and around an incident. What is needed is a holistic, organic integration. What is perhaps needed is a new form of analysis and synthesis, sensitive to the total situation of organized sociotechnical activity. But that is for later (Chapters 5 and 11, respectively).

STUDY QUESTIONS

1. Why has Heinrich's idea about industrial accident prevention been dubbed 'the broken windows theory' of safety?
2. What are the three central ideas in Heinrich's work?
3. Heinrich initially emphasized improving physical conditions and physical safeguards of work. Later he shifted to eliminating unsafe acts by workers as the most important intervention. What, do you think, accounts for that shift? And what are the consequences for how we manage safety even today?
4. What are the conceptual relationships between the domino theory and the Swiss cheese model of accident causation?
5. How did Frank Byrd try to enhance or improve on the results of Heinrich?
6. Explain the links between psychological behaviorism and behavioral safety.

7. What is the common cause hypothesis, and why do we need to separate it from any claims about a proportional relationship between incidents of different severity?

8. In which safety practices and models of accident causation since Heinrich can you recognize the common cause hypothesis?

9. Can you identify 'human error' as (1) an attribution, as (2) a political act and as (3) a consequence of trouble further upstream in how the concept is being treated in your own organization (or elsewhere)?

10. What are some of the consequences of using productivity measures as safety measures?

REFERENCES AND FURTHER READING

Amalberti, R. (2001). The paradoxes of almost totally safe transportation systems. *Safety Science, 37*(2–3), 109–126.

Arendt, H. (1967). *The origins of totalitarianism* (3rd ed.). London, UK: George Allen & Unwin Ltd.

Barnett, A., & Wang, A. (2000). Passenger mortality risk estimates provide perspectives about flight safety. *Flight Safety Digest, 19*(4), 1–12.

Besnard, D., & Hollnagel, E. (2014). I want to believe: Some myths about the management of industrial safety. *Cognition, Technology and Work, 16*(1), 13–23.

Bird, R. E., & Germain, G. L. (1985). *Practical loss control leadership.* Loganville, GA: International Loss Control Institute.

BP. (2010). *Deepwater horizon accident investigation report.* London, UK: British Petroleum.

Byrne, G. (2002). *Flight 427: Anatomy of an air disaster.* New York, NY: Springer-Verlag.

CSB. (2007). *Investigation report: Refinery explosion and fire, BP, Texas City, Texas, March 23, 2005* (Report No. 2005-04-I-TX). Washington, DC: U.S. Chemical Safety and Hazard Investigation Board.

Dekker, S. W. A., Long, R., & Wybo, J. L. (2016). Zero vision and a Western salvation narrative. *Safety Science, 88,* 219–223.

DePasquale, J. P., & Geller, E. S. (1999). Critical success factors for behavior-based safety: A study of twenty industry-wide applications. *Journal of Safety Research, 30*(4), 237–249.

DOE. (2002). *The department of energy behavior based safety process: Volume 1, summary of behavior based safety* (DOE Handbook 11/05/02). Washington, DC: Department of Energy.

GAO. (2012). *Workplace safety and health: Better OSHA guidance needed on safety incentive programs* (Report to congressional requesters, GAO-12-329). Washington, DC: Government Accountability Office.

Geller, E. S. (2001). *Working safe: How to help people actively care for health and safety.* Boca Raton, FL: CRC Press.

Graham, B., Reilly, W. K., Beinecke, F., Boesch, D. F., Garcia, T. D., Murray, C. A., & Ulmer, F. (2011). *Deep water: The Gulf oil disaster and the future of offshore drilling (Report to the President).* Washington, DC: National Commission on the BP Deepwater Horizon Oil Spill and Offshore Drilling.

Heinrich, H. W., Petersen, D., & Roos, N. (1980). *Industrial accident prevention* (5th ed.). New York, NY: McGraw-Hill.

Hollnagel, E. (2004). *Barriers and accident prevention.* Aldershot, UK: Ashgate.

Hollnagel, E. (2014). *Safety I and safety II: The past and future of safety management.* Farnham, UK: Ashgate Publishing Co.

Hopkins, A. (2001). *Lessons from Esso's gas plant explosion at Longford.* Canberra, Australia: Australian National University.

Hopkins, A. (2006). What are we to make of safe behaviour programs? *Safety Science, 44*, 583–597.

Hopkins, A. (2015). *Risky rewards: How company bonuses affect safety*. Farnham, UK: Ashgate Publishing Co.

HSA. (2013). *Behavior based safety guide*. Dublin: Health and Safety Authority.

Jensen, C. (1996). *No downlink: A dramatic narrative about the challenger accident and our time* (1st ed.). New York, NY: Farrar, Straus, Giroux.

JICOSH. (1964). *Concept of "Zero-accident Total Participation Campaign"*. Tokyo: Japan International Centre for Occupational Safety and Health, Ministry of Labor.

Kern, T. (1998). *Flight discipline*. New York, NY: McGraw-Hill.

Kerridge, I., Lowe, M., & McPhee, J. (2005). *Ethics and law for the health professions*. Sydney, Australia: Federation Press.

Krause, T. R. (1997). *The behavior-based safety process: Managing involvement for an injury-free culture*. New York, NY: Van Nostrand Reinhold.

Krause, T. R., & Seymour, K. J. (1999). Long-term evaluation of a behavior based method for improving safety performance: A meta-analysis of 73 interrupted time-series replications. *Safety Science, 32*, 1–18.

Krokos, K. J., & Baker, D. P. (2007). Preface to the special section on classifying and understanding human error. *Human Factors, 49*(2), 175–177.

Leveson, N. G. (2012). *Engineering a safer world: Systems thinking applied to safety*. Cambridge, MA: MIT Press.

Manuele, F. A. (2011, October). Reviewing Heinrich: Dislodging two myths from the practice of safety. *Professional Safety*, 56(10), 52–61.

Marsh, T. (2013). *Talking safety: A user's guide to world class safety conversation*. Farnham, UK: Gower Publishing.

Merlin, P. W., Bendrick, G. A., & Holland, D. A. (2012). *Breaking the mishap chain: Human factors lessons learned from aerospace accidents and incidents in research, flight test, and development*. Washington, DC: National Aeronautics and Space Administration.

Newlan, C. J. (1990). *Late capitalism and industrial psychology: A Marxian critique* (Master of Arts), San Jose State University, San Jose, CA. (1340534).

O'Neill, S., McDonald, G., & Deegan, C. M. (2015). Lost in translation: Institutionalised logic and the problematisation of accounting. *Accounting, Auditing & Accountability Journal, 28*(2), 180–209.

Perrow, C. (1984). *Normal accidents: Living with high-risk technologies*. New York, NY: Basic Books.

Rankin, W. (2007). Maintenance error decision aid investigation process. *Aero Magazine, 7*, 14–21.

Salas, E., & Maurino, D. E. (Eds.). (2010). *Human factors in aviation*. San Diego, CA: Academic Press.

Saloniemi, A., & Oksanen, H. (1998). Accidents and fatal accidents: Some paradoxes. *Safety Science, 29*, 59–66.

Shappell, S. A., & Wiegmann, D. A. (2001). Applying reason: The human factors analysis and classification system. *Human Factors and Aerospace Safety, 1*, 59–86.

Sharman, A. (2014). *From accidents to zero: A practical guide to improving your workplace safety culture*. London, UK: Maverick Eagle Press.

Sheratt, F. (2014). Exploring 'Zero Target' safety programmes in the UK construction industry. *Construction Management and Economics, 32*(7–8), 737–748.

Sheratt, F., & Dainty, A. R. J. (2017). UK construction safety: A zero paradox. *Policy and practice in health and safety, 15*(2), 1–9.

Snook, S. A. (2000). *Friendly fire: The accidental shootdown of US Black Hawks over Northern Iraq*. Princeton, NJ: Princeton University Press.

Storbakken, R. (2002). An incident investigation procedure for use in industry. *Masters of Science*, University of Wisconsin-Stout, Menomonie, WI.

Vette, G. (1984). *Impact Erebus*. New York, NY: Sheridan House Inc.

Warr, P. (Ed.). (1987). *Psychology at work* (3rd ed.). London, UK: Penguin.

Watson, R. I. (1978). *The great psychologists* (4th ed.). Philadelphia, PA: Lippincott.

Weick, K. E. (1995). *Sensemaking in organizations*. Thousand Oaks, CA: Sage Publications.

Woods, D. D., Dekker, S. W. A., Cook, R. I., Johannesen, L. J., & Sarter, N. B. (2010). *Behind human error*. Aldershot, UK: Ashgate Publishing Co.

Wright, L., & van der Schaaf, T. (2004). Accident versus near miss causation: A critical review of the literature, an empirical test in the UK railway domain, and their implications for other sectors. *Journal of Hazardous Materials, 111*(1–3), 105–110.

Yorio, P. L., & Moore, S. M. (2018). Examining factors that influence the existence of Heinrich's safety triangle using site-specific H&S data from more than 25,000 establishments. *Risk analysis, 38*(4), 839–852.

Young, S. (2014). From zero to hero: A case study of industrial injury reduction: New Zealand Aluminium Smelters Limited. *Safety Science, 64*, 99–108.

Zwetsloot, G. I. J. M., Kines, P., Wybo, J. L., Ruotsala, R., Drupsteen, L., & Bezemer, R. A. (2017). Zero accident vision based strategies in organizations: Innovative perspectives. *Safety Science, 91*, 260–268.

5 The 1940s and Onward
Human Factors and Cognitive Systems Engineering

KEY POINTS

- In the course of the 20th century, the human became increasingly acknowledged to be a recipient of safety trouble—trouble that was created upstream and then handed down to them by their tools, technologies, organizations, working environments, or tasks.
- Human factors was a field that grew from the insights of engineering psychologists in the 1940s. Confronted with the role of the human in increasingly complex technological systems, it represented an important hinge in this thinking about the relationship between humans, systems, and safety.
- Unlike behaviorism, human factors did not see technology, tools, or working environments as fixed or predetermined by technological capabilities alone. Instead, they were considered malleable, and they should be adapted to human strengths and limitations.
- Individual differences were less important than devising technologies and systems that would resist or tolerate the actions of individuals, independent of their differences. Safety problems had to be addressed by controlling the technology.
- Mental and social phenomena became important for understanding how best to design and engineer technologies that fit the strengths and limitations of human perception, memory, attention, collaboration, communication, and decision-making. This was the first cognitive revolution (1940s and 1950s), taking safety research away from behaviorism.
- The second cognitive revolution (1980s and 1990s) departed from the overly technicalized, individualist, laboratory task based and mentalist information processing paradigm of the first. It took the study of cognition 'into the wild' and set out to understand people's collaborative sense-making in their interaction with actual complex, safety-critical technologies.
- Cognitive systems engineering has taken the ideas of human factors and the second cognitive revolution into complex sociotechnical systems, where a multitude of humans and technological artifacts

manage safety-critical operations. Cognitive systems engineering does not divide systems up into human and machine components. Instead, it studies the joint cognitive system, with its functions and goals, as its unit of analysis. It has contributed to Safety Science with concepts such as mode error, automation surprise, and data overload.

5.1 INTRODUCTION

5.1.1 THE PLACE OF HUMAN FACTORS IN THE 20TH CENTURY

In the course of the 20th century, safety research increasingly revealed systematic, predictable organizational, operational, and design factors at work behind the creation of mishaps—not simply erratic individuals. The human became increasingly acknowledged as a recipient of safety trouble—trouble that was created upstream and then handed down to them by their tools, technologies, organizations, working environments, or tasks. As safety researchers Fleming and Lardner said about the popularity of behavior-based safety interventions:

> Whilst a focus on changing unsafe behaviour into safe behaviour is appropriate, this should not deflect attention from analyzing why people behave unsafely. To focus solely on changing individual behaviour without considering necessary changes to how people are organized, managed, motivated, rewarded and their physical work environment, tools and equipment can result in treating the symptoms only, without addressing the root causes of unsafe behavior.
>
> **Hopkins (2006, p. 588)**

This was not entirely new, of course. Even Taylorism contributed to the groundwork. For sure, it was preoccupied with fitting people to the task and to existing technologies. It engaged in measurement of people and their tasks, timing the execution of work, studying the motions of people, recording, administering, and developing the supervisory and bureaucratic machinery to make it all possible. But Taylorism also looked critically at the system—at plant and production-line design, planning, supervision—as it realized that just asking people to work harder with imperfect technologies was not very useful.

Heinrich, too, had seen how the potential for failure comes from conditions outside the person (recall Figure 4.2. in the previous chapter) and tried to trace how these conditions influence and constrain what makes sense to the person once on the job. As you probably also recall from Chapter 4, his insights were subsequently taken as a license to zero in on the fallible person and to target their imperfections with behavior modification programs and other interventions focused on the erratic individual. So, even as Heinrich's legacy ended up focusing on workers and their behaviors after all, he began with a focus on the system as the source of trouble for workers.

Human factors more firmly pulled the focus of Safety Science back to the antecedent system that in certain cases set people up for failure—almost independent of their personal characteristics. Why did it make sense for people to do what they did? What was the contribution of their physical work environment, tools, and equipment? Whereas behavior change toward a greater likelihood of safe outcomes might still be the goal, it did not mean that behavior itself should be the target. In fact, human factors argued we needed to target *other* things—people's working conditions and environment—to drive their behavior change:

> There is a basic fallacy in concluding that because the great majority of accidents are the result of human factors, in particular unsafe behaviour, the solution is to try to modify this behaviour directly.... It is noteworthy that 'human factors' specialists generally do not commit the fallacy identified above. It has long been recognized in the airline industry, for instance, that errors and violations are involved in the great majority of aircraft accidents, but the response of the industry has not been to promote safe behaviour programs; rather it has been to identity the factors that have contributed to these errors and violations such as cockpit layout, inadequate crew resource management and so on, and to work at changing these.
>
> **Hopkins (2006, pp. 585–587)**

Paying attention to what people need to do (or avoid) in order to create success at the sharp end despite the operational and organizational vulnerabilities and error traps yields important insights about what we might be able to do with the system they work in, so as to avoid them from getting in that situation in the first place. Human factors was the field that pioneered some of these insights and put them on the safety-scientific radar.

5.1.2 HUMAN FACTORS CHANGE BEHAVIOR, BUT NOT BY TARGETING BEHAVIOR

Human factors grew from the exploits of engineering psychologists in the 1940s. It represents an important hinge in our thinking about the relationship between humans, systems, and safety:

- Technology, tools, or working environments were not seen as fixed or simply predetermined by technological capabilities alone;
- Rather, they were considered malleable, and they should be adapted to human strengths and limitations;
- Individual differences were less important than devising technologies and systems that would resist or tolerate the actions of individuals, independent of their differences;
- Safety problems had to be addressed by controlling the technology.

As you will see in this chapter, human factors both needed and helped drive a change in psychological theorizing. Behaviorism was the school of psychology that dominated the first half of the 20th century, as you recall from the previous chapter:

- Behaviorism was focused on shaping people's behavior to existing environments, and did not consider changing the environment in order to change people's behavior.

TABLE 5.1

The Shift in Safety Focus between Behaviorism and Human Factors

Behaviorist Assumptions about Safety and People	Human Factors Assumptions about Safety and People
Human is cause of trouble	Human is recipient of trouble
Safety interventions target the human through selection, training, sanctions, and rewards	Safety interventions target the organizational and technological environment
Technology and tasks are fixed, the human has to be picked for them and adapted to them	Technology and tasks are malleable, and should be adapted to human strengths and limitations
Individual differences are key to fitting the right human to the task	Technologies and tasks should be devised to be error-resistant and error-tolerant, independent of individual differences
Safety problems addressed by controlling the human	Safety problems addressed by controlling the technology
Psychology is useful for influencing people's behavior, it allows us to engineer people to fit our systems	Psychology is useful for understanding perception, attention, memory, and decision-making, so that we can engineer systems fit for people

- Rather, it aimed to change behavior by focusing on behavior, using clever systems of selection, rewards, and sanctions.
- Investigating the mind to understand why people did what they did was not as important as working with their behavior to get them to do the right thing. It was a psychology that was not interested in mental phenomena.

The second half of the 20th century, in contrast, saw the growth of cognitive psychology, social psychology, and engineering psychology. Mental and social phenomena would once again become important for understanding how best to design and engineer technologies that fit the strengths and limitations of human perception, memory, attention, collaboration, communication, and decision-making. It would have far-reaching consequences for where we would seek safety improvements: inside people, or outside people? (Table 5.1).

In this chapter, you will find more detail about the emergence of 'human factors' as a discipline around the middle of the 20th century. It represents a strong transformation in our thinking about the sources of safety and risk—and where we should be directing our efforts to control them.

5.1.3 THE EMERGENCE OF 'HUMAN FACTORS'

We made a normal takeoff in a heavily loaded F-13. After leaving the ground, I gave the signal to the copilot for gear up. The aircraft began to settle back toward the ground as the flaps were retracted instead of the gear. Elevator control was sufficient to prevent contacting the ground again and flying speed was maintained. The copilot, in a scramble to get the gear up, was unable to operate the two safety latches on the gear switch and I was forced to operate this switch to raise the gear. Everything would have been all right had not the number-one engine quit at this point. Full rudder and aileron would not maintain directional control and the airspeed and altitude would not permit any retarding of power on the opposite side. As the gear retraced, the airspeed built up sufficiently to enable me to maintain directional control and finally to gain a few feet of altitude. The combination of difficulties almost caused a crash. I believe that the error of raising the flaps instead of the flaps instead of the gear was caused by inexperience of the copilot and the location of the two switches. Numerous instances of this error were reported by members of my squadron while overseas, although no accidents were known to be caused by it. However, it could result in very dangerous situations especially when combined with some other failure".

Fitts and Jones (1947, p. 350)

Thus, reports of the commander of a World War II (WWII) era aircraft about his narrow escape from a crash. The narrative comes from one of the first systematic investigations that went beyond the label of 'pilot error.' In 1947, Paul Fitts and Richard Jones, building on pioneering work by people like Alphonse Chapanis, wanted to get a better understanding of how the design and location of controls influenced the kinds of errors that pilots made. Using recorded interviews and written reports, they built up a large corpus of accounts of errors in using aircraft controls. The question asked of pilots from the Air Materiel Command, the Air Training Command, and the Army Air Force Institute of Technology, as well as former pilots, was this:

Describe in detail an error in the operation of a cockpit control, flight control, engine control, toggle switch, selector switch, trim tab, etc. which was made by yourself or by another person whom you were watching at the time.

(p. 332)

Practically all Army Air Force pilots, they found, regardless of experience and skill, reported that they sometimes made errors in using cockpit controls. Others, not involved in these studies, had started to notice the same problem. It was a confluence of evidence and insight that formed the beginning of a remarkable transformation in how we think about the relationship between safety and the human factor:

It happened this way. In 1943, Lt. Alphonse Chapanis was called on to figure out why pilots and copilots of P-47s, B-17s, and B-25s frequently retracted the wheels instead of the flaps after landing. Chapanis, who was the only psychologist at Wright Field until the end of the war, was not involved in the ongoing studies of human factors in equipment design. Still, he immediately noticed that the side-by-side wheel and flap controls—in most cases identical toggle switches or nearly identical levers—could easily be confused. He also noted that the corresponding controls on the C-47 were not adjacent and their methods of actuation were quite different; hence C-47 copilots never pulled up the wheels after landing. Chapanis realized that the so-called pilot errors were really cockpit design errors and that by coding the shapes and modes of operation of controls, the problem could be solved. As an immediate wartime fix, a small, rubber-tired wheel was attached to the end of the wheel control and a small wedge-shaped end to the flap control on several types of airplanes, and the pilots and copilots of the modified planes stopped retracting their wheels after landing. When the war was over, these mnemonically shape-coded wheel and flap controls were standardized worldwide, as were the tactually discriminable heads of the power control levers found in conventional airplanes today.

Roscoe (1997, p. 2)

A study like that of Fitts and Jones confirmed these insights, demonstrating how features of WWII airplane cockpits systematically influenced the way in which pilots made errors. Human factors was born out of an effort to make the work of operators more manageable; to make it more successful and less arduous. 'Pilot error' was put in quotation marks—denoting the researchers' suspicion of the term. The point was not the pilot error. That was just the symptom of trouble, not the cause of trouble. The remedy did not lie in telling pilots not to make errors. Rather, Fitts and Jones argued, change the tools and you eliminate the errors of people who work with those tools. Skill and experience, after all, had very little influence on error rates: getting people trained better would not have much impact. Rather, you change the environment, and by doing so you change the behavior that goes on inside of it.

> What happened was that in the last decades of the twentieth century, experts dealing with mechanization in many settings all moved away from a focus on the careless or cursed individual who caused accidents. Instead, they now concentrated, to an extent that is remarkable, on devising technologies that would prevent damage no matter how wrongheaded the actions of an individual person, such as a worker or a driver. In the withering away of the idea of accident proneness, it is possible to see the growing dominance of a new social strategy, solving safety problems by controlling the technological environment.

Burnham (2009, p. 5)

The insight would accelerate changes in how we studied and tried to improve human performance. Up to WWII, and for a decade or so beyond, the study of human behavior and cognition was governed by a fundamentally different assumption: the world was a fixed place. The only way to change behavior in it was to change something about the people. Through an intricate system of selection and training, rewards and sanctions, the right sort of behavior could be picked out and shaped so that it became adapted to its environment. Initially, the only way to fit the human to the machine was by trial and error. Either the human got the machine under his or her control, and was accepted, or did not and was rejected. Increasingly clever and intricate devices (e.g., simulators, particularly part-task simulators) had been developed to make such selections more systematic. Selection and training remained the major psychological preoccupation—until WWII. The reasons were largely pragmatic:

- WWII brought such a pace of technological development that behaviorism was no longer an adequate answer to the questions of safely and effectively matching humans and systems. Think, for example, about the problems of operator vigilance or decision-making. These were immune against typical behaviorist interventions. Up to that point, psychology had largely assumed that the world was fixed, and that humans had to adapt to its demands through selection and training. Human factors showed that the world was not fixed. Changes in the environment could easily lead to performance increments not achievable through behaviorist interventions. In a behaviorist intervention, performance had to be shaped after features of the world. In a human factors intervention, features of the world were shaped after the limits and capabilities of performance.

• The WWII was a total war, involving great masses of men and women. It was no longer possible to adopt Tayloristic principles of selecting a few specialized individuals to match a preexisting job or system. There was too much work to be done, and too large a population to do it with. Before WWII, it was possible to try to match and shape a carefully selected human being to a fixed task or system. Now the physical characteristics of people's operating environment had to become more widely accommodating of general human limitations and capabilities.

As usual, some conditions of possibility were already in place before WWII. Human factors research, for example, was able to borrow from aeromedical research done during the interbellum, or the period between the two world wars (Meister, 1999). This research tested the limits of human tolerance and endurance in environmental extremes. It established concrete limits of human performance in these environments, *and* gave credence to the overall idea that there are general rules and limits to human performance. It would not matter who would be put in that environment, all people would at some point suffer from the same issues.

This was the premise that Fitts and Jones took into their study of cockpit control design too. They opened their 1947 paper with an insight that could hardly be simpler, and hardly more profound. "It should be possible to eliminate a large proportion of so-called 'pilot-error' accidents by designing equipment in accordance with human requirements" (Fitts & Jones, 1947, p. 332). An example in the empirical data from their paper illustrated the need:

> I was acting as control for basic students shooting a stage. At completion of the day's flying and after all students had been dispatched to home base, I started up my airplane, BT-13, taxied out to the takeoff strip, ran through pre-takeoff checks and proceeded to advance the throttle. I held the plane down to pick up excess airspeed and as it left the ground, proceeded to pull back the propeller control to low RPM. Immediately, the engine cut out and I could see nothing but fence posts at the end of the field staring me in the face. Luckily, I immediately pushed the prop control forward to high RPM's and the engine caught just in time to keep from plowing into the ground. You guessed it. It wasn't the prop control at all. It was the mixture control[1].

Fitts and Jones (1947, p. 349)

Indeed, WWII pilots mixed up throttle, mixture, and propeller controls because the locations of levers kept changing across different cockpits. The study of Fitts and Jones showed that such errors were not surprising, random degradations of human

---

[1] The mixture control on this sort of piston engine controls the inflow of the air/fuel mixture into the engine's cylinders. At the backstop of the mixture control, the air/fuel is cut off altogether, sending nothing to the cylinders.

performance. Rather, they were actions and assessments that made sense once researchers understood features of the world in which people worked, once they had analyzed the situation surrounding the operator. Human errors are systematically connected to features of people's tools and tasks. It may be difficult to predict when or how often errors will occur (though human reliability techniques have certainly be tried: see Chapter 7). With a critical examination of the system in which people work, however, it is possible to anticipate *where* errors will occur. Human factors have worked off this premise ever since: the notion of designing error-tolerant and error-resistant systems is founded on it:

- **Error-resistant systems** oppose, retard, or do not even invite errors. Chapanis' flap-formed flap handle and wheel-formed wheel handle together are an error-resistant design. An Automatic Teller Machine that will not give you your money until you have pulled your bank card out of it is resistant to you walking away with your cash but forgetting your card.
- **Error-tolerant systems** are those that do not necessarily oppose, retard errors, but are relatively forgiving when they occur. Examples include the sorts of (sometimes annoying) questions asked by an interface (are you sure you want to do this?), but also the insertion of a squat (or Weight-On-Wheels, or WOW) switch in the undercarriage of an aircraft. Even if a pilot were to retract the wheels, if the switch has detected weight on the wheels, it will lock out the instructions from the retraction handle.

5.1.4　Work Inside and Outside the Research Laboratory

Human factors research expanded during the years following WWII. The largest scale of this expansion was in the United States (and also the UK), where the Cold War fueled a growth in Department of Defense-sponsored research laboratories. Some of these were exclusively devoted to human performance research, and those that were not, did contribute. The U.S. Office of Naval Research, for example, sponsored a textbook on Human Engineering as early as 1947, which evolved into the first textbook of the newly emerging discipline. Chapanis was its first author. The Aeromedical Research Unit at Wright Field in Ohio went on to determine human tolerance limits for high-altitude environments, among many other projects. And the United States was of course far from alone. By 1969, the Royal Air Force Institute of Aviation Medicine at Farnborough in the United Kingdom counted 50 human factors professionals out of a staff of 200.

Human factors also expanded outside of the military and its industrial complex. Bell Labs, for example, established a human factors group as early as 1946. Its purpose was, in part, to advise on the design of keyboards and numeric entry devices. The kind of research that was conducted across these places forced psychologists to work in close collaboration with designers, engineers, and medical scientists. It also forced engineers to listen to psychologists. As a result, experimental psychology was morphing away from pure laboratory studies, into an engineering or human factors psychology that pursued and tested practical answers to the encountered problems.

No tradition took such bold early steps out of the laboratory as the Francophone one did (Hoffman & Deffenbacher, 1993). A defining feature of this tradition was its emphasis on field studies, which involved in-depth observations and analyses of work in context (De Keyser & Woods, 1990). The aim was to contribute to the design of workplaces and technologies and avoid leaving a gap between the envisioned system tested in a lab and the design-in-practice. The question that animated this tradition was 'what actually happens' in a workplace. Francophone human factors, as a result, embraced a much broader notion of human-machine systems earlier on than Anglophone research did. The study of organization and social context as 'constraints' or 'work conditions' showed up early in the work of Jacques LePlat before they were popularized in, for example, the Anglophone Swiss cheese model (see Chapter 9) (Rasmussen, Duncan, & Leplat, 1987).

Those who were there in the early days would describe it as a heady time (Meister, 1999; Roscoe, 1997). Technological developments continued unabated, the research needs were great, and there was generally enough support for the research that needed to be done. The safety consequences for not considering the 'human factor' were deemed unacceptable, and there was enough evidence that our understanding of human capabilities and limitations were not keeping up with the pace of technology:

> By the 1950's and 1960's, it appeared as though aircraft designs were reaching the limit of human performance in terms of pilot workload and task saturation. Such aircraft as the F-4 Phantom and B-52 Stratofortress featured classic examples of steam gauge-type cockpits, with their confusing array of dials and switches. Pilots often found instrument faces too small to see clearly (especially during violent maneuvers), or instruments and switches were inconveniently placed, sometimes hidden behind other equipment, and difficult to reach.
>
> **Merlin, Bendrick, and Holland (2012, p. x)**

The gradual digitization of systems that people had to work with, driven in part by the 1960s Apollo program to put a man on the moon (Mindell, 2008), provided yet more acceleration for human factors as a research and engineering discipline. The need for it remained. In 1981, the U.S. Government Accountability Office warned that weapon systems procured by the Department of Defense often could "not be adequately operated, maintained or supported" (Meister, 1999, p. 157). The role of the human in interaction with these systems was fundamentally a safety-critical one, and not just in relation to their operation. Maintenance and support were just as vulnerable to mismatches between system design and human capabilities.

The widespread introduction of computerization and automation in fields ranging from nuclear power to aviation to healthcare would raise similar issues in the 1990s. It gave rise to a whole new take on human factors as a field, and created the need for what we now know as cognitive systems engineering. This new discipline no longer saw systems as separated into human and machine bits—connected through perceptions and actions—but rather as a joint cognitive system. The unit of analysis was the cognitive system, whose processes and functions should be described at that joint level, not at the level of the individual components constituting it (e.g., humans, engineered machines, computers).

In the meantime, the message of human factors had become quite empowering. The premise was that by designing things more smartly, human factors can make bad things go away. But there is a growing literature on how particular designs not only make us less dumb or less error-prone, but amplify capabilities and capacities, making us considerably smarter (Norman, 1993). It has been augmented by the pursuit of pleasurable products. This deploys knowledge of human factors and ergonomics to actively *add* to the quality and enjoyment of life, not just to reduce the lack of it (Jordan, 1998).

In safety, too, safe systems defined merely by the absence of negative events are making room for the idea that safe, or resilient, systems are capable of recognizing and adapting to challenges and disruptions—even those that fall outside the system's original design base. Rather than safety as an absence of negatives, it becomes defined by the presence of capacities, capabilities, and competencies (Hollnagel, Woods, & Leveson, 2006). Yet the basic commitment of human factors and safety could be seen as this: to make the world a better place, a more workable and survivable place, a more enjoyable place and perhaps, most recently, a more sustainable place.

But we are getting ahead of ourselves. Let's turn the clock back a little bit and first look at what human factors did to psychology, and vice versa. The relationship has had important consequences for Safety Science. It has fundamentally altered the questions we ask, the terms we use, and the models we have come to take for granted. Let's turn to that now.

5.2   HUMAN FACTORS AND CHANGES IN PSYCHOLOGY

5.2.1   BEHAVIORISM: CHANGING THE LEGACY

Remember from the previous chapter how James Watson wanted to instill in psychology a renewed sense of scientific credibility. This was during the first decades of the 20th century. Psychology could never be the science of the mind, he claimed, because the mind could not be studied scientifically. The mind was a black box, and we would better leave it closed. What mattered was putting things into the black box (stimuli) and studying what came out of it (responses). Introspection as a method to find out what happened in the mind was useless, unreliable, and unscientific. Behaviorism did not want any references to or investigations of consciousness. Psychologists had to turn their focus exclusively to phenomena that could be registered and described objectively by independent observers. This meant tightly controlled experiments that varied subtle combinations of rewards and sanctions in order to condition organisms (anything from mice to pigeons to humans) into particular behaviors. The outcome of such experiments was there for all to see, with no need for introspection. It all revolved around the stimulus and the response (or the so-called S-R complex). It allowed psychology to build a science for how to understand, manipulate, and control human behavior.

This logic, however, ran into trouble from the 1940s onward. The main driver was rapid technological development. It overtook the usefulness of the S-R framework. It called for the black box of the mind to be opened and to begin to understand what was actually going on inside of it. Just imagine the problem of an operator who has been watching a primitive (or, as it is known, primary) radar screen, on which she or he must spot incoming enemy aircraft. Imagine the issues that confront anybody (including the operator) who is concerned with the effectiveness and safety of that work:

- Vigilance (sustained attention);
- Long-term memory images of what to look for and operator expectancies of what to see;
- Perception and discrimination of stimuli;
- Things to keep in short-term memory while assembling a mental picture of what is going on;
- Prospective memory (remembering to do something when already engaged in some other task);
- Decision-making when stimuli are detected;
- Response selection;
- Criterion to be applied to respond to a stimulus;
- Clarity or sensitivity of the data generated by the stimulus (e.g., speed, heading, size, timing).

None of these issues can be studied very fruitfully, let alone influenced meaningfully, by behaviorism. What can behaviorism say, understand, or do about problems of sustained attention? Or working memory? The problems that human factors was confronted with, and helped make intelligible, created a need for understanding what went on in the minds of operators. This had essentially been forbidden territory. Questions about consciousness were ruled out. Introspection or other methods for creating access to internal phenomena were not allowed under behaviorism. Now the need for understanding mental processes had come back. The need was met by cognitive psychology, and at its very heart was the idea of the human as an information processor.

5.2.2   The First Cognitive Revolution: Information Processing

How are we to imagine or visualize what happens inside of the head? People have always used metaphors available in the world around them to model the workings of the mind. For instance, the processes of memory (encoding information, mixing it up, forgetting it) have historically been modeled on contemporary technologies for storing information, from wax tablets to paper to photography to gramophones to conveyor belts. In the 19th century, prominent U.S. psychologist William James was the first to distinguish between primary memory, with a limited capacity, and long-term memory. There was no good, practical simile in the real world for this at the time, but the postwar period would supply one: the computer. In one model, working memory was imagined as a visuospatial scratchpad for holding visual information, and a phonological loop that was used to keep auditory information "alive" for a short while. A central executive (or central processing unit) would then direct attention and coordinate processes, deciding what would make it into long-term memory and what would be discarded.

The first cognitive revolution reintroduced 'mind' as a legitimate object of study. Rather than manipulating the effect of stimuli on overt responses, it concerned itself with 'meaning' as a central concept. What did the stimulus mean (i.e., was it an enemy aircraft)? The aim of a renewed cognitive psychology was to discover and describe meanings that people created out of their encounters with the world, and then to propose hypotheses for what meaning-making processes were involved (Bruner, 1990). The very metaphors, however, that legitimized the reintroduction of the study of mind, also created significant problems for our subsequent understanding. The first cognitive revolution turned the mind into something overly technicalized.

The radio and the computer, two technologies accelerated by developments during the World War II, quickly captured the imagination of those once again studying mental processes. These were great similes of mind, able to mechanistically fill the black box (which behaviorism had kept shut) between stimulus and response:

- The innards of a radio showed filters, channels, and limited capacities through which information flowed. Not much later, all those words appeared in cognitive psychology. Now the mind had filters, channels, and limited capacities too.
- The computer was of course even better, containing a working memory, a long-term memory, various forms of storage, input and output, decision modules, and central processors. It did not take long for these terms to appear in the psychological lexicon.

What mattered was the ability to quantify and compute mental functioning. Information theory, for example, could explain how elementary stimuli (bits) would flow through processing channels to produce responses. It was not interested in meaning. In fact, information theory does not even contain the concept of 'meaning.' A processed stimulus was deemed informative if it reduced alternative choices, no matter whether the stimulus had to do with Faust or a set of digits from the periodic table. Here are the basic assumptions that most information processing theories share (Eysenck, 1993):

- Information made available by the environment is processed by a series of processing systems (e.g., attention, perception, short-term memory);
- These processing systems transform or alter the information in various systematic ways (e.g., three connected lines are presented to our eyes, but we see a triangle);
- The aim of cognitive psychology is to specify the processes and structures (e.g., long-term memory) that underlie cognitive functioning;
- Information processing in people resembles that in computers.

In its simplest form, an information processing model is stimulus-driven, linear, and serial. It is the outside world that triggers the series of processes, and one process is completed before the next one in the line begins (Figure 5.1).

The basic idea is the human (mind) is an information processor that takes in stimuli from the outside world, and gradually makes sense of those stimuli by combining

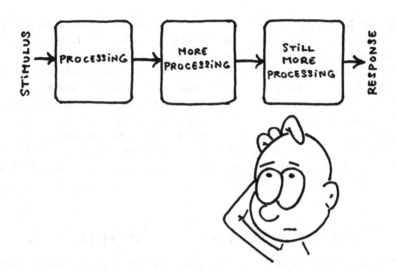

FIGURE 5.1 A human makes sense of a stimulus, finally leading to a response, through various stages of internal mental processing. Information processing model spoof inspired by Neisser (1976). (Drawing by author.)

them with things already stored in the mind. The information processing pathway of typical models mimics this flow, by taking a stimulus and pushing it through various stages of processing, adding more meaning the entire time. Once the processing system has understood what the stimulus means (or stimuli mean) can an appropriate response be generated (through a backflow from center to periphery; brain to limbs) which in turn creates more (new) stimuli to process.

It was obvious, however, that stimulus-driven (or bottom-up) processing did not tell the whole story. People brought expectations and past experience to the process-ing of information. This constituted top-down or conceptually-driven processing. And what if the processes for a cognitive task happen simultaneously? This has been referred to as parallel processing, often presumed to be used in thinking or problem solving. In all of these cases, the similarities between what a brain did and what a computer could be built or programmed to do, were irresistible. Whether this was because we built computers in a likeness of ourselves (i.e., we make computers look the way we think we internally do) or because we actually happen to think like com-puters (more so than that we think like a wax tablet, for instance) is open to debate. But in psychology, computability became the necessary and sufficient criterion for cognitive theories. Mind was equated to program. Through these metaphors, the 'construction of meaning' became the 'processing of information.'

People, as far as information processing psychology is concerned, have no goals of their own. They are merely the recipients of the information the world throws at them. At most, people's behavior gets adjusted in response to feedback generated by actions in the environment. The information processing model might work well for the laboratory experiments that supported it and helped bring it to life. Laboratory studies of perception and decision-making and reaction time reduced stimuli to single snapshots, fired off at the human processing mechanism as one-stop triggers. The information processing model may be helpful, but often only for the constrained, Spartan laboratory settings that keep cognition in captivity.

And yet, it is still being used for safety training for real-world settings today (see below). The simplicity and componential nature of the information processing model both make it attractive for getting ideas across to workers and practitioners. The strengths and weaknesses of the various processing steps can be discussed separately, all the time appealing to a model of cognitive functioning that is easy to understand. Aviation is one field where information processing models have made a significant impact on human factors and safety training (e.g., O'Hare & Roscoe, 1990).

INFORMATION PROCESSING AND SAFETY TRAINING

Here are some excerpts of one such training package (Eurocontrol, 2015). Working in aviation requires practitioners (pilots, controllers, air-side safety, etc.) to take in information from multiple sources, assess this information, prioritize it, and use it to make decisions and take actions. This complete process from sensing information (whether it is aural, visual, mental, kinesthetic, gustative, or olfactory) through to taking action is referred to as 'human information processing,' or information processing. The operator needs to process the type and amount of information within the required timeframe, and do so in an effective manner that leads to suitable responses.

There are at least four stages in information processing:

• sensing
• perceiving
• decision-making
• motor action, or performing

Information processing models can also contain a central executive, an attention-directing mechanism, and various parts of memory (sensory memory, working or short-term memory, long-term memory and motor memory). There is no claim that these are distinct sections of the brain, but rather a reminder of the usefulness to refer to the distinct functions, their goals and vulnerabilities. Most models also suggest at least one feedback loop, where the information processor senses changes that occur due to its own actions in the world. This is how we measure and correct our progress in achieving a task.

The model allows for individual differences and other variations in performance. Information processing capabilities vary from person to person, day to day, place to place, and task to task. Knowing how our information processing capabilities can be limited is important in designing and delegating tasks to ensure that the information processing requirements fall within the capabilities of employees and colleagues (i.e., within their memory, attention and decision-making capabilities) such that the following are minimized:

- Failure to see information;
- Misunderstanding information;
- Handling the information incorrectly;
- Forgetting the information;
- Reacting inappropriately.

Perhaps more importantly, we should understand our own limitations, especially during periods of high workload and/or when particular illusions may go unnoticed."

Through information processing, the psychology of mind was reduced to mechanical components and linkages. These components exchanged information between themselves, along a linear pathway and in proportional fashion. Cognitive and experimental psychology went to the laboratory. The various components of the human information processing system (sensory store, memory, decision-making) were tested in endless series of fractionalized laboratory experiments. The field split into specializations around each of the components of information processing. One such experiment, for instance, famously demonstrated the limits of working memory: humans could keep seven plus or minus two 'elements' stable in working memory for a period of time, but no more (Miller, 1956). With these experiments, the hope in the field was that more of the same would eventually add up to something different: that profound insight into the workings of the mind would emerge from the study of its invoked components. The assumption is that the functioning of the whole can eventually be explained once the functioning of its constituent components has been understood.

When information processing psychology is used as a basis for understanding safety, the focus necessarily becomes the individual. You can see this in the example above, and also in the one that follows below. Information processing, after all, puts the individual at the center. The environment merely serves new inputs to the individual and runs a feedback loop. Failure modes are limited to those inside of an individual: failing to see information, handling information incorrectly, forgetting information, and reacting inappropriately. There is an irony in this. Human factors never intended to see a 'human error' as the internal misfiring of an information processing system. Rather, it reminded us that 'human error' is a label that we put on other people's (or our own) behavior after the fact. This behavior, when studied in relation to its context, makes sense given the tools and tasks that people work with. The 'error' was systematically linked to those tools and tasks. In the mid-1990s, attributing safety problems to internal processing limitations would resurface with the popularization of 'situation awareness.'

5.2.3 Losing Situation Awareness

One term that has retained the information processing model to the present day is 'situation awareness.' Situation awareness (SA) had long been a common sense term in the fighter pilot community. One source sometimes credited is WWI German fighter ace, who reportedly realized the importance of gaining an awareness of the enemy before the enemy gained a similar awareness (Endsley, 1995). The concept of SA was adopted by human factors and safety researchers during the late 1980s, culminating in a special issue of the journal *Human Factors* (van Winsen, Henriqson, Schuler, & Dekker, 2015). Its popularity as an explanation for 'human error,' and a shorthand way of guiding systems designers to assist user 'awareness' seemed unstoppable. It is a "convenient explanation that [people] easily grasp and embrace" (Flach, 1995, p. 155).

CONVICTED FOR LOSING SA

Not long ago a criminal court case was concluded against a professional operator who, in the words of the prosecution (the Crown in this case), had "lost situation awareness" and collided his vehicle with a structure. He had been carrying out the duties of his job as normal that evening, but a whole array of circumstances—weather, equipment distractions, other vehicles, darkness, fatigue—came into play to cause a collision. As the Crown alleged, it was his "loss of situation awareness" that caused an accident. It killed two people. The Crown charged the operator with criminal negligence. The man had already been fired and was now sentenced to 4 years in jail. A man was accused and convicted of something that did not even exist 20 years before.

How do we help defend the practitioner who is accused (implicitly or explicitly) of losing SA? It would be quite difficult (Breakey, van Winsen, & Dekker, 2015). Consider the first sentence of a recent article that introduces a particular model of SA to anesthesia (Schulz, Endsley, Kochs, Gelb, & Wagner, 2013, p. 729): "Accurate situation awareness (SA) ... is integral for providing optimal performance..." But what does 'accurate awareness' actually mean? Accurate relative to what? And what is that 'awareness' or 'consciousness'? Ulrich Neisser warned in 1976 that psychology was not quite ready for grand questions about consciousness. Neisser feared that models of cognition would treat consciousness as if it were just a particular stage of processing in a mechanical flow of information. His fears were justified in the mid-1970s, as many psychological models did exactly that. SA did it again in the 1990s. Awareness, or consciousness, is equated to a stage of processing along an intrapsychic pathway (through 'levels of SA,' such as perception of elements, understanding their meaning, and then projecting their future state). As Neisser pointed out, this is an old idea in psychology. The three levels of SA popular today were anticipated by Freud, who also supplied flow charts and boxes in his Interpretation of Dreams to map the movements from unconscious (level 1) to preconscious (level 2) to conscious (level 3). The popularity of finding a home, a place, a structure for consciousness in the head is irresistible, says Neisser, as it allows psychology to nail down its most elusive target (consciousness) to a box in a flow chart.

Looking at events after the fact, outside-in, we can always show that there was more in the world (that we now know) than there was in somebody else's mind, because in hindsight anybody can show that. And we can then call that difference the practitioner's 'loss of situation awareness.' Look at it like this (see Equation 5.1):

$$\text{Loss of SA} = f\big(\text{what I know now} - \text{what you knew then}\big) \qquad (5.1)$$

Loss of SA, in other words, is the difference between what I know about a situation now (especially the bits highlighted by hindsight) and what somebody else apparently knew about that situation then. Interestingly, SA is nothing by itself, then. It can only be expressed as a relative, normativist function: for example, the difference between what people apparently knew back then and what they could or should have known (or what we know now).

This is of course a normativist. Normativism is the premise that it is possible to generate a 'true' and 'objective' characterization of the practitioner's experience. For example, Parasuraman and colleagues said (Parasuraman, Sheridan, & Wickens, 2008, p. 141) about SA that "there is a 'ground truth' against which its accuracy can be assessed (e.g., the objective state of the world or the objective unfolding of events that are predicted)." This idea, common in much SA work, says that the operator's understanding of the world can be contrasted against (and found more or less deficient relative to) an objectively available state of that world. It rests on the belief that the world is objectively available and apprehensible; that there is such a thing as a-perspectival objectivity. It requires us to supposedly take a 'view from nowhere' (Nagel, 1992), a value-free, background-free, position-free view that is true (the 'ground truth'). But does that even exist? And if it did, could we know it? Taking this normativist position comes at a price. Researchers or investigators learn less about *why* people saw what they did and why that made sense to them, because what they saw was 'wrong' relative to what was decided or predicted to be the 'ground truth.'

If this is how we see things, then we can even blame an operator's complacency for causing the loss of SA. If you are complacent, you pay attention to some things but not to others—which means you lose SA. You should have seen things, yet you did not. But, as Moray and Inagaki (2000) warned:

This point leads to a consideration of the problem of 'situation awareness'. It…is poorly specified. What is the situation of which the operator is required to be aware? In most of the experimental or field studies, it is not well defined or specific. Rather, and for obvious reasons, it is almost as though the investigators want the operator to be aware of 'anything in the environment which might be of importance if it should change unexpectedly'. This is, of course, not the phrase used in research. However, clearly, it is at the back of investigators' minds. The pilot of an aircraft needs a keen situation awareness because if anything abnormal occurs he or she must notice it and respond appropriately. However, there are, logically, an infinite number of events which may occur. Perhaps one, therefore, expects the operator to be aware of the status of just those variables which, if ever they change, represent significant events. But, what are those? If one does not define exactly what the set of events is that the operator must monitor, how can he or she devise an optimal or eutactic monitoring strategy? It is logically quite unreasonable to ask merely for 'situation awareness' as such, since if the set of events of which operators are to be aware is not defined, it is unreasonable to expect them to monitor the members of an undefined set; whilst if one defines a set, there is always the possibility that the dangerous event is not a member of the set, in which case the operators should not have been monitoring it.

(p. 360)

Loss of SA and complacency are concepts that assume the ability to provide a full accounting of goal-directed actions in the pursuit of tasks. That ability can be claimed only by an omniscient, normative arbiter who knows completely and accurately the values and interdependencies of all contextually dependent variables (K. Smith & Hancock, 1995). Such an arbiter would not only have to have privileged access to the activity of practitioners' minds, but also to a complete and infallible view of the unfolding of events (also known as 'the big picture'). Such arbiters (e.g., SA researchers) apply norms, implicitly asserting that their view of the world is 'correct' (or the ground truth) and their subjects' view is deficient. John Flach parodied the circularity that easily comes from this underspecification (Flach, 1995):

- Why did the operators lose SA?
- Because they were complacent.
- How do we know they were complacent?
- Because they lost SA.

This may explain in part why few researchers venture beyond claims of a 'loss of situation awareness.' There is little there. It sheds little light on the actual processes of attentional dynamics, or an understanding of 'awareness' or sensemaking in complex, dynamic situations. It does not help us understand the situation from the perspective of those who were inside it.

NAVY PILOT FAILS TO SEE INFORMATION

Neither Lieutenant Nathan Poloski's body nor his F/A-18 Hornet was ever found in waters almost three miles deep. All that was located in the Western Pacific after his fighter jet collided with another from the same aircraft carrier were his helmet and some pieces of debris. The pilot of the other jet ejected safely and was rescued shortly after.

The Navy accident report, all of eight pages long, was acquired by *The New York Times* under a Freedom of Information Act request (Schmitt, 2015). What remains a mystery is what exactly caused the accident, the report suggests. It was a clear afternoon with good visibility. Both pilots were healthy, properly rested, and under no unusual stress. There were no mechanical problems with either aircraft.

And so what is left, you wonder?

The two pilots involved in the midair, a Vice Admiral opined in reflections on the report when closing the investigation on April 20 this year, should have exercised more of what his military calls SA. In this case, it would have meant not relying only on cockpit instruments but looking outside to spot a looming catastrophe.

The eight-page report had originally only admonished (the dead) Poloski for losing SA. On September 12, 2014, Poloski had been on a practice bombing mission. With 221 h on the Hornet, he was less experienced than the other pilot, a Navy commander, who had taken off from the same deck a minute before to test repairs that had been made on the plane. At 7,000 feet, Poloski turned West and slowed to about 300 miles an hour. Poloski's jet caught up with the other plane and impacted its bottom left rear.

Did anyone ask sufficiently probing questions about the context in which his actions made sense? Poloski had told his mother shortly before deployment that he was looking forward to the mission. He did not go on the mission to do a bad job. It turned out that he was not aware that the other pilot had chosen the same route. And that controllers on the carrier were occupied with landing aircraft. But 'while there is no definitive evidence to suggest either pilot's S.A. or lack thereof directly contributed to this incident, greater S.A. by all parties may have prevented the collision,' the Vice Admiral concluded.

In the report, the focus was firmly on what might have been (or not been) in the heads of the pilots, rather than in their world. The focus was foremost on one of the pilots, Poloski, not on the context. The 'catastrophe,' after all, was 'looming.' All he needed to do was look up.

Based on the helmet investigators found—which had a big crack in it, extending from the bottom right side up to the crown with a hole halfway— they concluded that Poloski must have suffered massive, fatal head trauma. We may surely hope it was swift. But to then admonish someone for a 'lack of S.A.' whose perspective we never shared because we were not there, and with whom we will never be able to talk again?

When looked at from the position of retrospective outsider, a 'loss of SA' can look so very real, so compelling. They failed to notice, they did not know, they should have done this or that. But from the point of view of people inside the situation, as well as potential other observers, these deficiencies do not exist in and of themselves; they are artifacts of hindsight, 'elements' removed retrospectively from a stream of action and experience. To people on the inside, it is often nothing more than normal work. If we want to begin to understand why it made sense for people to do what they did, we have to put ourselves in their shoes. What did they know? What was their understanding of the situation? Rather than construing the case as a "loss of SA" (which simply judges other people for not seeing what we, in our retrospective omniscience, would have seen), there is more explanatory leverage in seeing the crew's actions as normal processes of sensemaking—of transactions between goals, observations, and actions. As Weick (1995) points out, sensemaking is:

> ...something that preserves plausibility and coherence, something that is reasonable and memorable, something that embodies past experience and expectations, something that resonates with other people, something that can be constructed retrospectively but also can be used prospectively, something that captures both feeling and thought... In short, what is necessary in sensemaking is a good story. A good story holds disparate elements together long enough to energize and guide action, plausibly enough to allow people to make retrospective sense of whatever happens, and engagingly enough that others will contribute their own inputs in the interest of sensemaking.

> *(p. 61)*

Even if one does make the concessions to the existence of 'elements,' as Weick does in the above quote, it is only for the role they play in constructing a plausible story of what is going on. A 'plausible story' is hardly something that fits inside one of the boxes of the information processing paradigm, or one of the levels of SA. A second cognitive revolution occurred in the 1990s, taking cognition outside of the laboratory, is understanding.

5.2.4 THE SECOND COGNITIVE REVOLUTION

In his reflections on the popularity of SA, Flach noted that researchers should not focus on the 'awareness' part, but rather on the 'situation' (Flach, 1995). The *situation* holds the constraints and opportunities for behavior, and *it* should be the focus:

- Information processing models (including that of SA) want to point to internal limitations and weaknesses that lead to decreases in human performance;
- What we now want to understand is how the world around people imposes constraints on their goal-directed behavior.

This notion can be traced to Gibson's research on visual perception (Gibson, 1979). The environment, Gibson argued, offers constraints and affordances to behavior. Ask not what is inside your head, he might have said, but what your head is inside of. Gibson was much more interested in *what* was perceived than in the imagined internal mechanisms by which sensations became perceptions. It also understood that people, or organisms were active in their environment. Their moving around, their acting and behaving changes things in the environment, and changes what is visible and what is not. This led Neisser to the idea of the perceptual cycle, a radical departure from the linear information processing model that started with a stimulus on one end and resulted in a percept on the other (Figure 5.2).

Consistent with the idea of this perceptual cycle, Weick would later say that sensemaking never starts: it is ongoing (Weick, 1995). People always are in the middle of things. While we may look back on our own experience as consisting of discrete "events," the only way to get this impression is to step out of that stream of

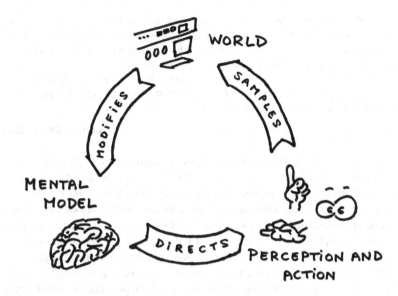

FIGURE 5.2 Sensemaking is ongoing, as people engage with the world around them. They perceive, take samples, and make changes in the world. This modifies and updates their understanding, which in turn directs further perception and action, and so on. Inspired by Neisser's 1976 perceptual cycle. (Drawing by author.)

experience and look down on it from a position of outsider, or retrospective outsider. It is only possible, really, to pay direct attention to what already exists (that which has already passed). "Whatever is now, at the present moment, under way will determine the meaning of whatever has just occurred" (Weick, 1995, p. 27).

There are even earlier traces of this. Psychologist William James (1842–1910) took what became known as a 'functionalist' approach to the field. He pointed out that people are integrated, living organisms engaged in goal-directed activities. Hence, the functionalism: what people do in their environments serves a function, a goal. They are not passive element-processors locked into laboratory headrests, buffeted about by one-shot stimuli from an experimental apparatus. James argued that the environment in which real activities play out shapes our responses. Psychological functioning is adaptive: it helps the human survive and adjust, by incrementally modifying and tweaking its composition or its behavior to generate greater gains on whatever dimension is relevant. He once illustrated our lack of attention to people's existence in an environment with a metaphor. It applies to information processing and SA as much as it did to the traditional psychology of his time:

> What must be admitted is that the definite images of traditional psychology form but the very smallest part of our minds as they actually live. The traditional psychology talks like one who should say a river consists of nothing but pailsful, spoonsful, quartpotsful, barrelsful and other moulded forms of water. Even were the pails and the pots all actually standing in the stream, still between them the free water would continue to flow. It is just this free water of consciousness that psychologists resolutely overlook.

Every definite image in the mind is steeped and dyed in the free water that flows around it. With it goes the sense of its relations, near and remote, the dying echo of whence it came to us, the dawning sense of whither it is to lead. The significance, the value, of the image is all in this halo or penumbra that surrounds and escorts it, or rather that is fused into one with it and has become bone of its bone and flesh of its flesh; leaving it, it is true, an image of the same thing it was before, but making it an image of that thing newly taken and freshly understood.

James (1890, p. 255)

The dynamics of experience, or of awareness, as captured by William James' metaphor, still represents a significant problem for psychology. Snapshots of short-term memory that are taken in a popular SA measurement technique are the 'spoonsful' of James' metaphor. These free the researcher to resolutely overlook the context of awareness, of consciousness. James tried to convey how the river of consciousness, of awareness, is complex, active, adaptive, and self-organizing. The separation between observer and observed makes little sense in his metaphor: space and time are intrinsic properties of experience. They are not 'out there' in a situation waiting for a mind to become aware of them. The mind is not an observer at all, but rather a central player in, and a creator of a perceptual cycle (Neisser, 1976). The perception of the environment is continually being created by cycles of expectation and action, by the questions asked of it. The dynamic transaction, or conversation, between situation and awareness is the only relevant reality for ecological psychology (Flach, Dekker, & Stappers, 2008).

As a result of these insights, a growing number of researchers and practitioners started questioning the continued usefulness of modeling the human as an information-processing system (Hollnagel & Woods, 2005). It tended to focus on the inner mechanisms of human perception, decision-making, and action outputs. More was needed. The second cognitive revolution in the 1990s recaptured and rehabilitated the impulses that brought to life the first that occurred in the 1950s. How do people make meaning? How can they make sense of ever more complex and technology-saturated work? In order to answer such questions, it is not only justifiable and necessary to throw the human factors net wider than information processing psychology ever did. Other forms of inquiry can shed more light on how we are goal-driven creatures in actual, dynamic environments, not passive recipients of snapshot stimuli in a sterile laboratory.

5.3 COGNITIVE SYSTEMS ENGINEERING

5.3.1 HUMAN ERROR (AGAIN)

The verdict of 'human error' (by whatever name) has remained as stubbornly seductive as ever. "In fact, in the first half of the 20th century, some estimated that 88% of accidents were caused by the unsafe acts of workers" (Holden, 2009):

- A full 85% of reports produced by the Australian Transportation Safety Bureau in 1 year (ATSB, 1996) contained references to a 'loss of situation awareness.'

- Investigators at the U.S. National Transportation Safety Board too, used 'loss of situation awareness.' For example, it allowed them to 'explain' why a regional airline crew took off from the wrong runway at Lexington airport in Kentucky in 2006, resulting in the deaths of 49 people including the captain (NTSB, 2007). Between 1999 and 2006, 96% of U.S. aviation accidents were attributed in large part to the flight crew. In 81%, people were the sole reported cause (Holden, 2009).
- The patient safety literature, too, is dominated by 'human error.' A recent review showed that although some of the studies actively looked at teams (17%) or organizations (14%), 98 of the 360 articles reviewed addressed the individual level of analysis, focusing on human error in all its forms (Waterson, 2009).

There is an almost irresistible belief that we are custodians of already safe systems that need protection from unreliable, erratic human beings. For example, in the words of two human factors researchers: "…the elimination of human error is of particular importance in high-risk industries that demand reliability" (Krokos & Baker, 2007, p. 175). The drive to 'eliminate' error can get converted into more rigid rules, tighter monitoring of people, sanctions, awareness campaigns, more automation, and computer technology to standardize practices. As human factors showed, however, the idea that any of this work is problematic. It is unsupported by empirical evidence when you examine how complex systems work. Real work by actual people is constrained by all kinds of design- and organizational and operational features. Designing things in certain ways and some errors become almost inevitable: human factors made that amply clear.

> The misconceptions and controversies on human error in all kinds of industries are rooted in the collision of two mutually exclusive world views. One view is that erratic people degrade an otherwise safe system. In this view, work on safety means protecting the system (us as managers, regulators and consumers) from unreliable people. We could call this the Ptolemaeic world view (the sun goes around the earth). The other world view is that people create safety at all levels of the socio-technical system by learning and adapting to information about how we all can contribute to success and failure. This, then, is a Copernican world view (the earth goes around the sun). Progress comes from helping people create safety. This is what the science says: help people cope with complexity to achieve success. This is the basic lesson from what is now called 'New Look' research about error that began in the early 1980s, particularly with one of its founders, Jens Rasmussen.
>
> **Woods, Dekker, Cook, Johannesen, and Sarter (2010, p. 6)**

5.3.2 JENS RASMUSSEN'S FOUNDATIONAL WORK

In the early 1970s, Rasmussen was studying electronics technicians who were working on the machinery of the Risø Lab in Denmark where he worked (see also Chapter 11). He and his colleague Jensen found that these technicians did not seem to 'rationally' trouble-shoot broken equipment. Instead, they engaged in rapid sequences of simple decisions, using what Rasmussen called 'informationally redundant' observations. The observations the technicians made seemed inefficient in terms of information economy (i.e., why check the same thing three times from different angles or approaches?) and thus not at all rational. They were also not driven by a handbook or procedure for the broken machinery.

But seen from the inside of the job, these informal procedures turned out to be highly locally rational. The technicians had goals such as minimizing time on diagnosing the problem, and reducing task demands or mental load (Rasmussen & Jensen, 1974). Following their eyes and hands along informationally redundant observations, and going along on a cascade of triggered decisions about where to look next, technicians were able to rapidly use previous patterns and stored mental models to cognitively simulate what might be ailing within the machinery. This kind of decision-making was later further formalized in what was called 'recognition-primed decision making' (Klein, 1993), an important contributor to cognitive systems engineering and an elaboration of which people actually make diagnoses and decisions in real-world situations.

The distinction between two images of work, error and expertise (called the Ptolomaeic and Copernican in the quote above), was inspired and given additional theoretical and empirical weight by Rasmussen's work and many those who followed it:

- In one image, there is a set of rational choices to be made by the worker in order to reach a goal. These decisions are supported or directed by written procedures that the worker should follow (remember from Chapter 2 that this is of course the Tayloristic ideal: the rules are right because the people who wrote them are smart. The people who follow them are dumber and simply need to follow the rules). If things go wrong, it is because the worker unwisely departed from the procedures and routines that should have been followed.
- In the other image of work, there is not a one-to-one mapping from written guidance to task, but rather a many-to-many mapping between means and ends present in the work situation as perceived by the individual.

As explained by Rasmussen:

> There are many degrees of freedom which have to be resolved at the worker's discretion. ... Know-how will develop..., 'normal ways' of doing things will emerge. ... Effective work performance includes continuous exploration of the available degrees of freedom, together with effective strategies for making choices, in addition to the task of controlling the chosen path to the goal".
>
> **Rasmussen (1990a, pp. 454–455)**

When you study the conditions under which people perform work, you can quickly find cognitive and resource limitations—as well as uncertainty and the sometimes sheer dynamics of unfolding situations. These all constrain the choices open to them. Hoven (2001) called this "the pressure condition," (p. 3) where practitioners are embedded in a narrow 'epistemic niche.' They can only know so much about their world at that time. All rationality is local: based on people's knowledge, understanding, and goals at the time. It is not based on some universal ideal overview of all the possible pathways and risks associated with them. Based on their understanding, people develop the most sensible or local way to work—even if that is not the rationality of the designer, the manager, or the procedure:

> Humans tend to resolve their degrees of freedom to get rid of choice and decision during normal work and errors are a necessary part of this adaptation, the trick in the design of reliable systems is to make sure that human actors maintain sufficient flexibility to cope with system aberrations, i.e., not to constrain them by an inadequate rule system.

> **Rasmussen (1990a, p. 458)**

Through his considerable influence on the field, Rasmussen helped the world look at work and incidents differently (Rasmussen, 1990b; Rasmussen et al., 1987). Bad outcomes are not the result of human stupidity or immoral choices. They are the product of normal, locally rational interactions between people and systems through which control is often maintained, and sometimes lost. He also showed that efforts by others to contain and combat 'the human error problem' ironically have unintended consequences that make systems more brittle and hide the sources of resilience that make systems work despite complications, gaps, bottlenecks, goal conflicts, and complexity. Safety rules that get in the way of doing work efficiently sometimes do precisely that, as do various kinds of warning systems or technological aids.

Rasmussen was able to step back from what seemed like an irrational act or 'error' on part of the worker. Labeling actions and assessments as 'errors' merely identified a symptom, not a cause. The symptom should invoke a more in-depth investigation of how a system comprising people, organizations, and technologies both functions and malfunctions. Because, "system breakdown and accidents are the reflections of loss of control of the work environment in some way or another" (Rasmussen, 1990a, p. 458).

And the way to prevent these, Rasmussen argued, is not by making it impossible for operators to do the wrong thing. For that would once again be organized around human limitations rather than strengths. It would see people as a problem to control—to be kept away from dangerous parts or processes as much as possible. It would suggest that the purpose of design is to lead people into safety, and to make them risk averse. It assumes, though, that the designers have been able to figure out, in advance, exactly what safety means and how operators can be led to it. Operators do not need to think, they just need to follow the path already laid out for them.

Rasmussen would have argued not for risk aversion in design, but for risk competence. Risk is dynamic, and it needs to be managed as such by those who see it change shape as they work (Rasmussen, 1997). In many ways and places, workers manage risk dynamically by 'finishing the design' of technologies that are not exactly up to the task of supporting them in the execution of their work (Wynne, 1988). Placing little reminders on controls (post-it notes, paper cups) is just one example of this. It plugs error traps, makes things work and brings risks under control in practice. We should design to support adaptation, Rasmussen argued (2000). Operators are not the problem. They are the solution that engineers and designers can harness. Make it possible for operators to finish the design, for them to adapt it, so that the design not only fits the complexities and dynamics of real-world process applications, but that it can absorb the inherent imperfections in what designers and planners predict might be useful in a field of practice.

5.3.3 TWO STORIES OF ERROR

Human factors, and later cognitive systems engineering, represent the most intense examination of the human contribution to safety. They both show that the story of 'human error' is remarkably complex. The way the fields have typically made their arguments is by referring to:

- a 'first story,' where human error is the cause of deviations, accidents, and breakdowns.
- a second, deeper story, in which the normal, predictable actions and assessments (which some call 'human error' after the fact) are the product of systematic processes inside of the cognitive, operational, and organizational world in which people work (e.g., AMA, 1998; Dekker, 2014).

Second stories show that doing things safely—under the pressure of limited resources and while trying to meet other (sometimes conflicting) goals—is always part of people's operational practice. People, in their different roles, are aware of potential paths to failure, and develop failure-sensitive strategies to forestall these possibilities. People are a source of adaptability required to cope with the variation inherent in a field of activity (see Chapter 11). A correlated argument (which can be traced back even to Taylor and Heinrich) of the second story is the idea that complex systems have a sharp end and a blunt end:

- At the sharp end, practitioners directly interact with the hazardous process;
- At the blunt end, regulators, planners, supervisors, administrators, economic policy makers, and technology suppliers control the resources, tools, technologies, constraints, and multiple incentives and demands that sharp-end practitioners must integrate and balance (Woods et al., 2010).

Small failures, problems, inconsistencies, and vulnerabilities are present in the organization or operational system long before an incident is triggered. Much of it germinates and collects at the blunt end of the organization, and trickles down to affect the many operational trade-offs and decisions that need to be made at the sharp end. All complex systems contain such conditions or problems, but only rarely do they combine at the sharp end in such a way as to create a mishap:

> The story of both success and failure is (a) how sharp end practice adapts to cope with the complexities of the processes they monitor, manage and control and (b) how the

strategies of the people at the sharp end are shaped by the resources and constraints provided by the blunt end of the system.

Woods et al. (2010, p. 7)

Doing things safely against the constant background of these systematic imperfections, resource limitations and goal conflicts, is *always* part of people's operational practice. Doing things safely needs to be done in the course of meeting other (often conflicting goals), while working around 'unfinished' and imperfect designs, and making up for operational and organizational deficiencies and complexities. In their different roles, people learn about the error traps and potential pitfalls and the routes to failure. They typically develop failure-sensitive strategies to forestall these possibilities (Figure 5.3).

This presents quite a different understanding of work than that of the Tayloristic model. Remember (e.g., from Figure 2.1.), Taylor's managers and planners were smart. They had to spell out to dumb workers what to do—every step of the way. The image of sharp versus blunt end suggests that workers are the smart ones, and managers and planners the dumb ones. They are 'dumb' in the sense that they are not

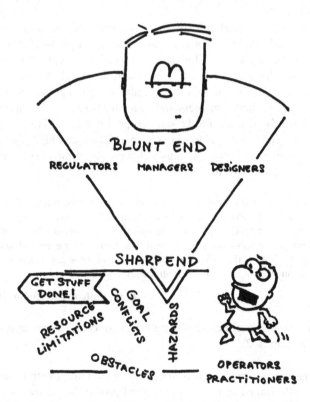

FIGURE 5.3   At the sharp end, practitioners directly interact with the hazardous process. There they have to safely manage the goal conflicts, resource limitations, and other problems that come down from blunt end designs, resource constraints, and multiple interacting demands. (Drawing by author.)

aware of the messy details and nuances of what it means to get stuff done. They are unlikely to know about the many subtle strategies and tactics that operators need to learn and develop, so that they can work safely under the pressure of multiple conflicting goals, design issues, resource limitations, and procedural obstacles (many of which come down from the blunt end).

5.3.4 INCREASED SOCIO-TECHNOLOGICAL COMPLEXITY

Just as human factors had been driven in large part by technological demands that could no longer be met with behaviorist logic, cognitive systems engineering was a response to the fact that systems had become ever more complex. Human factors could work well for the human-machine interface between an operator and a machine, but what about the control room of a nuclear power plant? Or mission control for space flight? These were systems that contained many people and teams, a whole host of interfaces and computers, various levels of automation, a multitude of procedures, checklists, screens and interfaces, and controls. In part, the nature of human work had started shifting dramatically: more supervisory control and less direct manipulation; more cognitive work; more collaborative decision-making.

Computers play a crucial role in the emergence of cognitive systems engineering. The digital revolution, or the massive takeover of work by computerization, is also known as the Third Industrial Revolution. It represents the rapid and almost wholesale change from analog, mechanical, and electronic technology to digital technology. The proliferation of digital computing, communication, and record-keeping has pushed computers and computer-based devices into virtually all safety-critical worlds—from control rooms to operating theaters to cockpits to traditional industries that now use tablets or smartphones for checklists and procedures. Such technological interventions obviously change the cognitive system in dramatic ways. It called for something different:

> The demands from fielding complex sociotechnical systems, often with significant safety ramifications, forced recognition that overall work system performance emerged from interactions of human operators with highly coupled technological components. This focus on human-technology integration was then further extended to recognize that a focus at the level of the individual operator was not sufficient. Teamwork matters, as do interactions within broader distributed work systems.
>
> **P. J. Smith and Hoffman (2018, p. 3)**

Hollnagel and Woods describe the driving forces for cognitive systems engineering as follows (2005, pp. 1–2):

- *Growing complexity of socio-technical systems.* This growth in complexity was directly linked to the unprecedented growth in the use of computer technology. By the 1970s, computers (and increasing automation) were poised to become the dominant force in how safety-critical technologies were operated;

- *Clumsy use of computer technologies* created problems that had not been seen before. Operators already had to adjust to greater complexity, but then also had to cope with data overload, getting lost in display page architectures, responding to cryptic computer warnings, finding patterns of meaning among seemingly meaningless data. Escalation of problems as a result of tight coupling in the underlying process and features of the interfaces that presented this to humans, had become a serious issue;
- *The information processing metaphor* which had mechanized our understanding of the mind was no longer adequate to explain what was going on between humans and their increasingly complex systems. Not only were new terms needed, a whole new unit of analysis (the joint cognitive system) was called for.

In the founding paper for the discipline, Hollnagel and Woods described cognitive systems engineering as more than just a reformulation of old ideas (Hollnagel & Woods, 1983). They argued that:

> Technological developments alone have changed the nature of the human-machine interface from emphasizing the human's physical tasks to emphasizing cognitive tasks, and thereby made a purely technological approach to human-machine systems obsolete. The costs and consequences of ignoring the cognitive functions of human-machine systems are noted in technological failures daily (we need only mention the Three Mile Island accident), and on a less dramatic scale in the small annoyances we all encounter when we use computers. An alternative approach is therefore necessary which will apply and further develop the techniques and knowledge base of cognitive psychology to the design of human-machine systems. This area of human-machine studies we call Cognitive Systems Engineering (CSE).

> *(p. 585)*

What human factors as a field could no longer cope with was an intertwined set of developments that significantly increased the complexity of work itself, particularly of work that was connected with computers and safety-critical processes. A cognitive system, Hollnagel and Woods explained, needs to be in control of these processes. There is a lot, however, that challenges the system's capacity to remain in control. Among them:

- Striving for efficiency brings a cognitive system closer to the limits of safe performance (you will learn more about this in Chapter 7);
- Operating a cognitive system in a complex world can lead to novel, unforeseen, and unpredictable situations in which maintaining control is difficult or impossible (Dekker, 2006);
- Automation that helps keep the system away from those limits of safe performance (overflow alarms, stall warnings, etc.) can actually serve to increase the overall complexity of the cognitive system and create a sense of some of its constituents being 'out of the loop' (Bainbridge, 1987);

- With cognitive systems now responsible for almost everything that matters in an organization or even society, the consequences of failure can be huge (Lee, 1992; Perry, Wears, & Cook, 2005). The design, implementation, management, governance, and maintenance of these systems is as important as operating them, but those aspects are seldom under the control of those who actually operate these systems;
- The amount of data has been increasing considerably since the adoption of computers. Computers can help in storing, transforming, sorting, transmitting, and presenting data, for sure, but these capacities have barely kept up with their sheer ability to produce such data. Data overload is thus a significant concern (Woods, Patterson, & Roth, 2002).

The stakes were high (P. J. Smith & Hoffman, 2018). Cognitive systems engineering spread from the United States to other countries and spawned a network of research centers and dedicated field researchers. Rediscovering the Francophone tradition of work studies, researchers used observations and interviews, did scaled simulation investigations, and conducted process-tracing studies (Woods, 1993). They embedded among those who did operational work, and sometimes even 'went native' themselves. Among the research outcomes relevant to safety are a number of recognizable patterns in the functioning of cognitive systems that have since made it into the safety literature as well as popular imagination.

5.3.5 JOINT COGNITIVE SYSTEMS

In order to make sense of its findings, and to be able to communicate what it saw, the field needed a new unit of analysis. This unit of analysis needed to capture the goals, processes, representations, interactions, and activities of a host of people and cognitive devices. Looking at how complex systems actually functioned in the world, it was obvious that this unit was able to adapt to the way it approached a problem; or who or what was even part of it. It was clearly goal-oriented, wanting to achieve something in the world. It operated using knowledge of itself and the environment, and could plan and modify plans to adjust for changes. The new unit of analysis became known as the 'joint cognitive system':

- The unit of analysis for understanding cognitive functioning should be the joint cognitive system, not the individual human information processor;
- A cognitive system consists of people and cognitive artifacts;
- Activities within this joint cognitive system, and between it and its environment, can be described using information processing terms. The way information is represented and changes representation, however, is often directly observable and does not need to be inferred from what goes on in the head of an individual;
- Outcomes of interest (like a safe landing) depend on the functioning of the cognitive system as a whole.

So are safe outcomes actually mostly determined by single people and their individual cognitive processes? This question was raised in a landmark paper published in Cognitive Science in 1995 (Hutchins, 1995b). Ed Hutchins, himself a pilot and cognitive scientist, wrote *How a Cockpit Remembers Its Speeds*, based on his field studies and own experiences. Cognitive Science, Hutchins proposed, normally takes the individual as its unit of analysis. But in many endeavors, including safety-critical ones, the outcome is determined not entirely or not even largely by the information processing properties of individuals. The successful completion of a task in a real-world safety-critical system, he argued, depends on the interactions between not only multiple people, but between those people and a suite of technological (and less high-tech) devices.

Hutchins called these devices 'cognitive artifacts.' He took the theoretical approach of the cognitive system as unit of analysis literally 'into the wild' (Hutchins, 1995a) by considering cognition situated and distributed across humans and artifacts. By tracing how information gets represented and processed through a cognitive system (which consists of both people and artifacts), he showed how memory for airspeeds during an approach to landing is a distributed cognitive property of an airliner cockpit. Memory for speed was not just a number in the head of a single pilot, but rather a construct that emerged from the interactions between pilots, booklets, speed bugs, pointers, dials, and other artifacts. Such a take on the issue offers new leverage points for carefully and meticulously improving the safety of the operation—from enhancing problem representations in checklists or displays, to changing designs, adding reminders, taking away unnecessary steps or amending collaboration through callouts, and other procedures.

As you read the study below, you will quickly understand how you cannot meaningfully examine any aspect of this problem in a laboratory setting. You have to go 'into the wild' to study the joint and distributed cognitive activities and learn to speak the language of the domain. As Ed Hutchins used to say, you can only recognize the universal concepts of joint cognitive performance in the particulars of an actual joint cognitive system. At the same time, the blooming confusion of the particulars only makes sense in reference to those universal concepts. The ability to translate back and forth between concept-dependent and context-specific descriptions of what goes on has become a hallmark of cognitive systems engineering research (Woods & Hollnagel, 2006).

HOW A COCKPIT CALCULATES ITS SPEEDS: AN EXAMPLE OF A JOINT COGNITIVE SYSTEM

The calculation of airspeeds in an airliner cockpit for takeoff is not something that Hutchins studied, but his approach to investigate the workings of a joint cognitive system applies nicely. This was done in a study a few years back (Henriqson, van Winsen, Saurin, & Dekker, 2011). Calculating the correct speeds for takeoff in an airliner is a collaborative activity that is vulnerable to interruptions and miscalculations. It takes place at a point in the departure when there is actually very little time, and when almost everything for an on-time departure now hinges on the cockpit crew's ability to quickly calculate the correct takeoff data. The issue is that the activity cannot be started until everybody is onboard, and all the baggage and cargo has been loaded, and it is known exactly where everybody sits and where the bags and cargo are placed. Of course, once everybody is onboard and everything is loaded, the aircraft is also ready to be pushed back from the gate. But first, the correct takeoff data need to be calculated. The door of the aircraft is still open at this point, because the dispatcher or ground agent still needs completed and signed paperwork from the pilots. Often they wait near the cockpit as the pilots complete these calculations. This is also typically a time that radio interactions with cabin staff, with air traffic control, and with the driver of the pushback truck are necessary.

The aim of analyzing the calculation of takeoff speeds in a modern airline cockpit as a distributed cognitive activity is to better identify the possible vulnerabilities that it is exposed to, particularly given the introduction of new technologies supporting this task.

The study considered the production of takeoff speeds as a 'calculation' because it involves the manipulation of several input parameters in order to get the output values required for operation. Today's activities surrounding the calculation of airspeeds involve computer technology of a generation beyond what Hutchins described in his 1995 paper, particularly the flight management computer (FMC).

A joint cognitive system involving pilots and flight dispatchers, as well as many technological artifacts, is responsible for determining the correct takeoff Gross Weight (GW) of the airplanes and its related takeoff speeds, the so-called V-speeds: V1 (decision speed to keep going), Vr (rotation speed to get airborne), and V2 (safety speed to maintain once airborne). Each of these V-speeds has technical and procedural consequences for a safe takeoff and climb of the airplane. If the calculation of the correct V-speeds fails, two consequences are likely: (1) too high takeoff speed values will result in more runway distance for takeoff, but often with little implications for safety; (2) too low takeoff speed values will lead pilots to prematurely commence liftoff, which will also increase runway distance for takeoff, and might even cause an excursion, due to the incapacity to get airborne.

Another typical outcome in the situation of too low V-speed values is a tail strike. Tail strikes occur when the airplane's tail touches the runway during takeoff. It can badly damage the airplane and therefore be costly for the airline. In a recent case in New Zealand, a tail strike occurred because the Vr was set 33 knots below the 163 knots required for a safe takeoff, due to an incorrect assumption of the GW to be 100 tons less than the airplane actually was. A takeoff weight transcription failure led to the miscalculation of the takeoff data using runway analysis charts, which in turn resulted in a low power setting and excessively slow takeoff reference speeds. In another event, the same type of failure caused a fatal accident with a cargo airplane in Canada. In this episode, however, instead of using runway analysis charts, the crew was performing calculations with a cockpit laptop (Henriqson et al., 2011).

Indeed, different artifacts are used to perform the calculations (e.g., runway analysis charts or cockpit computers) and they involve multiple agent interactions (e.g., pilots, flight dispatchers, and cockpit computers).

In a typical jet transport airplane, the calculation of takeoff speeds involves cooperative work between aircrew and several other agents (such as ground staff) when performing procedures required to conduct this activity. Even though the sequence of actions may vary according to airplane type/manufacturer, airline, or contingencies of the natural work (such as interruptions and last minute changes), four general processes of takeoff speeds calculation involve:

a. definition of the airplane's operating weights;
b. computation of the airplane's operating parameters;
c. FMC programming;
d. verifications and checklist readings.

As a rule, the definition of the airplane's operating weight is related to the determination of the Zero Fuel Weight (ZFW), the Takeoff Fuel Weight (TFW), and the GW. The GW of the airplane is the sum of the ZFW and the TFW (which usually includes some extra weight for burn-off during taxi-out). The ZFW is determined by totaling the aircraft's empty weight (without fuel), the passengers, and the cargo load. The TFW is chosen by the aircrew considering the flight plan, en route weather forecasts, and company standards, and refers to the total fuel on board. Performance implications, such as field limit (e.g., runway length, runway slope, wind, pressure altitude, and temperature), climb limit (e.g., rate of climb during departure and obstacle clearance), or landing weight limit, may influence the ZFW and/or TFW and consequently the GW. These weight determinations should be considered prior to the calculation of takeoff speeds and are usually provided to the aircrew by means of a load sheet document.

The load sheet contains the list of operating weights, taking into account previous performance computations, and is typically created by company flight dispatchers. As the final ZFW of the airplane depends on the actual number of passengers and load on board, preliminary values of ZFW, TFW, and GW are provided to the aircrew, in order for them to execute flight deck preparations, route planning, and FMC programming. Minutes before the departure time, the final load sheet is delivered to the aircrew, for them to do the final calculation of V-speeds and other performance parameters.

The calculation of the airplane's operating parameters involves manipulation of a variety of data and can be carried out in two ways: using cockpit computers or using runway analysis charts. When using laptops, the aircrew first copies airdrome information (such as wind, temperature, and altitude) into the proper field of the flight plan paper document or data card. Even though some airlines do not use data cards anymore, it is a sheet provided to register important takeoff information, still in use by many companies. Thereafter, the aircrew inserts airdrome and load sheet data in the laptop, using dropdown menus. The software then calculates takeoff parameters for engine power settings, optimum wing flap settings, and V-speeds, according to the information entered by the pilots. Parameters for takeoff with 'full' and 'reduced' power are generated depending on the operating circumstances (e.g., runway braking action, weather conditions). The output values presented on the laptop then need to be transcribed onto the data card (when it is in use) or directly inserted into the FMC.

Instead of using cockpit computers (such as laptops), another manner to compute the airplane's operating parameters is by means of runway analysis charts. This involves the selection of the proper runway chart and the manual manipulation of numeric values of flap setting, weight, temperature, and wind. From these charts, the pilots should extract takeoff parameters for maximum GW, assumed temperature for takeoff power setting, and V-speeds. The resulting values should then be transcribed onto the data card, and from this, into the FMC. Because the use of runway analysis charts involves the mental operation of numeric values by the pilots, data cards are more often used (compared with the airlines that use the laptop tools to determine these parameters) as they assist pilots' memory in these cases.

After computation of the takeoff parameters, using either laptops or runway analysis charts, the FMC must be programmed. This programming requires several steps, such as the performance and engine power settings and the determination of takeoff speeds. Typical performance requires the pilots to enter the previously defined parameters of ZFW and TFW into the FMC. The FMC then automatically generates the GW value, which is the basic reference for V-speeds calculation at this point. Some aircraft, equipped with later versions of FMC software, may require the pilots to input the GW value themselves.

The next step involves engine power settings. Engine power is determined by inserting the related (assumed) temperature in accordance with prior performance analysis, in order to save the engine from harm caused by excessively high temperatures. Still, other methods can be combined to reduce takeoff power settings (to save fuel and reduce wear on the engines) such as 'de-rated' takeoff or climb.

Finally, takeoff speeds need to be linked to flap position for takeoff and CG (the airplane's Center of Gravity) into the FMC, based on information from the data card or directly from the laptop. The FMC requires that aircrew validate V1-, Vr-, and V2-FMC-generated speeds in the appropriate fields of the FMC (depending on the FMC version, pilots should insert speed values, rather than validate them). By validating or inserting these V-speeds on the

FMC, reference marks of these speeds are then displayed on the cockpit's air-speed indicators. In older airplanes with analogical or hybrid (analogical/digital) displays, pilots may have to adjust the speed bugs (reference markers on the airspeed indicators). This is what was still done in Hutchins' 1995 study. A further validation of the FMC's V-speeds is also performed by cross-checking the V2 speed generated, with the GW V2 chart, which is usually a plastic card with a table of GW and related V2 values.

After the FMC programming, the aircrew will perform departure checklist readings, in order to establish a common ground for the operational actions during the takeoff and initial climb. When doing this, the crew cross-checks weight and speed parameters for takeoff, according to the available data in the cockpit. These data are now supported by multiple (technical) artifacts, such as the load sheet, data card, laptop, FMC, and the cockpit's speed indicators.

The above outlined determination of the typical course of actions for the calculation of takeoff speeds (with and without cockpit computers) shows that:

a. the process involves entering a set of values in the cockpit (weights and performance parameters), processing of these values, and ends with specific marks on the cockpit's speed indicators;
b. the process uses preliminary and final information about estimated operating weights, in order to conduct the calculation;
c. the process requires cooperation of several agents (computers, pilots, and ground staff—flight dispatchers, loading- and fueling agents); and
d. the process connects people and artifacts following an algorithmic logic of procedure (e.g., first, entering values on a runway analysis chart; second, computing operational parameters; third, transcribing these parameters onto the data card; and fourth, inserting these values into the FMC).

The cockpit is the place where the calculation of the takeoff speeds occurs. Several agents and artifacts interact in order to produce these takeoff speeds. These interactions involve the performance of many subtasks, distributed across the agents. Cognitive agents are single players, human or machine, able to execute actions based on different inputs. The main human players in the cockpit are the pilots (captain and first officer). In some situations, there may be extra aircrew members, such as the flight engineer (required in some types of airplane), or a third pilot. Air traffic controllers, company staff, or minor agents, such as fueling and loading people, are considered elements of the environment here, not of the joint cognitive system under study, because they cannot control the cockpit calculation, despite their influence on the outcome of the calculation.

The main machine players in the cockpit are the laptop and the FMC. These are considered cognitive agents because these machines perform actions related to the joint functions in the system. These machines have cognitive capabilities such as control authority, information processing, and decision-making. According to one of the pilots interviewed, " [the FMC] can be seen as another pilot inside the cabin. It is somebody else to share tasks with, but it is also somebody else who you will need to be monitoring and synchronize with during the operation" (Henriqson et al., 2011, p. 222). Moreover, it is not unusual to hear pilots state things like "what is it doing now?" when programming or monitoring automation.

Each set of goal-oriented interactions of people and cognitive artifacts (machines) related to the calculation of takeoff speeds is part of the (joint) cognitive system. Together, they function as the unit of analysis. The joint cognitive system is driven by internal representations (and the transformations these undergo as they travel through the system) that create a shared representation. An example could be the captain calling out the operating weight values to the first officer by means of reading them from the load sheet. At the same time, the first officer is entering these numbers into the laptop. Multiple transformations of representation take place: (1) the captain reading the load sheet values; (2) the first officer entering these values into the laptop; and (3) the information transfer of symbolic values between pilots, speed bugs, and FMC. During the definition of the takeoff operating weights, external representations provided by several artifacts in the cockpit (e.g., the load sheet, the flight plan) constitute the pilots' representation of the working situation. This constitution produces local cognitive system representations; at the level of the captain, when reading the load sheet; at the level of the aircrew, when the captain is calling out weight values to the first officer; and at the level of the first officer, when he hears the captain call out the weights. At these levels, internal (the pilots' minds) and external (e.g., entering weight values into the laptop) transformations are temporary forming the representations. Unique, but partially (always incomplete), the cockpit representation is created like a joint system's representation. As a result, cockpit performance of takeoff speeds calculation depends on both the performance of single individual cognitive systems, as well as the joint cognitive systems' interactions. This, again, is the reason why the joint cognitive system (not the individual human or machine agents) is the unit of analysis.

Ephemeral representations are susceptible to interferences of the pilots' prospective memory (Loukopoulos, Dismukes, & Barshi, 2009). Prospective memory is self-initiated (not directly based on external stimuli) remembering, in order to perform an intended action. Verbal communications are among the most common ephemeral representations. For example, the verbalization of the ground staff of preliminary values of operating weights communications with refueling agents, and minor load agents, as well as radio-communication with air traffic control. These ephemeral representations occur between pilots, mostly when one is calling out operating values to another or when they are performing checklists. They are susceptible to various constraints of the working phase, such as multitasking management, interruptions, lack of sequence and orderliness, and the use of estimated versus actual values.

In sharp contrast with ephemeral representations, some artifacts produce strong and durable representations constituting the cockpit's temporary representation, such as the data card or the default values of the laptop. But it is no guarantee of accuracy. In one incident, the data card values from the takeoff of the previous flight were mistakenly used to carry out the calculation of operating takeoff parameters. In another incident, the laptop default value of takeoff GW from the previous flight was used (Henriqson et al., 2011).

Another computational problem happens when the joint cognitive system matches the first digit of the GW for an immediate lower level. In one recent case, a GW of 347,000 lb was computed as 247,000 lb. Once such (possibly incorrect) values are validated on the takeoff data card during the ephemeral processes of transcriptions from one media to another, and inserted in the computers, the entire system is provided with this new representation. Double-checking of the procedures that leads to V-speeds will probably not reveal the inaccuracy of these values and the consequent representations.

Failures in this process are mostly related to the creation of meaning. If meaning is something considered plausible or reasonable for the situation, the representation will likely be accepted as valid. For example, a GW above the maximum takeoff structural weight should be refused as valid meaning by both the computer and the crew. On the other hand, values floating between the maximum and empty weight are possible to be considered and accepted across different representations in the cockpit. Pilots, especially when transitioning from one type of aircraft to another (e.g., from Airbus A330 to Boeing B747), are not able to identify orders of magnitude values related to weight or speeds, due to its constant fluctuation during natural operating contexts.

In three recent incidents, the ground staff provided the aircrew with incorrect operating weight values of either ZFW or GW. The weight values upon which the V-speeds are calculated are virtual because there is no possibility for the cockpit (pilots or computers) to check the received values, as there is no system available to actually weigh the airplane. In this way, the received weight values are just 'virtual' values that create a cockpit representation of a presumed airplane weight. Additionally, the actual values of ZFW, TFW, and GW are always changing, because during the calculations, the airplane is being refueled and loaded. As a consequence, both, the preliminary and the final values are responsible for the construction of the cockpit's weight representation.

The richness of representations, and the calculations and transformations of these representations, as a cockpit prepares its data for takeoff during a highly pressurized, time-limited portion of flight preparation, shows that simply blaming 'human error' for a mistake is not very useful. A careful study of the joint cognitive system can instead reveal the chafing points and vulnerabilities that put the eventual product of the joint cognitive system (takeoff speeds in this case) at risk.

5.3.6 Patterns in Cognitive Systems Engineering

Whereas a keen eye for detail is necessary to begin to understand the workings of a joint cognitive system, the lessons that can be drawn from it (and the leverage for changing the cognitive system to assure safer outcomes) are not really visible at the level of those context-rich details. This is where stepping up to a concept-based language is necessary. The context-specific discoveries themselves are not exportable (or externally valid) by themselves either. The discoveries that we make in complex fields of practice need to be transformed into a conceptual language, so that we can take those patterns and regularities and see whether they exist in other fields of practice as well (and whether they are subtly or starkly different there). Through a range of studies across complex, safety-critical domains, cognitive systems engineering has generated a number of these patterns and concepts. Here are just some:

- *Clumsy use of technology*, which leads to new kinds of cognitive work instead of freeing up resources. Clumsy technology makes it difficult to enter into team play with computers and automated systems because people do not know what the systems are doing. Technology always changes the work it was supposed to replace, and thus calls on new kinds of human judgments, expertise, and knowledge. Clumsy technology leads to new error opportunities and pathways to system failure (Woods et al., 2010);
- *Data overload*, which is one consequence of ubiquitous computerization. This has tremendously advanced our ability to collect, transmit, and transform data. Operators are typically bombarded with computer-processed data, especially when anomalies occur. Our ability to digest and interpret data, or present it in ways that help people make sense and meaning, has failed to keep pace with our abilities to generate and manipulate greater and greater amounts of data (Woods et al., 2002).
- *Mode error*, which occurs when an operator executes an intention in a way that would be appropriate if the device were in one configuration (one mode) when it is in fact in another configuration. Mode errors are a joint cognitive system breakdown, involving users who lose track of current system configuration, and a system which interprets user input differently depending on the current mode of operation. The potential for mode error increases as a consequence of a proliferation of modes and interactions across modes without changes to improve the feedback to users about system state and activities (Sarter & Woods, 1995).
- *Automation surprise*, where automated systems act on their own without immediately preceding inputs from the user(s). Gaps in users' mental models of how the automation works and weak feedback about the activities and future behavior of the automation are important factors (Sarter, Woods, & Billings, 1997).

To illustrate the contribution of cognitive systems engineering to our understanding of safety, and to highlight the contrast to information processing-based approaches that focus mostly on the human operator, let's look at automation surprises in the last part of this chapter. Automation surprises continue to be a significant safety concern in aviation and beyond. The system safety community's search for effective strategies to mitigate them is ongoing. Guidance for such mitigation efforts come not only from operational experiences, but from our understanding of the cognitive processes that govern human interaction with automation. This helps us better explain the kinds of breakdowns that can occur, and know what exactly it is that we need to prevent (Figure 5.4).

So far, however, the literature has offered divergent directions because they are based on two fundamentally divergent assumptions about the nature of cognition and collaboration with automation—a difference that reaches back beyond the so-called 'cognitive revolutions' of the 1950s and 1990s (Dekker, 2015; Hutchins, 1995b). One sees cognition as individual information processing, the other as collaborative and distributed across human and machine agents. They have produced different diagnoses, understandings and definitions, and concomitantly different prescriptions for mitigating automation surprises. Below you will see the results of a field study that empirically compares and contrasts two models of automation surprises that have been proposed in the literature based on this divergent understanding: a normative model and a sensemaking model:

- The first (normative) model focuses on suboptimal human performance (Parasuraman & Manzey, 2010);
- The second (sensemaking) focuses on the complexity of the context (Dekker & Lützhöft, 2004).

FIGURE 5.4 Automation surprise: the automated system acts on its own without immediately preceding user input. Automation surprises are linked to higher workload phases, gaps in users' models of how the automation works, and weak feedback the system offers about its behavior. (Drawing by author.)

This difference is relevant for our understanding of the cognitive processes that govern the human interaction with automation and measures that are to be taken to reduce the frequency of automation-induced safety events.

The first research tradition on cognition in human factors fits within a larger metatheoretical perspective that takes the individual as its central focus. Recall how, for human factors, this has meant that the processes worth studying take place within the boundaries of the individual (or her/his mind), as epitomized by the mentalist focus on information processing and various cognitive biases. In this tradition, automation surprises are ultimately the result of mental proclivities that can be attached to an individual. One such tendency has been defined as 'automation bias,' which is "the tendency to over-rely on automation and exhibit complacency because the highly (but not perfectly) reliable automation functioned properly for an extended period prior to this first failure" (Parasuraman & Manzey, 2010, p. 382). Remedies rely on intervention at the level of the individual as well, particularly training. Some innovations in training are already being developed, such as the nurturing of flight path management skills using manual flight from the outset and then introducing increasing levels of autoflight systems to achieve the same flight path tasks on the Airbus A350 (FAA, 2013). There is consensus, however, that training and a focus on individual knowledge is a partial solution at best. As the FAA study observed about flight management system and automation surprises:

> the errors are noteworthy, and it has not been possible to mitigate them completely through training (although training could be improved). This reflects that these are complex systems and that other mitigations are necessary.
>
> *(2013, p. 55)*

The realization that the problem of automation surprises might rather be defined, studied, and understood as the product of complex systems is found in cognitive systems engineering, and in the study of distributed cognition, which renewed the status of the environment as active, constituent participant in cognitive processes. This forms the basis for the second model. The cognitive systems synthesis invokes a new unit of analysis: all of the devices that do or shape cognitive work should be grouped with human cognitive work as part of a single joint human-machine cognitive system. Automation surprises in this conception are not the result of either pilot error or a cockpit designer's over-automation. Instead, they exhibit characteristics of a human-machine coordination breakdown—a kind of weakness in a distributed cognitive system (Woods, 1996). There is still a placeholder for individual cognition in this model too, under the label of mental model. When conceptualized in this way, automation surprises reveal the following pattern (Sarter and Woods, 1995):

- Automated systems act on their own without immediately preceding directions, inputs, or commands from human(s);
- There are gaps in users' mental models of how the automation works;
- Feedback about the activities and future behavior of the automation is weak.

TABLE 5.2

Two Different Takes on 'Automation Surprise'

Automation Surprise According to Information Processing	Automation Surprise According to Cognitive Systems Engineering
Automation surprise is due to human error or designer over-automation of the system	Automation surprise represents the breakdown of coordination in a joint cognitive system
Cognition is a matter of human information processing. Deficiencies in this process can lead to automation surprises	Cognition is distributed across human agents and technological artifacts which jointly produce outcomes
An automation surprise is the result of human complacency in monitoring the automation, leading to a loss of SA. It resembles 'learned carelessness'	An automation surprise is the end result of a deviation between human expectation and actual system behavior
Automation surprise is addressed by controlling the human through more training and reminders	Safety problems addressed by controlling the joint cognitive system through better team play, e.g., focusing on feedback and directability

An automation surprise can then be defined as the end result of a deviation between expectation and actual system behavior that is only discovered after the crew notices strange or unexpected behavior and that may already have led to serious consequences by that time (Dekker, 2014) (Table 5.2).

The two divergent models of human–automation interaction that have been proposed in the literature (the integrated model of complacency and automation bias (a normative model), and the crew-aircraft contextual control loop (a sensemaking model)) make fundamentally different assumptions about the nature of human cognition. The first falls within the premises of the first cognitive revolution—particularly that all relevant processes occur within an individual (e.g., in the form of biases or motivational short-comings such as complacency). The second is premised on the second cognitive revolution: pertinent cognitive processes are situated in a work environment and distributed across both human and machine partners. As you will see below, support for the distributed cognitive, or sensemaking model was found in a lack of reduction in trust in the automated system following an automation surprise, in the discovery of the automation surprise by the respondent himself/herself and in the attribution of the cause of the automation surprise to a lack of knowledge about the automation.

What are the implications for improving safety, in this case? The sensemaking model suggests that our understanding of the interaction between humans and automation can be improved by taking into account systemic factors and the complexity of the operational context, rather than focusing on suboptimal human performance. Automation surprise seems to be a manifestation of the system complexity and interface design choices in aviation today, and rarely the result of individual under-performance. Further research into individual automation surprise experiences can substantiate whether a 'buggy' mental model was underlying the events, and if so whether redesign of the automation to facilitate the generation of an appropriate mental model, supported by sufficient training and the buildup of experience is warranted.

AUTOMATION SURPRISE: TAKING
CONCEPTS BACK TO CONTEXT

One aspect of the translation between context-specific and concept-dependent is that the concepts that are developed need to make sense to the community of practitioners from which the initial data came. In other words, concepts developed in this sort of research need to be calibrated with the context-specific settings that helped form them. Parasuraman and Manzey (2010) suggested that automation surprise is a result of 'complacency' and 'attentional bias,' which is induced by overtrust in the proper function of an automated system. Their 'integrated model of complacency and automation bias' shows 'complacency bias' leading to 'attentional bias in information processing' and then loss of SA. A lack of contradictory feedback induces a cognitive process that resembles what has been referred to as 'learned carelessness.' Even a single instance of contradictory feedback may lead to a considerable reduction in trust in the automated system. The authors suggest that both conscious and unconscious responses of the human operator can induce these attentional effects.

The integrated model of complacency and automation bias is a normative model because the authors suggest that there is "some normative model of 'optimal attention allocation'" by which the performance of the human operator can be evaluated (p. 405). This supposedly enables the identification of 'complacency,' 'automation bias,' 'learned carelessness,' and 'lack of Situational Awareness'.

In contrast, and more recently, Rankin, Woltjer, and Field (2016), following cognitive systems engineering work, propose the 'crew-aircraft contextual control loop.' The element of surprise marks the cognitive realization that what is observed does not fit the current frame of thinking. Commencing with automation in aviation, they extend their model to include other operational issues between the crew and the aircraft. Other literature outside aviation supports their suggestion that cues are ignored due to a previously existing frame or mental model, until a sudden awareness occurs of the mismatch between what is observed and what is expected.

The contextual loop is a sensemaking model in that it assumes:

> ...the retrospective and prospective processes of data framing, re-framing and anticipatory thinking. Sensemaking involves the continuous process of fitting the data (what we observe) into a frame and fitting a frame around the data (what we expect). [...] Recognising and preparing for difficult challenges is directed by where we focus our attention, which is based on our expectations. This is a highly contextualised process where responses are simultaneously being identified based on the constraints of the situation.

> (p. 624)

The normative character of the 'integrated model of complacency and automation bias' indicates that negative performance consequences due to automation failure are perceived as a result of 'a conscious or unconscious response of the human operator,' i.e., human error. Numerous personal, situational, and automation-related factors play a role, but the main finding is that we need "to make users of automation more resilient to complacency and automation-bias effects" (Parasuraman & Manzey, 2010, p. 406).

Rankin et al.'s 'crew-aircraft contextual control loop' on the other hand suggests that we need to "adequately prepare pilots to cope with surprise, such as by using scenarios with ambiguous and potentially conflicting information." Human operators are submerged in "ambiguities and trade-offs regarding manual control, procedure applicability, system knowledge and training" and balancing different goals and strategies while trying to make sense of possibly conflicting inputs. Under these circumstances we need to better "prepare crews for the unexpected" (p. 641).

Although the two models share many common factors that affect the prevalence of automation surprise (such as personality, fatigue, and automation design) they each predict different outcomes:

1. According to the integrated model even a single instance of contradictory feedback may lead to a considerable reduction in trust in the automated system. In case of the contextual control loop, the trust in the automation is not predicted to be reduced through contradictory feedback, although the mental model may be updated.
2. In the integrated model the complacency potential increases over time as long as contradictory feedback is lacking. This suggests that alerting systems (ATCs) or a second crew member will be instrumental in alleviating the mismatch between what is observed and what is expected. In contrast, reframing in the contextual control loop may occur within the individual without external trigger as a function of time.
3. The integrated model implies that the cause of automation surprise is attributed to too much trust in automation and a lack of SA, whereas in the case of the contextual control loop, the predominant cause is expected to be lack of knowledge about the automation in relation to the current operational context.

The objective of the study laid out below was to empirically validate either the integrated model or the contextual control loop model. The statistical analyses done for this study are not reported here, but are of course available from the original (de Boer & Dekker, 2017).

A questionnaire was developed that contained 20 questions of which 14 were multiple choice, four were open questions and two that were based on 6-point Likert scales. There were three parts to the questionnaire: (1) respondent demographics and flight experience, (2) specific details about the last automation surprise experience that can be recalled, and (3) experiences with automation surprise in general. Respondents were directed to a website that hosted the questionnaire or filled out the questionnaire on paper. The questionnaire was anonymous and it was assured that data would remain confidential.

Pilots were prompted to describe a recent case of automation surprise. "For this research, we are specifically interested in the last time you experienced automation surprise. The following questions [...] are aimed at the last time you exclaimed something like: 'What is it doing now?' or 'How did it get into this mode?.'" The change in trustworthiness of the automation following the most recent automation surprise experience was determined on a 6-point Likert scale (1 = trust in automation did not change – 6 = much less trust in automation) in response to the question: "Automation surprises can cause the automation to be perceived as less trustworthy. As a result of your last automation surprise, did your trust in the system change?" It also asked: "How was this last automation surprise discovered?" Four mutually exclusive responses were possible: the respondent him/herself, a colleague pointed it out, by an ATC discovered it. It also asked about the causes of automation surprise: "Below is a list of causes for Automation Surprises. Please state which ones are applicable to your last Automation Surprise. Multiple options are possible." We also asked the respondents "...to indicate how often [the same] factors are involved in Automation Surprise in general" on a 6-point Likert scale (1 = never – 6 = very often). We included nine possible preprinted responses and a field for "other."

During the study, 200 questionnaires were filled in and returned. The respondents were predominantly male (96%), 54% were in the rank of captain, and 42% were first officers (the balance is in the rank of second officer). With regard to aircraft type currently operated, respondents listed (in order of frequency): Boeing 737NG, Airbus A330, Boeing 777, Embraer 170/190, and Fokker 70/100. The age of the respondents ranged from 23 to 58 years (median 38 years). The flying experience varied from 750 to 27,500 h (median 7,500 h). The median number of flights executed per month was 28, with a range from 3 to 43 flights per month.

In 90% of the automation surprise events an undesired aircraft state was not induced, and only in one case of automation surprise (0.5%) was consequential damage to the aircraft reported. Based on the severity and prevalence of automation surprise, each pilot is expected to experience a reportable automation surprise event about once every 3 years.

Out of the 180 respondents who had reported an automation surprise, 106 of them (59%) reported that their trust had not changed following the automation surprise event. Only 17 respondents (9%) reported a reduction in trust of score 4 or higher (i.e., over the halfway mark). The distribution of respondents across the various rating categories was not random. These findings are in support of the contextual control loop: in the integrated model even a single instance of contradictory feedback is supposed to lead to a considerable reduction in trust in the automated system.

The discovery of the last automation surprise was predominantly by the respondent. Of the 176 respondents that had reported an automation surprise event and responded to this question, 89% reported that he or she noticed it themselves. 7% were warned by an ATC, and 5% were notified by the fellow pilot. Air traffic control did not contribute to the discovery of an automation surprise event in our sample. The distribution of respondents across the rating categories was not random. This finding supports the contextual control loop model, which suggests that reframing may occur as a function of time within the individual without external trigger. The integrated model suggests that ATCs or a second crew member will be instrumental in alleviating the mismatch between what is observed and what is expected, because the complacency potential increases over time as long as contradictory feedback is lacking.

In answer to what the cause was of the last automation surprise event that was experienced (multiple answers permitted), a total of 315 reasons were given for the 180 automation surprise events that were reported. In nearly two-thirds of the automation surprise events (63%) one of the relevant causes was claimed to be system malfunction. We interpret this as a lack of understanding of the system, because of the high reliability of aviation systems, the tight coupling of the modules of cockpit automation, the interface design choices and the use of automation in aviation that is above the critical boundary of action selection. Other frequently reported causes include wrong manual input (24%) and lack of knowledge concerning aircraft systems (19%). Too much trust in the proper operation of systems and/or complacency was only mentioned in 38 cases (12%) and lack of SA 25 times (8%). The study also asked how often specific factors are involved in automation surprise in general (6-point Likert scale, average response across all factors is 2.9). There were relatively high average scores for wrong manual input (3.4) and lack of knowledge concerning aircraft systems (3.1). Too much trust in the proper operation of automated systems scored equal to average (2.9) and lack of SA scored lower than average (2.6). Both the results about the last automation surprise event and about automation surprise in general lend credibility to the contextual control loop, because this model suggests that a lack of knowledge about the automation in the context of the operation will be predominant cause of automation surprise. In contrast, the integrated model implies that the cause of automation surprise is attributed to too much trust in automation and a lack of SA, which scored lower in both questions on the cause of automation surprise in our survey.

The data found in this study were a good fit with the contextual control loop on all three points:

1. Despite experiencing an automation surprise, more than half of the respondents did not report a reduction in their trust of the automation. Only a small minority reported a strong reduction in trust. This is inconsistent with the integrated model, which suggests that even a

single instance of contradictory feedback may lead to a considerable reduction in trust in the automated system;

2. The integrated model predicts that 'complacency potential' increases over time as long as contradictory feedback is lacking. This suggests that ATCs or a second crew member will be instrumental in alleviating the mismatch between what is observed and what is expected rather than the operator himself/herself. However, our data shows that the discovery of the last automation surprise was predominantly (89%) by the respondent himself/herself. This finding supports the contextual control loop model which suggests that reframing may occur as a function of time without external trigger;

3. The integrated model implies that the cause of automation surprise is attributed to too much trust in automation (or complacency) and a lack of SA. However, the data indicate that a lack of understanding of the system, manual input issues, and buggy knowledge concerning aircraft systems were the predominant causes of automation surprise—as predicted by the contextual control loop.

Parasuraman and colleagues have documented some of the problems associated with measuring 'trust,' and the discrepancy that typically occurs between objective versus subjective measures of trust. No strong link, for example, has yet been established between low sampling rates (or poor automation monitoring) and high self-reported trust in automation (Parasuraman & Manzey, 2010). As a result, previous studies have typically measured trust subjectively.

The study reported here is consistent with that, so as to be as fair as possible in our empirical evaluation. This, of course, means that an alternative, but slightly far-fetched, explanation could be that the way that the automation surprise is discovered is by a higher memory retention for positive events (though the literature is ambiguous on this), or by cognitive dissonance (or another self-centered outcome perception bias). Experiencing automation surprise might challenge one's self-esteem, which is restored if the event is perceived to be resolved by the self. Similarly, a low reduction in trust that was identified might be a result of the self-reporting nature of this study. As in many surveys, we acknowledge that the external validity and reliability of the collected data may be limited due to the self-report nature of the survey, and the post-hoc nature of having to rate 'trust' in the wake of an event. As part of this, we also recognize that the voluntary participation may also have biased the results, particularly on the prevalence of automation surprises.

Nonetheless, the fit between the data and the crew-aircraft contextual control loop suggests that the complexity of the context needs to be addressed to resolve issues in human–automation interaction. This will likely produce better safety outcomes than a focus on suboptimal human performance, at least within routine airline operations.

Support for the need to take a sensemaking (rather than a normative) approach can further be found in the study, where a surprisingly high number of respondents (63%) indicated that a system malfunction was one of the causes of the most recent automation surprise event. This may be the result of the ambiguous nature of a 'system malfunction.' The term can be used to denote:

1. that there is a discrepancy between what the designer intended and what the crew thinks it should be doing—therefore, not an actual malfunction but a 'buggy' mental model; or
2. that the system is functioning correctly (i.e., logically), but not doing what the designer intended (i.e., poor execution of the design); or
3. that the system is not functioning correctly (i.e., 'broken').

We propose that due to the stringent certification requirements in aviation that design errors (2) and malfunctions (3) will be less likely and the former (i.e., (1), the 'buggy' mental model) will be predominant. This is supported by the relatively high score in our survey for "lack of knowledge concerning aircraft systems" and the higher prevalence of automation surprise events associated with higher degrees of automation, in which an appropriate mental model of how the system works (as opposed to 'how to work the system') is more difficult to achieve. The high number of reports of 'system malfunction' in the survey lends support to the need to understand how pilots make sense of the automation.

The sensemaking model suggests that our understanding of the interaction between humans and automation can be improved by taking into account systemic factors and the complexity of the operational context, rather than focusing on suboptimal human performance. Automation surprise seems to be a manifestation of system complexity and interface design choices, and rarely the result of individual underperformance.

STUDY QUESTIONS

1. How can the field of human factors endeavor to change behavior, but not by changing behavior itself? Why was this a significant insight into the management of safety, and what sorts of discoveries brought that insight about?
2. What is the difference between error-resistant and error-tolerant design?
3. Describe the first cognitive revolution and how it supported the adoption and development of human factors.
4. Describe the second cognitive revolution and its role in shaping our understanding of human interaction with technology, as well as in how we should study it.
5. What are some of the problems you might encounter when you use 'loss of situation awareness' as an explanation of why things went wrong?

6. How is Gibson's aphorism 'Ask not what is inside your head, but what your head is inside of' a critique of information processing psychology as well as of more recent three-level theories of situation awareness?

7. Why, in cognitive systems engineering research, do you need to be able to study the particulars of an actual joint cognitive system but can only make sense of those particulars in reference to the universal concepts or patterns of joint cognitive work?

8. Have you come across cases of 'automation surprise'? What would the conventional explanation of 'human error' say about how to solve automation surprises?

REFERENCES AND FURTHER READING

AMA. (1998). *A tale of two stories: Constrasting views of patient safety*. Chicago, IL: American Medical Association.

ATSB. (1996). *Human factors in fatal aircraft accidents*. Canberra, Australia: ACT.

Bainbridge, L. (1987). Ironies of automation. In J. Rasmussen, K. Duncan, & J. Leplat (Eds.), *New technology and human error* (pp. 271–283). Chichester, UK: Wiley.

Breakey, H., van Winsen, R. D., & Dekker, S. W. A. (2015). "Loss of situation awareness" by medical staff: Reflecting on the moral and legal status of a psychological concept. *Journal of Law and Medicine, 22*(3), 632–637.

Bruner, J. (1990). *Acts of meaning*. Cambridge, MA: Harvard University Press.

Burnham, J. C. (2009). *Accident prone: A history of technology, psychology and misfits of the machine age*. Chicago, IL: The University of Chicago Press.

de Boer, R. J., & Dekker, S. W. A. (2017). Models of automation surprise: Results of a field survey in aviation. *Safety, 3*(20), 1–11.

De Keyser, V., & Woods, D. D. (1990). Fixation errors: Failures to revise situation assessment in dynamic and risky systems. In A. G. Colombo & A. Saiz de Bustamante (Eds.), *System reliability assessment* (pp. 231–251). Dordrecht, The Netherlands: Kluwer Academic.

Dekker, S. W. A. (2006). Past the edge of chaos. *Human Factors & Aerospace Safety, 6*(3), 235–246.

Dekker, S. W. A. (2014). *The field guide to understanding 'human error'*. Farnham, UK: Ashgate Publishing Co.

Dekker, S. W. A. (2015). *Safety differently: Human factors for a new era*. Boca Raton, FL: CRC Press/Taylor & Francis Group.

Dekker, S. W. A., & Lützhöft, M. (2004). Correspondence, cognition and sensemaking: A radical empiricist approach to situation awareness. In S. Banbury & S. Tremblay (Eds.), *Situation awareness: A cognitive approach* (pp. 22–42). Aldershot, UK: Ashgate Publishing Co.

Endsley, M. R. (1995). Toward a theory of situation awareness in dynamic systems. *Human Factors, 37*(1), 32–64.

Eurocontrol. (2015). Information processing. From Skybrary, Eurocontrol. http://www.skybrary.aero/index.php/Information_Processing

Eysenck, M. (1993). *Principles of cognitive psychology*. Hillsdale, NJ: Lawrence Erlbaum Associates.

FAA. (2013). *Operational use of flight path management systems: Final report of the performance-based operations aviation rulemaking committee/commercial aviation safety team flight deck automation working group*. Washington, DC: Federal Aviation Administration.

Fitts, P. M., & Jones, R. E. (1947). Analysis of factors contributing to 460 "pilot error" experiences in operating aircraft controls (TSEAA94-12). Dayton, OH: Aero Medical Laboratory, Air Material Command, Wright-Patterson Air Force Base.

Flach, J. M. (1995). Situation awareness: Proceed with caution. *Human Factors, 37*(1), 149–157.

Flach, J. M., Dekker, S. W. A., & Stappers, P. J. (2008). Playing twenty questions with nature: Reflections on the dynamics of experience. *Theoretical Issues in Ergonomics Science, 9*(2), 125–155.

Gibson, J. J. (1979). *The ecological approach to visual perception.* Boston, MA: Houghton Mifflin.

Henriqson, E., van Winsen, R. D., Saurin, T. A., & Dekker, S. W. A. (2011). How a cockpit calculates its speeds and why errors while doing this are so hard to detect. *Cognition, Technology and Work, 13*(4), 217–231.

Hoffman, R. R., & Deffenbacher, K. A. (1993). An analysis of the relations between basic and applied psychology. *Ecological Psychology, 5*(4), 315–353.

Holden, R. J. (2009). People or systems: To blame is human. The fix is to engineer. *Professional Safety, 54*(12), 34–41.

Hollnagel, E., & Woods, D. D. (1983). Cognitive systems engineering: New wine in new bottles. *International Journal of Man-Machine Studies, 18*(6), 583–600.

Hollnagel, E., & Woods, D. D. (2005). *Joint cognitive systems: Foundations of cognitive systems engineering.* Boca Raton, FL: Taylor & Francis Group.

Hollnagel, E., Woods, D. D., & Leveson, N. G. (2006). *Resilience engineering: Concepts and precepts.* Aldershot, UK: Ashgate Publishing Co.

Hopkins, A. (2006). What are we to make of safe behaviour programs? *Safety Science, 44*, 583–597.

Hoven, M. J. V. (2001). *Moral responsibility and information and communication technology.* Rotterdam, NL: Delft University.

Hutchins, E. L. (1995a). *Cognition in the wild.* Cambridge, MA: MIT Press.

Hutchins, E. L. (1995b). How a cockpit remembers its speeds. *Cognitive Science, 19*(3), 265–288.

James, W. (1890). *The principles of psychology* (Vol. 1). New York, NY: Henry Holt & Co.

Jordan, P. W. (1998). Human factors for pleasure in product use. *Applied Ergonomics, 29*(1), 25–33.

Klein, G. A. (1993). A recognition-primed decision (RPD) model of rapid decision making. In G. A. Klein, J. Orasanu, R. Calderwood, & C. E. Zsambok (Eds.), *Decision making in action: Models and methods* (pp. 138–147). Norwood, NJ: Ablex.

Krokos, K. J., & Baker, D. P. (2007). Preface to the special section on classifying and understanding human error. *Human Factors, 49*(2), 175–177.

Lee, L. (1992). *The day the phones stopped: How people get hurt when computers go wrong.* New York, NY: Donald I. Fine.

Loukopoulos, L. D., Dismukes, K., & Barshi, I. (2009). *The multitasking myth: Handling complexity in real-world operations.* Farnham, UK; Burlington, VT: Ashgate Publishing Co.

Meister, D. (1999). *The history of human factors and ergonomics.* Mahwah, NJ: Lawrence Erlbaum Associates.

Merlin, P. W., Bendrick, G. A., & Holland, D. A. (2012). Breaking the mishap chain: Human factors lessons learned from aerospace accidents and incidents in research, *flight test*, and development. Washington, DC: National Aeronautics and Space Administration.

Miller, G. A. (1956). The magical number seven, plus-or-minus two or some limits on our capacity for processing information. *Psychological Review, 63*, 81–97.

Mindell, D. A. (2008). *Digital Apollo: Human and machine in spaceflight.* Cambridge, MA: MIT Press.

Moray, N., & Inagaki, T. (2000). Attention and complacency. *Theoretical Issues in Ergonomics Science, 1*(4), 354–365.

Nagel, T. (1992). *The view from nowhere.* Oxford, UK: Oxford University Press.

Neisser, U. (1976). *Cognition and reality: Principles and implications of cognitive psychology.* San Francisco, CA: W. H. Freeman.

Norman, D. A. (1993). *Things that make us smart: Defending human attributes in the age of the machine.* Reading, MA: Addison-Wesley Pub. Co.

NTSB. (2007). *Attempted takeoff from wrong runway Comair flight 5191, Bombardier CL-600-2B19, N431CA, Lexington, Kentucky, August 27, 2006 (NTSB/AAR-07/05).* Springfield, VA: National Transportation Safety Board.

O'Hare, D., & Roscoe, S. N. (1990). *Flightdeck performance: The human factor* (1st ed.). Ames, IA: Iowa State University Press.

Parasuraman, R., & Manzey, D. H. (2010). Complacency and bias in human use of automation: An attentional integration. *Human Factors, 52*(3), 381–410.

Parasuraman, R., Sheridan, T. B., & Wickens, C. D. (2008). Situation awareness, mental workload and trust in automation: Viable, empirically supported cognitive engineering constructs. *Journal of Cognitive Engineering and Decision Making, 2*(2), 140–160.

Perry, S. J., Wears, R. L., & Cook, R. I. (2005). The role of automation in complex system failures. *Journal of Patient Safety, 1*(1), 56–61.

Rankin, A., Woltjer, R., & Field, J. (2016). Sensemaking following surprise in the cockpit: A re-framing. *Cognition, Technology and Work, 18*, 623–642.

Rasmussen, J. (1990a). Human error and the problem of causality in analysis of accidents. *Philosophical Transactions of the Royal Society of London, B, 327*(1241), 449–462.

Rasmussen, J. (1990b). The role of error in organizing behavior. *Ergonomics, 33*(10–11), 1185–1199.

Rasmussen, J. (1997). Risk management in a dynamic society: A modelling problem. *Safety Science, 27*(2–3), 183–213.

Rasmussen, J. (2000). Designing to support adaptation. *Ergonomics and Ergonomics Society Annual Meeting Proceedings, 44*, 554–560.

Rasmussen, J., Duncan, K., & Leplat, J. (1987). *New technology and human error.* Chichester, UK; New York, NY: Wiley.

Rasmussen, J., & Jensen, A. (1974). Mental procedures in real-life tasks: A case study of electronic trouble shooting. *Ergonomics, 17*(3), 293–307.

Roscoe, S. N. (1997). The adolescence of engineering psychology. In S. M. Casey (Ed.), *Human factors history monograph series* (Vol. 1, pp. 1–9). Santa Monica, CA: Human Factors and Ergonomics Society.

Sarter, N. B., & Woods, D. D. (1995). "How in the world did we get into that mode?" Mode error and awareness in supervisory control. *Human Factors, 37*(1), 5–19.

Sarter, N. B., Woods, D. D., & Billings, C. E. (1997). Automation surprises. In G. Salvendy (Ed.), *Handbook of human factors and ergonomics* (2nd ed., pp. 1926–1943). New York, NY: Wiley.

Schmitt, E. (2015, May 13). Navy pilot's death reflects hazards of job. *International New York Times*, New York, NY, p. 6.

Schulz, C. M., Endsley, M. R., Kochs, E. F., Gelb, A. W., & Wagner, K. J. (2013). Situation awareness in anesthesia: Concept and research. *Anesthesiology, 118*(3), 729–742.

Smith, K., & Hancock, P. A. (1995). Situation awareness is adaptive, externally-directed consciousness. *Human Factors, 27*, 137–148.

Smith, P. J., & Hoffman, R. R. (2018). *Cognitive systems engineering: The future for a changing world.* Boca Raton, FL: CRC Press.

van Winsen, R. D., Henriqson, E., Schuler, B., & Dekker, S. W. A. (2015). Situation awareness: Some conditions of possibility. *Theoretical Issues in Ergonomics Science, 16*(1), 53–68.

Waterson, P. (2009). A critical review of the systems approach within patient safety research. *Ergonomics, 52*(10), 1185–1195.

Weick, K. E. (1995). *Sensemaking in organizations.* Thousand Oaks, CA: Sage Publications.

Woods, D. D. (1993). Process-tracing methods for the study of cognition outside of the experimental laboratory. In G. A. Klein, J. M. Orasanu, R. Calderwood, & C. E. Zsambok (Eds.), *Decision making in action: Models and methods* (pp. 228–251). Norwood, NJ: Ablex.

Woods, D. D. (1996). Decomposing automation: Apparent simplicity, real complexity. In R. Parasuraman & M. Mouloua (Eds.), *Automation technology and human performance.* Mahwah, NJ: Lawrence Erlbaum Associates.

Woods, D. D., Dekker, S. W. A., Cook, R. I., Johannesen, L. J., & Sarter, N. B. (2010). *Behind human error.* Aldershot, UK: Ashgate Publishing Co.

Woods, D. D., & Hollnagel, E. (2006). *Joint cognitive systems: Patterns in cognitive systems engineering.* Boca Raton, FL: CRC/Taylor & Francis Group.

Woods, D. D., Patterson, E. S., & Roth, E. M. (2002). Can we ever escape from data overload? A cognitive systems diagnosis. *Cognition, Technology & Work, 4*(1), 22–36.

Wynne, B. (1988). Unruly technology: Practical rules, impractical discourses and public understanding. *Social Studies of Science, 18*(1), 147–167.

6 The 1950s, 1960s, and Onward
System Safety

Drew Rae and Sidney Dekker

KEY POINTS

- System safety was recognized as a formal discipline relatively recently. Safety should get built into the system from the beginning. Once a system is in operation, system safety specifies the requirements for effective and safe management of the system.
- This requires system safety to recognize the technical, human, and environmental contributors to the creation and erosion of safety, and to map and resolve (to the extent possible) the conflicts and trade-offs between safety and other factors in the design and operation of the system.
- Systems engineering for safety involves standardized process steps, with many variations in the detail of technique applied at each step. The steps are (semi-) formal modeling of the system under development; analysis of draft system designs; and analysis of the final design to demonstrate safety and to inform post-design safety efforts.
- From a design perspective, systems can be unsafe through requirements error (designing the wrong system), or implementation error (designing the system wrong). The aim is to prevent foreseeable events and minimize the consequences of unforeseen ones. The increasing complexity of automation and computerization (particularly when added to legacy systems) can make this very difficult.
- System safety, through its formal language and techniques, has defined safety as freedom from unwanted events, and protection against unwanted outcomes. As systems have become more complex and anticipating all pathways to failure has become virtually impossible, an emphasis is shifting to assuring the presence of capacity to handle unforeseen events, rather than assuring the absence of failure modes.

6.1 INTRODUCTION

Chapter 5 has shown how social choices in the framing and discussion of safety issues can lead to arbitrary and often misleading problem categories. 'Human error' and 'Poor interface design,' for example, are different ways to describe the same situation, but require the application of different principles to solve. 'Systems engineering' presents a new way of looking at safety—as a design problem. The field is variously known as:

- Systems engineering;
- System safety engineering;
- System safety;
- Safety engineering.

We will use the term 'system safety' throughout this chapter. Interestingly, system safety blends two seemingly divergent perspectives that come out of the foundations of safety science:

- The Tayloristic world—where safety is achieved through following defined procedures;
- The pre-Taylor focus on physically engineered solutions.

The product of this mix is a set of standardized processes to produce safe systems.

Safety is often concerned with negative properties—properties or behaviors that a system should *not* exhibit. One problem with that approach is that you are never done. You can never create an exhaustive list that has taken into account all the ways in which something could go wrong. Indeed, we can seldom even describe the list of undesirable properties comprehensively in anything but the most abstract terms, let alone prove their absence. This requires us to understand and acknowledge that (a lack of) safety is not a concept that can be formally defined or mathematically proven. All designs are fallible, as are the people and processes used to produce and operate those designs. Designs and processes can be safer, or less safe, but they cannot be definitively safe or unsafe.

System safety was recognized as a formal discipline relatively recently, and is less mature than disciplines such as construction engineering or mechanical design. This results in a wide spectrum of views on what it is. These views primarily disagree on where the emphasis should be placed. There are, however, some principles of system safety that are universal:

1. The primary aim of any system safety activity is to *save lives*. Whereas activities may have procedural or regulatory aspects, this should always be in support of their primary aim;
2. Effective management of safety is both *proactive* and *reactive*. It involves identification and treatment of hazards prospectively as well as responding to information, events, and concerns;
3. System safety is most effective if it is applied when and where it is most needed. Scaling safety activities in response to risk is often termed a *risk-based* approach to safety.
4. System safety is inherently *skeptical*. It requires positive evidence of safety, rather than relying on a lack of evidence of danger.

From a design perspective, systems can be unsafe through:

- requirements error (designing the wrong system), or
- implementation error (designing the system wrong).

The idea of system safety is that both types of error should be reduced through the application of appropriate design processes and principles:

- The ambition of system safety is that safety gets built into the system from the beginning, not just added to a completed design;
- The aim is to prevent foreseeable events and minimize the consequences of unforeseen ones;
- And once a system is in operation, system safety can also help specify the requirements for effective management of the system so that it continues to achieve desired levels of safety.

This requires system safety to recognize the technical, human, and environmental contributors to the creation and erosion of safety, and to map and resolve (to the extent possible) the conflicts and trade-offs between safety and other factors in the design and operation of the system (related to cost, efficiency, weight, and so forth).

System safety involves standardized process steps, with many variations in the detail of technique applied at each step. The steps are:

1. Representation of the system under development as a formal or semi-formal model;
2. Identification of 'hazards:' the unsafe states or events in the model;
3. Analysis of draft system designs to identify safety improvements;
4. Analysis of the final design to demonstrate safety and to inform post-design safety efforts.

These steps may be applied multiple times at different levels, using abstraction, refinement, and decomposition, in which:

- *Abstraction* is the removal of unnecessary detail;
- *Refinement* is the addition of extra detail, while preserving key system properties;
- *Decomposition* is the separation of a design into smaller, less complicated pieces.

The strengths and limitations of system safety are closely linked to the maturity of design processes for different applications. They are of course also linked to the technological possibilities available for the design and its alternatives. For low-complexity systems, usually involving well-understood technology, it is straightforward both to specify what 'safe' looks like, and to verify that a given design meets a safe specification. Extrapolating this to systems with higher complexity, particularly those with software and automation, is much more difficult. For these situations, there is often considerable debate about what constitutes a good design process (or even about what constitutes the 'system'). The existence of such debate is an indication that many decisions integral to the systems engineering approach are socially constructed rather than scientifically determined. Such decisions include:

- How systems should be modeled or represented;
- How 'safe' and 'unsafe' should be determined with respect to the system model;
- How abstraction, refinement, and decomposition should be used to support analysis;
- Which techniques should be applied for analyzing whether a system model is safe;
- Which techniques should be applied for developing a system to match the model, and for testing that the developed system matches the model.

Each of these decisions is made more difficult by the inclusion of humans and organizations as part of the system to be modeled and analyzed. System safety techniques have evolved to incorporate these sociotechnical aspects, as well as to deal with the increasing complexity of engineered artifacts involving software and integrated circuitry. This integration, however, is complex and will probably remain forever unfinished. The example toward the end of the chapter illustrates as much.

6.2 HISTORICAL BACKGROUND

6.2.1 FLY-FIX-FLY

Immediately before and during World War II (WWII), warring parties were concerned with maximizing the production, distribution, and sustainment of military goods. The ability to build and repair more planes, tanks, and ships would provide a decisive advantage over their opponents. Prior to 1940, the safety of these sorts of systems was typically achieved by attempting to control obvious hazards in the initial design. Other problems, once detected, were then corrected once the system was in use—or, in the best case, while it was still in a testing phase (Stephans, 2004). It was becoming obvious that this was no longer really acceptable or desirable. The 1950s and 1960s saw increasingly large systems being built and put into operation, such as nuclear missiles and fleets of nuclear power plants. These were systems whose failure would have huge and devastating consequences for many. Allowing significant failures to show up in a testing or operational phase through a 'fly-fix-fly' (or trial-and-error) regime of safety improvement was no longer deemed adequate. What used to work—you test what you fly, and you fly what you test—might still work, but the potential consequences would be too dire.

A first formal system safety effort grew out of WWII allies calling upon leading manufacturers to share and promote the 'science' of industrial success. One of the techniques that was heavily promoted during the war was statistical quality control. Quality control, as a general practice, involves preventing faulty components from becoming integrated into the product, and preventing products from leaving the factory and being distributed. Statistical quality control achieves this goal by identifying and eliminating sources of variation in the processes that produce the products. Rather than inspecting every product for defects, the new approach allowed factories to inspect only a small sample of products in each batch, and ultimately to reduce the total number of defects.

Statistical quality control originated with the work of Walter Shewhart, a physicist working in quality improvement for the Western Electric Company, a telephone equipment supplier for AT&T. Shewhart recognized that there were two different types of variation in a manufacturing process:

- Common causes—inherent variability that follows a statistical distribution (Shewhart called this 'chance' or 'noise' variation)
- Special causes—unpredictable variability from disturbances (Shewhart called this 'assignable cause' or 'signal')

Shewhart demonstrated, both mathematically and practically, that attempting to respond to common cause variation was counterproductive for quality, and that quality improvement required identifying and fixing special cause variation. He invented a simple graph—now known as a control chart—that showed variation of a quality variable over time, along with a set of rules for determining if a process was 'in statistical control' (subject to only common cause variation) or experiencing special cause variation. Joseph Juran and W. Edwards Deming applied Shewhart's principles to American manufacturing during WWII and to economic recovery in Japan after the war. In addition to its straightforward application to manufacturing, they extended quality control to cover management processes as well (Deming, 1982).

6.2.2 MISSILES, NUCLEAR, AND AEROSPACE

The U.S. Minuteman Intercontinental Ballistic Missile (ICBM) program is another important route into what we know as system safety today (Stephans, 2004). American military planners began developing ballistic missiles (using ex-Nazi know-how and experts) immediately after WWII. Driven by the accelerating arms race with the Soviet Union, the U.S. Air Force had mapped out a development plan for ICBMs, capable of carrying a nuclear warhead, by 1954. The first missile, the Atlas, was essentially a highly evolved German V-2 missile, which the Nazis had used against the allies in the waning days of WWII. It was powered by rocket engines that burned a mixture of liquid fuel and oxidizer.

By 1954, though, the Atlas was nowhere near production and had not even been flight tested. One of the engineering challenges was obviously to make it fly to a significantly greater distance than the V-2 ever had, which required both a lot more fuel but also lighter weight. In 1958, the air force contracted Boeing in Seattle to assemble and test the new missile, even though Boeing had no experience with missiles. The program created a new national industry, with tens of thousands of industrial and air force managers, engineers, and workers getting trained and converging on Seattle and northern Utah. The first Minuteman flew from Cape Canaveral in Florida in 1961.

For a few years, the air force had already been planning the (safe) deployment of Minuteman missiles. Wanting a gigantic 'missile farm' equipped with as many as 1,500 missiles, the air force ended up with language it already knew well. It divided the missiles across a couple of administrative 'wings,' then into 50-missile 'squadrons,' which were further divided into 'flights.' One site chosen was on the high plains around Great Falls, Montana. In case of attack or nuclear accident, the low population density would be an advantage. Based on access to previously classified military archives Sagan (1993) explains the many procedures and safeguards in place:

> The safety rules for the Minuteman ICBM force were designed to prevent both inadvertent and unauthorized arming or launching of the missiles. A complex set of personnel regulations, operational rules and redundant mechanical safety devices designed to achieve these objectives was prescribed. In each of the two-man underground launch control centers (LCCs) for the ICBMs, a set of ten switches (the Launch Enable System switches) had to be activated in order to arm the missiles, which connected the missiles ignition circuits. The two LCC officers had to turn two separate key switches on opposite sides of the room (the Launch Control Switch and the Cooperative Launch Control Switch) within two seconds of one another in order to launch the missile. A status panel existed in the LCC, which provided both a light and an audible alarm to inform the officers whether a missile under their control was on alert, was armed, was not functioning properly, was in the middle of a launch countdown, or had been fired. Each Minuteman missile was also equipped with a safety control switch on its motor ignition system, which could be manually locked in safe position during routine maintenance. All electrical circuits between the missile and the LCC were required to have electrical surge arrestors and to be checked prior to deployment to ensure that they could not be inadvertently activated by electrical tests or monitoring equipment.
>
> *(pp. 82–83)*

This complex system of checks and balances, however, was designed for peacetime. It was meant for normal conditions without tight coupling, without undue time pressure or the sudden need to bring systems to operational alert. When the pressure came on, the design of the elaborate system was stretched to (and beyond) its limits. It was a demonstration of the difference between work-as-imagined (as above) and work-as-done *avant la lettre* (Hollnagel, 2012b).

The first squadron of solid-fuel Minuteman I missiles was about to be deployed when the Cuba missile crisis rapidly escalated in late 1962. The first Minuteman was placed on operational alert at Malmstrom Air Force Base in Montana on the morning of October 26, and a total of five of the ICBM's were similarly ready 4 days later. Sagan (1993) found that the fly-fix-fly approach (even though the missiles weren't technically 'flying' but at least operationally ready to fly) had still very much to be used to prevent and fix continually emerging problems:

> To achieve this rapid surge of in readiness, many routine safety procedures used at other bases had to be adjusted. Throughout the ICBM complex, Air Force and contractor personnel jerry-rigged the missile systems with available material: according to the declassified history of the 341st Strategic Missile Wing's activities, 'lack of equipment, both standard and test required many work-arounds.' Individual missiles were

therefore repeatedly taken on and off alert as maintenance crews discovered problems and improvised fixes to permit maximum emergency readiness. Numerous unanticipated operational difficulties had to be surmounted under the tense pressure of the crisis alert. Officers quickly discovered, for example, that these Minuteman missiles might be inadvertently armed through miswiring, wire shorts, or burn-throughs, and each missile at Malmstrom was therefore tested for such faults when it was removed from alert status. Emergency communications and launch crew certification procedures also had to be improvised. It was discovered that the land lines to the Malmstrom Launch Control Center (LCC) intermittently failed, and because there wasn't time to test and fix the communication system as designed, a special 'walkie-talkie' system of transceiver radios was installed at the site. There was also insufficient time for regular launch crew certification procedures to be followed as the ICBMs were rushed onto alert status, and the crews taking control of the first operational Minuteman missiles therefore were *not* fully evaluated or certified when they accepted this awesome responsibility.

(pp. 81–82)

Based on its experience with the ICBM program and the still too frequent and devastating explosions of these liquid-propellant vessels, the U.S. Air Force Ballistic System Division realized that it needed more systematization of requirements, testing, quality control, verification, and all the associated documentation. As Leveson (2012) points out, although there were operators, there were actually no pilots to blame when things went wrong with a missile, because there were no pilots onboard. The assurance of safety had to be sought elsewhere, and much earlier on in the design and production process. Safety became a system problem. Early air force attempts at such documentation provided the basis for what is now known as MIL-STD-882, or the *System Safety Program for Systems and Associated Subsystems and Equipment*. The U.S. Department of Defense (DOD) has since turned it into the touchstone for its system safety efforts.

And missiles were not alone: the aerospace industry as a whole was an important contributor to the emergence of system safety in the 1950s and 1960s. During and after WWII, the air force lost 7,715 aircraft and killed 8,547 while doing so. Most of these accidents had been blamed on the pilots (see the previous chapter), but, like their human factor counterparts, aerospace engineers were not so easily fooled. Safety, they argued, had to be designed and built into aircraft in a much more systematic way—just like performance, stability, and structural integrity (Leveson, 2012). Also, whereas the National Aeronautical and Space Administration (NASA) developed its own system safety program and requirements, it was using much of the same missile technology, contractors, personnel (and sometimes missions) as the air force. So its developments closely paralleled the approach of MIL-STD-882 and the DOD.

In the late 1960s, the U.S. Atomic Energy Commission had become aware of system safety efforts in these aerospace and defense communities. It made the decision to hire a retired manager of the National Safety Council to develop an atomic energy system safety program. Although individual programs and contractors had adequate or even excellent safety programs in place, these varied widely. It hampered effective monitoring, evaluation, and control of safety efforts. The aim of the Atomic Energy Commission was to improve the overall safety effort with a new, standardized approach to system safety that incorporated the best features of existing system safety efforts (Stephans, 2004). In the mid-1970s Atomic Energy Commission was reorganized into the Department of Energy (DOE), which we have already encountered in other chapters.

A 1975 report labeled *WASH-1400, The Reactor Safety Study* (U.S. Nuclear Regulatory Commission, 1975) set out to calculate the risk associated with the operation of a nuclear power plant with the help of probabilistic risk assessment, which takes into account the likelihood as well as the severity of adverse events. To calculate the risk, the team modeled all kinds of possible accident scenarios in event trees. They concluded that the risk associated with the operation of 100 power plants was lower than the risk associated with other natural or man-made disasters such as airplane crashes (Keller & Modarres, 2005). After publication, the report was criticized for:

- The calculations;
- The handling of uncertainties;
- The risk comparisons which had been "prejudging an acceptable level of risk for nuclear energy" (p. 280).

The main critique went to the heart of the idea of system safety at the time: the calculated probabilities and consequences of adverse events. A few years after its publication, the Three Mile Island nuclear accident (see Chapter 8) showed just how far off the mark these calculations may indeed have been.

6.2.3 COMPLEXITY, CULTURE, AND COMPUTERS

As systems have grown in cost and complexity—also beyond aerospace, nuclear, and weapons—the recognition has grown as well that a more sophisticated, systematic upstream safety approach was needed. Some other fields had been looking at system safety past their noses, considering the up-front efforts too expensive without a good return on the investment. Notwithstanding, system safety tools and techniques gradually spread in the 1970s and 1980s, and became more accepted and more useable for others as well. The better designs they helped produce eventually convinced a growing number of fields that they could benefit from spending money for safety in advance of building and testing, for example:

- Naval facilities;
- The U.S. Army Corps of Engineers;
- The petrochemical industry;
- (Military) construction projects;
- Rail;
- The yet-to-be-built NASA Space Station.

These all became sites for the application and further development of system safety efforts. In parallel with reactive quality control programs, manufacturers began introducing predictive reliability programs. This period gave rise to what we might readily recognize as system safety techniques today, such as HazOp (Hazard and Operability programs), and fault trees, failure modes, effects, and criticality analyses.

They were all intended to help track, identify, and control (mostly mechanical and process) risks systematically.

System safety has since recognized the importance of human, organizational, and cultural contributions to the safety of the overall system. High-visibility accidents in the 1980s and 1990s have alerted it to the deeply intertwined aspects of management, resources, goal conflicts, and organizational culture that all impact on the design and use of safety-critical technologies. System safety has generally been ill-equipped to handle this enormous complexity and the harder-to-pin down aspects of human and organizational performance. Early attempts at Human Reliability Analysis (a quantification of human error potential) and later work on safety culture are examples of attempts to provide more formal approaches to those aspects of system safety.

In addition, the proliferation of software to control these systems has since added an entirely new layer of complexity and risk (Lee, 1992; Leveson, 2012; Leveson & Turner, 1992; Nemeth, Nunnally, O'Connor, Klock, & Cook, 2005; Norman, 1990). Today, researchers and practitioners are working hard to keep up with this changing face of hazard, with methods such as STAMP (Systems-Theoretic Accident Model and Process) (Leveson, Daouk, Dulac, & Marais, 2003) and those of Cognitive Systems Engineering (see the previous chapter) on the frontline of providing better tools to predict and prevent (Perry, Wears, & Cook, 2005).

6.3 FORMAL CONCEPTS OF SYSTEM SAFETY

One of the important contributions of system safety is its formalization of the language, as well as the tools and techniques for safety. Although some concepts are still more vague than others (hazards, for example), this approach to safety engineering has had an impact on how we talk about safety in other domains as well (including occupational health and safety with its risk assessments, hazards, hierarchy of controls, and more).

6.3.1 HAZARDS

The premise of system safety is that in order to prevent accidents, their potential causes must be managed in a systematic way. This still applies, even though we have increasingly recognized that accidents arise from complex webs of causes. System safety employs the concept of 'hazard' as a manageable division of the accident causes. Informally, a hazard is a potential accident, or an 'accident waiting to happen.' It is surprisingly difficult to make this concept more precise, and there is no consensus on the definition of a hazard. It could be applied to substances, events, states, behaviors, attitudes, properties, and more.

In order to describe a hazard, you may refer to specific states of the system or events within the control of the system that characterizes its undesirable behavior. The key points that system safety considers true about hazards are:

- Hazards are a management concept rather than a physical property of a system. The 'correctness' of a set of hazards arises from its usefulness in preventing accidents. Two different sets of hazards for the same system may be equally valid;

- A behavior can be a hazard, even if the system designers intended the behavior to occur, or the requirements allow the behavior to happen;
- If a hazard is conceptually possible, it presents risk, even if it is believed to be unlikely that the hazard will occur;
- There are some occasions (such as where an event outside the system presents danger, or causes the system to malfunction) where the basic definition of a hazard as a behavior of the system fails to capture the key event of interest. Any hazard identification process must consider these occasions.
- Hazards cause accidents; a hazard must occur for an accident to occur. If it is possible for an accident to occur without an identified hazard occurring, then not all hazards have been identified.

As with accidents, you can consider the severity and likelihood of a hazard. There is an issue, however, of how to evaluate the severity of a hazard. This must be done in terms of the ensuing accident, but where there is more than one possible outcome, which accident or accidents should you consider? Here are some of the options:

- The most severe potential accident?
- The most likely potential accident?
- The 'expected' outcome, factoring in the likelihood that the hazard will develop into each potential accident?

Hazard likelihood and risk are defined in ways analogous to accident probability and risk:

- Hazard likelihood is the combined likelihood of all the circumstances that are sufficient to cause that hazard.
- Hazard risk is the combination of the likelihood of the hazard and the consequences of any ensuing accident, taking into account the likelihood that the hazard evolves into the accident.

NO 'CORRECT' WAY TO EVALUATE A HAZARD

Consider the following scenario:

1. The hazard probability is $10-6$ per hour;
2. The most likely accident arising from the hazard involves 1 death;
3. Only 1 in 10 occurrences of the hazard actually result in a death.
4. Then, the risk is $10-6 \times 1 \times 10-2$ p giving $10-7$ deaths per hour on a "most likely outcome" basis.

There is no 'correct' way to evaluate hazard severities and risk, and practice varies between industry sectors. However, standards are often mute on this point, and this is an issue which needs to be defined, e.g., in a safety plan or safety management system. Clearly, the measured risk will vary depending on how hazard severity is defined. At minimum, within any one project, it is important to be consistent in the definition. We briefly consider the impact of the choice here. Taking the most severe potential accident (the 'worst case') may lead to overengineering if this outcome is very improbable. In particular, it makes it difficult to prioritize hazards, since a worst case of 'catastrophic' can be imagined for almost any hazard. It can also perversely lead to disregarding mitigations, which are of genuine benefit, but do not prevent the worst case. Using the most likely potential accident may be the most 'efficient' approach to design, eliminate, or control the issues which would cause most difficulty, and avoid putting effort into very improbable events. This strategy may result in under consideration of hazards where there is a credible outcome significantly worse than the most likely outcome. Using expected values is perhaps the most 'precise,' allowing for refined risk control, although this assumes that the underlying data on which the analysis is done is accurate. Evaluating hazard probability requires us to understand the causes of the hazards, and how likely they are to arise. Thus, our next step is to 'move back down the causal chain' and to consider what can occur to give rise to hazards.

Hazards can occur due to:

- Causes within a system (endogenous). Internal, or endogenous, causes of hazards are generally failures or malfunctions of structural components or platform systems, including those that arise because of design error;
- External threats (exogenous). Examples of exogenous causes are lightning strike, impact of another system, and electromagnetic interference. These are environmental factors that can affect system behavior;
- Or combinations of both.

It is common to use the terms fault and failure. Historically, and unfortunately, these terms have been used somewhat interchangeably. But you can make a useful, and necessary, distinction between them by using events and states:

- A failure is an unintended action or event;
- A fault is an unintended state of a system.

Here is an example of the difference between faults and failures:

- System failure—closing of road barrier at level crossing when heavy lorry passes;
- System fault—inadequate locks that release the barrier under vibration (from heavy vehicle passing).

As can be seen from these examples, faults give rise to failures. The fault–failure relationship occurs across levels in the design hierarchy, with a failure of a unit or subsystem giving rise to a fault at the system level, which in turn will give rise to a system failure, and so on. However, faults will not necessarily propagate in this way, in all circumstances. The system fault identified above is always present (until the design is changed) but is only triggered when a heavy lorry passes, i.e., the failure arises because of the combination of an internal fault and an external trigger. If no heavy vehicle ever passed the level crossing the failure would never arise. Thus, not all faults will necessarily cause system-level failures, and part of the aim in designing systems is to eliminate single points of failure that can give rise to a hazard. Thus, system safety refers to triggers: the set of conditions that activate a fault giving rise to a failure.

The definitions of both faults and failures refer to 'unintended,' not 'incorrect,' or 'unspecified.' Systems are not intended to cause accidents. There is considerable empirical evidence that many accidents arise due to flaws in requirements or specifications. You cannot say that the system behaved according to its specification, so the accident did not occur. That is why system safety uses the term 'unintended' and rely on the fact that the intent was to prevent or eliminate accidents rather than to talk in terms of specifications. The definitions of failure and fault cover all deviations from intent, so not all failures will have safety significance.

6.3.2 Risk Assessment

Remember that in the formal language of system safety, risk is the combination of the likelihood of the hazard and the consequences of any ensuing accident, taking into account the likelihood that the hazard evolves into the accident. Once you have understood the potential accidents that are associated with a particular hazard, you can try to determine the severity of the credible accidents that we have identified. You should also try to establish the probability of that hazard or that accident. You can do that in two different ways:

- You can estimate the probability of the accident, and hence get a measure of accident risk. Then you (or other stakeholders) have to decide whether the estimated risk is acceptable. This is done in many domains, including rail and military aviation.
- You can also establish an acceptable risk beforehand, and set probability targets that need to be met by the various parts of the system (including those associated with human and organizational performance that gets tricky very quickly).

Accident severity categories are qualitative, and may in some cases amount to little more than a 'best guess.' They typically are:

1. *Catastrophic*: fatalities or multiple severe injuries or major damage to the environment. Also, a loss of a major system;

2. *Critical*: single fatality or severe injury or significant damage to the environment. Includes a loss of a major system;
3. *Marginal*: minor injury or significant threat to the environment. Includes severe system damage;
4. *Insignificant*: possible minor injury, and only minor system damage.

The terms used in this process can make the discussions around a risk assessment—and the considerations and voices that go into those discussions—more interesting than the actual outcome. What, exactly, is the line between major damage to the environment, for example, and 'merely' significant damage? And who gets to say this, or from whose perspective? The outcome of a risk assessment always remains negotiable, precisely because of the qualitative nature of the categories and the way they are arrived at. This also goes for accident probability, if not more so. The typical categories there are:

- *Frequent*: likely to occur frequently; the hazard will continually be experienced;
- *Probable*: will likely occur several times or even often in the operation of the system;
- *Occasional*: will likely occur several times in the operation of the system;
- *Remote*: likely to occur somewhere in the lifecycle of the system, and it should be reasonably expected to occur at some point;
- *Improbable*: unlikely to occur;
- *Incredible*: extremely unlikely to occur.

This actually asks analysts to consider not only how likely a hazard is to arise, but also how likely it is to turn into an accident. Again, phrases such as 'it should be reasonably expected to occur' make an appeal to the expertise of those who are doing the risk assessment, and whose experience and professionalism can help determine what 'reasonable' actually means in that context. The assessment is thus heavily dependent on who participates, and even then there are no clearly determinable and nonnegotiable outcomes.

AN ACCIDENT STOPS BEING INCREDIBLE ONCE IT HAS HAPPENED

Another interesting note is that the 'incredible' nature of an accident changes dramatically after it has happened. What people often consider 'incredible' after an accident, is that nobody saw it coming—particularly not the regulators or operators. This is consistent with what is known as the 'hindsight bias.' It is interesting to note that, although he is generally credited with drawing attention to the phenomenon, Fischhoff did not use the term hindsight bias. Instead, he preferred the description 'unperceived creeping determinism' (Fischhoff, 1975, 2007). His study demonstrated that finding out an outcome that has occurred increases its perceived likelihood. In other words, a retrospective reviewer, knowing the outcome of an event, may have an exaggerated sense of their own probable ex ante ability to predict it (as in: I would have known this was going to happen). Paradoxically, and crucially for risk assessment, hindsight bias is, therefore, actually about the prediction of probabilities of future events. Fischhoff observed that people are usually unaware of this changed perception because they immediately assimilate the outcome knowledge with what they already know about the event in order to make sense of the past. He noted that "making sense out of what one is told about the past seems so natural and effortless a response that one may be unaware that outcome knowledge has had any effect at all" (1975, p. 288).

This takes us to one of the greatest conundrums in safety work: to explain why a slide into disaster, so easy to see and depict in retrospect, was missed by those who inflicted it on themselves or others. Judging, after the fact, that people suffered a failure of foresight (see also the next chapter) is easy. All you need to do is plot the numbers, trace out the trajectory, and spot the slide into disaster. But that is when you are standing amid the rubble and looking back. Then, it is easy to marvel at how misguided or misinformed people must have been. But why was it that the conditions conducive to an accident were never acknowledged or acted on by those on the inside of the system—those whose job it was to not have such accidents happen?

Fischhoff demonstrated how foresight is not hindsight. There is a profound revision of insight that turns on the present. It converts a once vague, unlikely future into an immediate, certain past. When we look back and we see drift into failure, it is, of course a by-product of that hindsight. It seems as if the organization was inescapably headed toward the failure, and we almost automatically reinterpret the availability of signs pointing in that direction as really hard to miss: "People who know the outcome of a complex prior history of tangled, indeterminate events, remember that history as being much more determinant, leading 'inevitably' to the outcome they already knew" (Weick, 1995, p. 28). But this certainty of trajectory and outcome is the product of hindsight and outcome knowledge. The vantage point of the retrospective outsider is what helps reveal the drift that occurred.

People's situations at the time would have looked very different. Of course, there may have been multiple possible pathways. Some of these pathways or options might have been proven, some recognizable and familiar, and many of them plausible. An additional image of certainty is created when outcome severity and probability are combined into a so-called risk-matrix, but this too is a human, and often collaborative, production. What this does do, however, is to offer an inroad into the next discussion, and that is whether the risk is tolerable or not—and to what extent (and how) it needs to be controlled. A pragmatic criterion here is the so-called ALARP (as low as reasonably practicable) principle. This means that a risk can be agreed to be taken only if the benefit is desired, or when risk reductions are so prohibitively expensive that they exceed any improvement they would offer. Of course, you can always spend more money to (try to) improve safety, but it is not always cost-effective. And as you will see in the next chapter, Perrow's work shows how the added complexity of safety devices and barriers can actually *increase* the risk that a system is exposed to.

6.3.3 Safety Cases

A safety case is a justification that a product or system is acceptably safe for its intended role. Implicit in the growth of safety regulation was the development of formal processes for demonstrating to a regulator that a system, installation, substance, or process was safe. In the nuclear industry, this was overseen by the Nuclear Installations Inspectorate (United Kingdom), and the Nuclear Regulatory Commission (United States). The United Kingdom established a system of operating licenses requiring reports justifying the safe design, construction, and operation of nuclear facilities. The United States commissioned the Reactor Safety Study, which used quantitative risk assessment to justify the safety of nuclear facilities. In Europe, the Seveso Directive for Technological Disaster Risk Reduction of 1982 included mandatory publication of safety reports to demonstrate compliance with the directive. It now applies to more than 12,000 industrial sites in the European Union where dangerous substances are used or stored in large quantities. This is mainly in the chemical and petrochemical industry, as well as in fuel wholesale and storage. In the United Kingdom, the report by Hidden into the 1988 Clapham Junction rail collision (Hidden, 1989), and the Cullen enquiry into the Piper Alpha disaster of the same year led to extension of this safety case approach into the rail and offshore oil industries.

In tightly regulated industries, individual organizations have little flexibility in what to present or how to present it: their safety case is largely built around compliance with what other stakeholders tell them to demonstrate. In cases of more flexibility about what standards or processes to follow, there is an obligation to explain and justify why those standards and processes are deemed appropriate (and why others are not). A safety case, in either case, presents a record of all the safety-related information concerning the product or system. This includes:

- an overall safety argument;
- safety evidence from analyses of the product or system;
- applicable design information;
- evidence from development, testing, and in-service experience.

The role of argument in safety cases is frequently neglected in favor of voluminous but unstructured (numeric) evidence. Of course, argument without evidence is unfounded and probably unconvincing. But evidence without argument, without contextualization and explanation, can lead to a bamboozling of stakeholders by sheer weight of numbers and the putative hard work behind generating them. The complex combination of environment and software, of people and interfaces, of organization and culture is still capable of producing vexing failures that would have been very difficult to foresee through any safety case.

6.3.4 RELIABILITY AND SAFETY

'*Safety*' is sometimes viewed as a special case of the more general property '*dependability*.' Dependability is a portmanteau term including the other 'ilities,' i.e., reliability, availability, maintainability, and security.

- *Reliability* is the probability that a part or a system can perform its required function, under given conditions, for a defined time interval;
- *Availability* is the probability that an item is performing its required function, under given conditions, at a defined time (regardless of the number of times it may have failed and been repaired in the interval);
- *Maintainability* is the probability that a failed item can be repaired in a defined time so it can perform its required function.

Safety is not synonymous with reliability (or availability or maintainability). Any failure may reduce reliability, but only failures that contribute to hazards affect safety. A part or system may also be reliable with respect to its specification, but unsafe in combination with other parts, systems, or the environment. Although improving reliability often improves safety, there are often conflicts between safety and availability: improving safety involves reductions in availability, and vice versa. In general, there is a need to make trade-offs between different dependability attributes, and this is a key aspect of systems engineering for safety-critical applications.

And there is more. The increased complexity and coupling of systems has given the relationship between safety and reliability a new twist. A primary focus on failure events in the sorts of analyses that are typical for system safety, may blind us to safety risks. Because the assumption that better reliability means more safety is largely false. The more complex a system, the less this actually applies. As Leveson argues, increasing safety by increasing system or component reliability, and thinking that if components or systems do not fail, then accidents would not happen,

> ...is one of the most pervasive assumptions in engineering and other fields. The problem is that it's not true. Safety and reliability are *different* properties. One does not imply nor require the other. A system can be reliable but unsafe. It can also be safe but unreliable. In some cases, these two properties even conflict. That is, making the system safer may decrease reliability and enhancing reliability may decrease safety.
>
> **Leveson (2012, p. 19. Italics in original)**

Component failure accidents, Leveson adds, have received most attention in engineering. But as systems have become more complex, and components more reliable, this has changed. More common now are component interaction accidents. In these, component failures are rare and not even necessary to have an accident. Accidents, rather, come from designs that are no longer intellectually or organizationally as manageable as they once were. Planning, understanding, anticipating, and guarding against foreseeable failure pathways becomes much more difficult. Even when all components operate reliably, an accident can happen.

RELIABLE, AND THEREFORE NOT SAFE

As an example of a system that behaved as designed but nonetheless became an accident, Leveson (2012) describes what happened to a batch chemical reactor in the United Kingdom. A batch reactor is a type of vessel widely used in the process industries. Reactants, or substances that take part in and undergo change during a reaction, are placed in the vessel and allowed to react. In the process, a product is formed and taken out of the vessel. The process can then be repeated (i.e., the next batch). There are typically valves for inlet and outlet, and perhaps a heater or stirring system to encourage the reaction. Some batch reactors are designed to keep a constant pressure by varying content volume, whereas others are the other way around: controlling volume by manipulating pressure.

In this case, a computer was responsible for controlling the flow of a catalyst into the vessel. A catalyst is a substance that increases the rate of a chemical reaction without itself undergoing any permanent chemical change. The computer was also responsible for the flow of water into the condenser that was designed to cool off the reaction. Sensor inputs in the computer would warn of any problems in various other parts of the plant that were connected to this batch reactor. Programmers were told that if a fault occurred elsewhere in the plant, then they could leave all their controls as they were, since the system would take care of itself.

On this occasion, the computer received a signal about a mechanical issue elsewhere in the plant: a gear box was reporting a low oil level. The requirements had the computer sound an alarm, but otherwise it left everything as it was, as it had been programmed to do. The problem was that a catalyst had just been added to the reactor, though the computer had only just started to increase the cooling water flow to the condenser. The flow did not keep pace with the chemical process and the reactor overheated. The relief valve did work as designed, venting the chemical into the factory. All components worked as the requirements had specified that they should, but together they created an accident. Prevention would have required identifying and eliminating or mitigating unsafe interactions among the system components and the software code driving them.

The example above is perhaps relatively straightforward, and you could argue that it is the kind of thing that people who build a plant like that might likely foresee. But what if things are no longer contained in a single plant, or if they are built and operated by different groups on different premises? What happens when those begin interacting? The tightly packed city of Amsterdam shows such interesting possibilities of interactions, with the potential for mass casualty events. In these cases, the reliability of each subsystem actually increases the safety risk for the entire system.

RELIABLE COMPONENTS, UNSAFE SYSTEM

In 2018, experts at the Amsterdam Fire Service argued that the multiple use of space, particularly underground, creates the possibility for risky interactions that fall outside the scope of the consideration of safety for each of the individual parts. Rules and regulations have fallen behind this multiple use, and fire safety gets evaluated for each of the components, not for the entire system. This leads to a situation where the assured and certified reliability of the fire suppressing system of each individual component actually leads to a greater safety risk for the whole. Amsterdam Central Station, for example, sports four layers of infrastructure—one above the other. The lowest is the North-South metro line. Then there is the Michiel de Ruijtertunnel for cars, then there's a shopping center, and then there's a bus station that is covered by a huge glass roof. The various levels, of course, are connected to each other by stairs and escalators.

A crash and fire in the car tunnel can produce significant issues at the other levels through various interactions. Since the tunnel is only 360 m long, there are no smoke suppression ventilators (they are required only for tunnels over 500 m long). Smoke can only leave through the mouths of the tunnel. The glass roof of the bus station (which sits on top of the infrastructure pile) adds to the problem: it actually covers the entire tunnel, including the two mouths. With smoke escaping from the tunnel mouths, it will get trapped under the glass dome of the bus station roof. The smoke then cools, and descends back down into the train station and shopping center—through the stairs and escalators (which are also the means of evacuation). Simulations show that within a few seconds, large numbers of people will be trapped and overcome by smoke.

Other metro stations show different kinds of interactions between the reliability of one part and the lack of safety of the whole. Stations such as Vijzelgracht and Rokin, where the subway tunnels are sunk relatively deep underground, are surrounded by subterranean parking garages. Permission for those garages was granted without taking into account the presence of the subway tunnels. Since the fire department does not enter such garages once there is a fire inside, they are equipped with automatic carbon dioxide suppression systems. These are rated based on the volume of an empty garage. This means that if the garage is full of cars, there will be excess CO_2. This is heavier than air, and will thus sink into the subway stations. It will make the air there unbreathable. The reliability of the suppression system in one part thus increases the safety risk of the whole.

As you will see, the work of Perrow (Chapter 8) tries to take these kinds of realizations into a semiformal language of interactive complexity and tight coupling. This is an attempt to take into account features of the human and organizational order, as well as the technological and engineered system and somehow begin to map out the potential for unforeseen interactions and rapid escalation of faults and failures.

6.3.5 System Safety and Understanding Complex System Breakdowns

Rapid escalation also occurred in the example below. It shows how the insights of Cognitive Systems Engineering (recall the previous chapter) can shed light on a situation in which the coordination inside a joint cognitive system quickly goes sour. This is a case of a part legacy system which has, in various versions, been operational for 50 years, but in which the successive addition of layers of computerization and automation have created new complexities and unforeseen pathways to system breakdown. Careful analysis with the help of system safety concepts (e.g., hazard probability, failures versus faults, singe-failure modes) could make a difference when taken together with the lexicon of Cognitive Systems Engineering (e.g., buggy mental models, problem representations, automation surprises, mode errors). It is at the intersection of these two where breakdowns occur, and this is also where we should look to increase the resilience of a joint cognitive system like it.

SYSTEM SAFETY, LEGACY DESIGNS, AND A JOINT COGNITIVE SYSTEM GOING SOUR

On the morning of February 25, 2009, a Boeing 737–800 was vectored by air traffic control for an instrument approach to runway 18R at Amsterdam Schiphol Airport (AMS) at 2,000 feet, less than 5.5 nm (nautical miles) from the runway threshold (DSB, 2010). The aircraft was flying on autopilot. The right autopilot (known as Autopilot B or CMD B) had been selected on, and the right Flight Control Computer (known as FCC B) was giving it all inputs. The short turn before the runway prompted the crew to use vertical speed mode to capture the glideslope from above (a necessity because of the close-in vector while being kept at 2000 feet). Air traffic controller workload had been mounting at the time and the approach sector was to be split shortly after this flight. The First Officer (F/O), a newly hired 42-year-old pilot (with 4,000 h of air force flight experience) undergoing line training, was pilot flying (PF).

Upon the crew's selection of vertical speed mode, and leaving 2,000 feet, the 737s autothrottle (A/T) retarded to idle, which was consistent with crew expectations (and, for all they knew) their instructions to the automation. Approaching a new flap setting, the airplane had to slow down and go down simultaneously, something that required idle power at that point. For the next 70 seconds, the automation behaved exactly as the crew would have expected.

The A/T, however, had automatically retarded in a mode (the so-called retard flare mode) that is not normal in this situation, but that was triggered by a faulty radar altitude reading by the left radar altimeter (RA) and other flight parameters after leaving 2,000 feet. There is no A/T indication in the cockpit that uniquely marks out the retard flare mode. The RA fault had not been reported to the crew, and there was no fault flag, no warning, no light nor any other direct annunciation about it in the cockpit.

Essentially, because of the faulty RA input, the A/T decided that it was time to land. It no longer tracked the selected speed, nor did it provide so-called flight envelope protection. The autopilot, however, was still flying the airplane, tracking the electronic glideslope down to the runway. In other words, the A/T was trying to land; the autopilot was trying to fly.

Earlier on, the crew had become aware of problems with the left radio altimeter during their descent toward Amsterdam, and they may have set up the cockpit so as to insulate the automation from those problems. They had FCC B selected as the so-called Master FCC, and had selected Autopilot B on. FCC B has its own independent RA. The training and documentation available to B737 pilots suggest that this would be sufficient for protecting the automation against left RA anomalies.

But it is not. What is not in Boeing 737 documentation and training available to pilots is that the A/T always gets its height information from the left RA, independent of which FCC has been selected as master and independent of which autopilot is selected on. The knowledge available through training and pilot documentation is so underspecified that it in fact can create a false or buggy mental model about the interrelationships between the various automated systems and their sensor input. This produced the automation surprise. The anatomy of this automation surprise is in Table 6.1.

Through training and documentation, the crew (as all B737 crews) could have been led to believe that a problem with the left RA had no consequences as long as FCC B was selected to control the automation that was flying the

TABLE 6.1
Anatomy of an Automation Surprise (DSB, 2010)

What is Trained	What a Crew Could Have Concluded	What is Actually Going On
One of the FCC's is specified as the master FCC	We select FCC B as master	FCC B controls Autopilot B. Autopilot B is following the glideslope
Each FCC continues to calculate thrust, pitch, and roll commands. The A/T adjusts the thrust levers with commands from the FCC	FCC calculates thrust commands. FCC B is going the A/T its commands	
Two independent radio altimeters provide radio altitude to the respective FCC's	FCC B has its own independent radio altimeter A faulty RA on the left has no impact, since the left RA goes to FCC A, not FCC B	The A/T gets its height information directly from the left radio altimeter, independent of which FCC has actually been selected master
FCC A controls autopilot A and FCC B controls autopilot B	We fly this approach on autopilot B, which gets input from FCC B	Flying the approach on FCC B does not protect you from a fault in the left RA

airplane. In fact, the left RA was still providing height data to the A/T, a fact that cannot be found in B737 pilot training or documentation.

Because of the tight vector onto final approach, the crew was completing the landing checklist during the 30 seconds when the airspeed decayed below the selected landing speed (as the A/Ts kept the thrust levers in the idle position). On an approach such as this one, the autopilot will keep the nose where it needs to be to stay on glideslope, not where it should go to recover lost speed (because, at this phase of flight, maintaining airspeed is the job of the A/T, after all).

The crew was surprised by the automation when it turned out that it had not been keeping the airspeed. When they did notice and tried to intervene, it was too late in this situation, for this crew, to recover.

The analysis of the accident (DSB, 2010):

- Shows the consequences of air traffic control (ATC) not previously announcing or coordinating a short and high turn-in for final approach, so that the flight crew has no chance to properly prepare;
- Exposes design shortcomings in the Boeing 737 New Generation systems that can lead to one part of the automation doing one thing (landing) based on corrupted input while the other is doing something else (flying);
- Highlights shortcomings in industry training standards for automated aircraft such as the Boeing 737 New Generation. Such training does not support crews in developing an appropriate mental model of how the automation actually functions and what effect subtle failures have.

Both the length and depth of type training, as well as procedural compliance at the airline matched industry standard. The Captain (also instructor), moreover, had close to 11,000 h on the Boeing 737. If such training, procedural standards, and line experience are not enough to insulate a flight crew from an automation surprise such as the one that happened on this flight, then few other airlines could feel safe that their training and procedures protect their flight crews from a similar system safety event. Postaccident manufacturer recommendations, however, in effect, tell flight crews to mistrust the machine, stare harder at it, and intervene earlier. They leave a single-failure pathway in place.

Instead, a system safety model that does accurate predictive work for an accident like this needs to take into account culture, collaborative cognition, complexity, and legacy systems, and see automation surprises as a breakdown of coordination and sensemaking. More context-specific details then start to make sense.

During the descent toward Amsterdam, the crew discussed and showed awareness of an anomaly of the (left) radio altimeter. The aircraft was kept at 2,000 feet while being vectored to capture the localizer at less than 5.5nm from the 18R runway threshold at AMS. This put it above the glideslope. The crew appeared aware of the tight vectoring given to them by ATC (the landing gear was already down and the flaps were at 15 even before localizer intercept). No prior warning or coordination from ATC occurred, which would be normal and desirable with such a tight vectoring for approach.

Upon the crew's selection of vertical speed mode to capture the glideslope from above (as a result of the tight vectoring received from air traffic control), the 737s A/T retarded to idle, consistent with crew expectations. The aircraft had to descend and simultaneously slow down to the next (flap 40) target speed. Upon selecting V/S (vertical speed) mode, the A/T window of the flight mode annunciator (FMA) on the primary flight display in the cockpit showed RETARD.

The B737 has two RETARD modes that combine A/T and autopilot functions: (1) *Retard flare* and (2) *Retard descent*. Retard descent commands the thrust levers to the aft stop to allow the autopilot to follow a planned descent path. Retard descent mode is normally followed by the Armed (ARM) mode, in which the A/T protects the flight envelope and maintains a selected speed.

ARM mode also allows crews to manually set the thrust levers forward again.

In contrast, the Retard flare mode is normally activated just prior to touchdown when an automatic landing is performed. The A/T does the retard part, the autopilot the flare part, so as to jointly make a smooth landing. In retard flare mode, the A/T no longer offers flight envelope protection, does not maintain any selected speed, and it will keep the thrust levers at the idle stop (or pull them back there if the crew would push them forward).

The A/T window on the FMA offers no way for a flight crew to distinguish one RETARD mode from the other. While the A/T window would normally have shown "MCP SPD" upon selecting V/S mode (knowledge that a PF has his 17th leg on a B737 is unlikely to have ready at hand), the RETARD mode made aircraft behavior insidiously consistent with crew expectations. They needed to go down and slow down so as to capture the glideslope (from above) and get the aircraft's speed down to the speed for the next flap setting. The A/T announced that it did what the crew commanded it to do: it retarded, and aircraft behavior matched crew expectations: the aircraft went down, slowed down, and then captured and started tracking the glideslope.

As it only showed RETARD (and not FLARE), the FMA annunciation gave the appearance as if the A/T went into RETARD descent mode. However, the A/T went automatically into the unexpected RETARD flare mode—not because the crew had selected V/S, but because a number of conditions had now been fulfilled, and the A/T was acting according to its own logic: the aircraft was going below 2,000 feet RA (in fact, it was at -7 feet RA, according to its only available (and corrupted) input to the A/T system), the wing flaps were more than 12.5 degrees out and the F/D mode was no longer in ALT HOLD.

While the A/T had, in effect, decided it was time to land, FCC B was still commanding the F/D and Autopilot B to stay on glideslope. One part of the automation was doing one thing (landing), while the other part was doing something else (flying). The part that was landing (the A/T) had control over the airspeed, the part that was flying (Autopilot B) did not; it only tracked the descent path on the glideslope.

Based on their training and documentation, the crew would have believed that they had protected their aircraft and its flight from any preexisting problems with the left RA. The right autopilot (known as Autopilot B or CMD B) had been selected on, and the right FCC B was giving it inputs.

Boeing pilot training materials and documentation do not reveal that the A/T always gets its height information from the left radio altimeter; that on pre-2005 737NG models, it does not cross-check its RA input data with other RA data; and that the right RA does not provide input to the A/T—even when FCC B has been selected as master and Autopilot B is flying.

The crew was completing their landing checklist during the 30 seconds when the airspeed decayed below the selected landing speed as a result of this automation mismatch. Interleaving task demands, speed tape design issues, and the erosion of cognitive work surrounding the calculation of final approach speeds in automated airliners, becoming visual with the runway, and landing checklist design could all have interacted with the crew's attention during the 30 seconds of speed decay below approach speed.

Believing, on the basis of their training, documentation, and experience, that they had insulated their cockpit setup from any problem with the left RA, the flight crew was surprised by the automation when it turned out that it had not been keeping the airspeed. When they did notice as a result of the stick shaker, and tried to intervene, it was too late in this situation, for this crew, to recover. As said, postaccident manufacturer recommendations that, in effect, telling flight crews to mistrust their machine and to stare harder at it, not only mismatches decades of Human Factors, system safety, and Cognitive Systems Engineering research, but also leave a single-failure pathway in place.

A broad issue in system safety and Cognitive Systems Engineering is at stake in a study like the one above. If we want to learn from practitioners' interactions with each other and technology, we need to study their practice—whether it goes well or not. Process tracing methods are a key approach developed in Cognitive Systems Engineering. They are part of a larger family of cognitive task analysis and cognitive task design (Flach, 2000; Hollnagel, 2003). They aim specifically to analyze how people's understanding evolved in parallel with the situation unfolding around them during a particular problem-solving episode (such as the automation surprises in the example above) and how failure modes multiply or get halted as a result. A scenario that leads to these develops both autonomously and as a result of practitioners' management of the situation, which gives it particular directions toward an outcome. Typically, process-tracing builds two parallel accounts of the problem-solving episode:

- one in the context-specific language of the domain;
- one in concept-dependent terms.

Recall from the previous chapter how it is in and through the latter that you can discover or recognize (even from other domains) regularities in the particular performance of practitioners during the episode (Hutchins, 1995; Woods, Dekker, Cook, Johannesen, & Sarter, 2010). What you then look at is not a meaningless mass of raw data, but a set of patterns of human performance. The context-specific or domain-dependent particulars make sense through the concepts applied to them. What looks like a flurry of changing display indications and confused questions about what the technology is doing, for example, can be made sense of by reading into them the conceptual regularities of automation surprise (Sarter, Woods, & Billings, 1997). At the same time, concepts can only be seen in the particulars of practitioners' performance; conceptual regularities begin to stand out across different instances of it. What happens if we take the highly context-dependent description of the accident above and try to turn it into a more conceptual cognitive engineering language so that we can draw broader lessons from it?

The aircraft in the example above had a flight crew of three—all of whom perished, in addition to six other occupants. There had been an initially inexplicable speed decay for more than a minute late on the approach, with the airplane stalling close to the ground, about a mile short of the runway. It plowed into a muddy field and broke apart. As it turned out, this airplane type is equipped with automation where one part is responsible for tracking the vertical and horizontal approach path to the runway (the autopilot) and the other part is responsible for maintaining (through engine A/T) the correct airspeed to do so. The latter part was getting readings from a faulty (but not flagged as "failed") RA on the captain's side that suggested the aircraft had already landed. The crew knew about the RA problem. The training package for this aircraft had implied that if the copilot's autopilot does the flying, it also uses the RA on that side to maintain correct thrust and thus airspeed. So, the crew had set up an approach in which the copilot's side was doing all the flying. Everything was good and manageable and they were cleared to land.

Yet the crash showed that it is, in fact, always the RA from the captain's side that supplies height information to the A/T system (the one that takes care of airspeed on an approach like this). This fact about the RA could be found in none of the manuals and none of the training for this aircraft. It probably resulted from legacy issues (this jet has, in one form or another, been flying since the 1960s, starting off with much fewer systems and much less automation and complexity). Let's limit ourselves to the pilots here (not the designers from way back). If no pilot could have known about this because it was never published or trained in any manual they ever saw, it is pretty safe to conclude that the pilots involved in this accident did not know it either. The global pilot community of this aircraft type could not have known it on the basis of what it was told and trained and given. 'What other booby traps does this airplane have in store for us?' was how the highly experienced and knowledgeable technical pilot of my own airline at the time put it.

Because of its intermittent RA fault (essentially believing the jet was already on the ground) that it was telling the engines to go to idle. This while the crew believed the A/T was getting faithful information from the copilot's RA, as that is what they thought they had set up. The copilot's autopilot, in other words, was diligently flying the approach track to the runway (or trying to keep the nose on it at least) while the A/T system had already decided it was time to land. As a result, speed bled away. The aircraft had been kept high by air traffic control and had had to go down and slow down at the same time—this is very difficult with newer jets, as they have highly efficient wings. Going down and slowing down at least requires the engines to go to idle, and remain there for a while. In other words, the action of the A/T system (while actually the result of a misinformed RA) was entirely consistent with crew intentions and expectations. It was only during the last seconds that the crew would have had to notice that the engines remained idle, rather than slowly bringing some life back so as to help them stay on speed. In other words, the crew would have had to suddenly notice the absence of change. That is not what human perceptual systems have evolved to be good at.

There is no reason to conclude that the approach was 'rushed.' The crew anticipated the late glideslope capture by lowering the gear and selecting flaps 15 even before capturing the localizer, and the only items to be completed after glideslope capture were final flap setting and the landing checklist. Landing clearance had already been obtained. The flight accurately reflects human factors research on plan continuation (Orasanu & Martin, 1998). Decisions to go around or intervene in a student pilot's actions involve the assessment and reassessment of the unfolding situation for its continued do-ability. The dynamic emergence of cues about the do-ability of the approach, suggested to the crew that continuing was possible and not problematic.

A breakdown in crew coordination cannot be substantiated either. There is little to no evidence in the primary data source (the CVR) for overlapping talk, for second-pair part silences, or for other-initiated repairs—three aspects of conversational interaction that have recently been implicated in CRM (crew resource management) breakdowns. The Captain was well liked, and a popular instructor.

The accident fits the substantial research base on automation surprises. For 70 seconds, automation and aircraft behavior were consistent with crew expectations (the A/T had insidiously reverted to an unexpected mode that seemed—and was annunciated—as if it followed crew instructions). After that period, the really difficult task for the crew was to discover that the automation was actually not (or no longer) following their instructions. This was discoverable not through a change in aircraft behavior (as it usually is during automation surprises), but through a lack of change of aircraft behavior (and a lack of mode change). The aircraft did not stop slowing down, and the automation did not change mode. The crew would have had to discover one or two nonevents, in other words. Discovering nonevents is very difficult, particularly when initial system behavior is consistent with expectations, when design does not show the behavior but only the status of the system, and when there is no basis in a crew's mental model to expect the nonevent(s) (De Keyser & Woods, 1990; Sarter & Woods, 1995).

6.4   SYSTEM SAFETY AS THE ABSENCE OF NEGATIVE EVENTS?

Ultimately, and you will see this in the final parts of the book, the question is how long it makes sense to define safety as freedom from unacceptable or intolerable risk (Hollnagel, 2006). Safety has long had us focus on prevention of unwanted events and protection against unwanted outcomes. There is a lot of mileage to be had from that approach. But recall the example above. The complexity of an accident like that, with an often modified and improved aircraft type that have been flying for many decades, suggests that there are limits on our ability to map out all failure pathways. In a safe, technologically advanced and partially or largely automated system, hazards never disappear entirely. Due to complexity, however, both hazard likelihood and hazard risk can get buried in the increasing opacity of the system—and hidden from view by many hours or years of problem-free performance.

So continuing to measure safety as the absence of negative events has a 'best before' date. It is literally 'best before' the risk of something negative has become too small as to become statistically meaningless, or noise. The less there is to measure (particularly once you arrive at the mythical 'zero harm'), the less there is to control. Indeed, once a very low level of risk has been attained, the menu of eliminating hazards, preventing initiating events, and protecting against consequences has little to offer anymore. Of course, as said before, more rules and protections can always be created. But they may serve to make things both more expensive and ultimately less safe.

With risk assessments as well as with safety cases, then, the key issue with complex systems may not be the freedom from unwanted events or the protection against unwanted outcomes. Safety cases are naturally built around a lot of arguments and data about how negative events are precluded, prevented, or mitigated. But given the complexity of some of the systems that they get applied to, and the fundamental unpredictability of emergent events, there is a lot of merit in explicitly building a safety case around the presence of positive capacities that make things go well (most of the time). Perhaps that would not be a 'safety case' in the narrow sense, but rather a 'resilience case.' And it wouldn't be a 'risk assessment,' but rather a 'resilience assessment' (see Chapter 11).

The other thing the logic and language of system safety does is divide the operation of systems into normal and abnormal (Hollnagel, 2009):

- *normal operations* ensure that the system works as it should and produces the intended outcomes;
- *abnormal operations* disrupt or disturb normal operations or otherwise render them ineffective.

The purpose of system safety is to maintain or, as far as possible (or practicable) assure normal operations by preventing disruptions or disturbances. Abnormal operations are putatively prevented by barriers, regulations, procedures, standardization, elimination, and other controls. But as we have already seen in the section on reliability and safety, the normal operation of some parts can actually contribute to the abnormal operation of the system. And not only that, but the distinction between 'normal' and 'abnormal' turns out not to be so clear at all. The next chapter will introduce you to the work of Barry Turner, who was the first to describe the banal and unexciting life of an organization that was about to have a spectacular accident: while unwittingly on the way to something very abnormal, organizations are pretty much tied up in their usual, utterly normal, everyday preoccupations. As Vaughan (1996) later described with respect to NASA in the wake of the 1986 Challenger Launch decision, the 'messy interior' of an organization is pretty much the same— whether it has had an accident or not (yet). Abnormal operations *are* normal operations. And normal operations always are somewhat abnormal.

This even goes for the engineering state of an operational system. Take the normal operation of a jetliner, for instance. There are always some things broken onboard. A radio might not work, or a coffee maker, or a particular part of the collision avoidance system. There is a process around these things, for sure (organized around a 'minimum equipment list' that allows pilots to fly the aircraft with certain things not installed or broken, if for limited periods). But it does show that there's always something 'abnormal,' to the point where the very word becomes problematic or meaningless.

This is why Hollnagel has preferred calling it 'varying conditions' (Hollnagel, 2014). The challenge for a system—to be resilient—is to be able to function under varying conditions. Things go right because people learn to overcome design flaws and shortcomings, or to work around broken or unfinished equipment (as the Minuteman teams did). People adjust their performance, and that of their system, to meet demands. They do active work to interpret and apply procedures and match them to conditions (Dekker, 2003); they manage the complexity and emergent events that come from underspecified and difficult-to-oversee systems and combinations of systems. Accidents, then, come in part from unexpected combinations of events and actions. Risks, in turn, can be represented by dynamic combinations of performance variability. Hollnagel's Functional Resonance Analysis Method (or FRAM) is a unique, if labor intensive, technique developed to map the possibility of such functional resonance in a complex system (Hollnagel, 2012a). You will learn more about related resilience concepts in Chapter 11.

STUDY QUESTIONS

1. Explain the fly-fix-fly approach to developing safe systems. Can you argue that the human factors pioneers (see the previous chapter) adopted such an approach?
2. What made the fly-fix-fly approach unacceptable?
3. Why are human, organizational, and cultural contributions to safety so difficult to formalize compared to those of an engineered system?

4. How has the growing integration of software in safety-critical systems of all kind created new challenges for system safety?
5. Complexity science offers its own new ways of modeling (e.g., agent-based systems), which have had some success outside of safety, and have found some applications in safety, dealing with systems such as power networks, autonomous vehicles, etc. Do these approaches offer a solution to complexity, or a desperate attempt to preserve the relevance of a reductionist approach as complexity becomes undeniable?
6. What are the links between System Safety, Human Factors, and Cognitive Engineering? Name at least three.
7. System safety, through its formal language and techniques, defines safety as freedom from unwanted events, and protection against unwanted outcomes. What are the limits of seeing safety that way (and what are the reasons for those limits)? And what is the alternative?

REFERENCES AND FURTHER READING

De Keyser, V., & Woods, D. D. (1990). Fixation errors: Failures to revise situation assessment in dynamic and risky systems. In A. G. Colombo & A. Saiz de Bustamante (Eds.), *System reliability assessment* (pp. 231–251). Dordrecht, The Netherlands: Kluwer Academic.

Dekker, S. W. A. (2003). Failure to adapt or adaptations that fail: Contrasting models on procedures and safety. *Applied Ergonomics, 34*(3), 233–238.

Deming, W. E. (1982). *Out of the crisis.* Cambridge, MA: MIT Press.

DSB. (2010). Turkish airlines, crashed during approach, Boeing 737–800, TC-JGE Amsterdam Schiphol airport. In D. S. Board (Ed.), *Aircraft Accident Report.* The Hague, The Netherlands: Dutch Safety Board.

Fischhoff, B. (1975). Hindsight ≠ foresight: The effect of outcome knowledge on judgment under uncertainty. *Journal of Experimental Psychology: Human Perception and Performance, 1*(3), 288–299.

Fischhoff, B. (2007). An early history of hindsight research. *Social Cognition, 25*(1), 10–13.

Flach, J. M. (2000). Discovering situated meaning: An ecological approach to task analysis. In J. M. Schraagen, S. F. Chipman, & V. L. Shalin (Eds.), *Cognitive task analysis* (pp. 87–100). Mahwah, NJ: Lawrence Erlbaum Associates.

Hidden, A. (1989). *Clapham junction accident investigation report.* London, UK: Her Majesty's Stationery Office.

Hollnagel, E. (2003). *Handbook of cognitive task design.* Mahwah, NJ: Lawrence Erlbaum Associates.

Hollnagel, E. (2006). Resilience: The challenge of the unstable. In E. Hollnagel, D. D. Woods, & N. G. Leveson (Eds.), *Resilience engineering: Concepts and recepts* (pp. 9–17). Aldershot, UK: Ashgate Publishing Co.

Hollnagel, E. (2009). The four cornerstones of resilience engineering. In C. P. Nemeth, E. Hollnagel, & S. W. A. Dekker (Eds.), *Resilience engineering perspectives, Volume 2: Preparation and restoration* (pp. 117–134). Aldershot, UK: Ashgate Publishing Co.

Hollnagel, E. (2012a). *FRAM the functional resonance analysis method: Modelling complex socio-technical systems.* Boca Raton, FL: CRC Press.

Hollnagel, E. (2012b, February, 22–24). *Resilience engineering and the systemic view of safety at work: Why work-as-done is not the same as work-as-imagined.* Paper presented at the Gestaltung nachhaltiger Arbeitssysteme, 58. Kongress der Gesellschaft für Arbeitswissenschaft, Universität Kassel, Fachbereich Maschinenbau, pp. 19–24.

Hollnagel, E. (2014). *Safety I and safety II: The past and future of safety management.* Farnham, UK: Ashgate Publishing Co.

Hutchins, E. L. (1995). *Cognition in the wild.* Cambridge, MA: MIT Press.

Keller, W., & Modarres, M. (2005). A historical overview of probabilistic risk assessment development and its use in the nuclear power industry: A tribute to the late Professor Norman Carl Rasmussen. *Reliability Engineering & System Safety, 89*(3), 271–285. doi:10.1016/j.ress.2004.08.022

Lee, L. (1992). *The day the phones stopped: How people get hurt when computers go wrong.* New York, NY: Donald I. Fine.

Leveson, N. G. (2012). *Engineering a safer world: Systems thinking applied to safety.* Cambridge, MA: MIT Press.

Leveson, N. G., Daouk, M., Dulac, N., & Marais, K. (2003). *Applying STAMP in accident analysis.* Cambridge, MA: Massachusetts Institute of Technology.

Leveson, N. G., & Turner, C. S. (1992). *An investigation of Therac-25 accidents* (UCI Technical Report No. 92–108). Retrieved from University of California at Irvine.

Nemeth, C. P., Nunnally, M., O'Connor, M., Klock, P. A., & Cook, R. I. (2005). Getting to the point: Developing IT for the sharp end of healthcare. *Journal of Biomedical Informatics, 38*(1), 18–25. doi:S1532-0464(04)00155-8 [pii] 10.1016/j.jbi.2004.11.002

Norman, D. A. (1990). Human error and the design of computer systems. *Communications of the ACM, 33*(1), 4–7.

Orasanu, J. M., & Martin, L. (1998). Errors in aviation decision making: A factor in accidents and incidents. Human Error, Safety and Systems Development Workshop (HESSD) 1998. Retrieved from http://www.dcs.gla.ac.uk/~johnson/papers/seattle_hessd/judithlynnep

Perry, S. J., Wears, R. L., & Cook, R. I. (2005). The role of automation in complex system failures. *Journal of Patient Safety, 1*(1), 56–61.

Sagan, S. D. (1993). *The limits of safety: Organizations, accidents, and nuclear weapons.* Princeton, NJ: Princeton University Press.

Sarter, N. B., & Woods, D. D. (1995). "How in the world did we get into that mode?" Mode error and awareness in supervisory control. *Human Factors, 37*(1), 5–19.

Sarter, N. B., Woods, D. D., & Billings, C. E. (1997). Automation surprises. In G. Salvendy (Ed.), *Handbook of human factors and ergonomics* (2nd ed., pp. 1926–1943). New York, NY: Wiley.

Stephans, R. A. (2004). *System safety for the 21st century: The updated and revised edition of system safety 2000.* London, UK: John Wiley & Sons.

U.S. Nuclear Regulatory Commission. (1975). *Reactor safety study. WASH-1400.* Washington, DC: Nuclear Regulatory Commission.

Vaughan, D. (1996). *The challenger launch decision: Risky technology, culture, and deviance at NASA.* Chicago, IL: University of Chicago Press.

Weick, K. E. (1995). *Sensemaking in organizations.* Thousand Oaks, CA: Sage Publications.

Woods, D. D., Dekker, S. W. A., Cook, R. I., Johannesen, L. J., & Sarter, N. B. (2010). *Behind human error.* Aldershot, UK: Ashgate Publishing Co.

7 The 1970s and Onward
Man-Made Disasters

KEY POINTS

- In the 1970s, the size and complexity of many of society's safety-critical systems were becoming apparent to many—and in certain cases alarmingly so. Large disasters with socio-technical systems, and many near-disasters, brought safety and accidents back to center stage.
- The Tenerife collision between two Boeing 747 airliners in 1977, which killed 583 people, and the 1978 Three Mile Island nuclear plant accident in the United States, brought the risk of system accidents out of the shadows of system safety calculations, and into the societal and political arena.
- This greater visibility helped give rise to two decades of productive scholarship, and set the stage for a lot of the conversation about safety, accidents, and disasters we are having to this day.
- Accidents were increasingly understood as social and organizational phenomena, rather than just as engineering problems. Man-made disaster theory was the first to theorize this, closely followed by high reliability theory and normal accident theory.
- Disasters and accidents are preceded by sometimes lengthy periods of gradually increasing risk, according to man-made disasters theory. This buildup of risk goes unnoticed or unrecognized. Turner referred to this as the incubation period. During this period, he suggested, latent problems and events accumulate that are culturally taken for granted or go unnoticed because of a collective failure of organizational intelligence.
- The accumulation of these events can produce a gradual drift toward failure, the gradual erosion of safety margins under the pressure of resource limitations, competition, and goal conflicts. Understanding drift, and finding ways to intervene and prevent it, has become a practical and theoretical preoccupation.

7.1 MAN-MADE DISASTER THEORY

The book *Man-Made Disasters* first appeared in 1978. It was the transformation of a doctoral dissertation written at the University of Exeter by Barry Turner. Why was it so prescient, and so foundational? Turner, in his own words, had realized "early in the 1970s that there was a potential store of information about administrative failures and shortcomings in reports of public inquiries into large-scale accidents" (Turner & Pidgeon, 1997, p. xii). Strangely enough, as far as the published literature is concerned, he was, and remained, in sparse company at the time. Understanding the risk of an accident or disaster, as you might have seen from the previous two chapters, was not something that social science busied itself with. It was a problem—if at all—that was reserved for hard scientists and engineers. And their concerns were expressed in numbers, not in reflections on the social order of an organization or industry.

7.1.1 SAFETY AND SOCIAL SCIENCE

Writing in the 1970s, Turner observed how there was not much coherent research on disasters at all:

> ...a large body of information about the preconditions of disasters has not been built up, partly because of over-specialization amongst the various disciplines studying different features of disasters, but also because of the prevalence of a belief that all disasters are different. If those trying to examine and understand disasters are really trying to study a group of phenomena which have no common characteristics, at least in their early phases, they are clearly in a situation where science cannot operate, for no generalizations are possible, and they are dealing with a unique and unprecedented class of phenomena whose properties can never be predicted from past experience. However, it seems highly doubtful that this is, in fact, the case.
>
> **Turner (1978, p. 38)**

Turner did identify shared properties of disasters. His original data came from mining, railways, medicine manufacturing, an oil storage explosion, shipping accidents, a smallpox outbreak, and a building fire. The patterns and commonalities he found across disasters in these different worlds were *sociological*, not necessarily technical. Turner broke new ground in safety science because of:

- Identifying and describing the commonalities and patterns observable across different technical worlds in which disasters happened;
- Studying and explaining disasters as sociological phenomena.

Given the major involvement of organizational life in the incubation of disaster, Turner said, "anyone concerned to understand more about the origins of disaster will need to study the forces which affect the manner in which organizations handle communications about [risk] of all kinds—whether these communications are within the organizations concerned, between organizations, or between an organization and the [environment]" (1978, p. 199). Safety science has been building on that new ground to this day. You will recognize in Turner's man-made disaster theory the basis for many ideas that are now in use in:

- Defenses-in-depth thinking (e.g., latent errors or resident pathogens that are already present and help incubate disaster (Reason, 1990));
- High reliability theory (e.g., weak signals that do not get communicated or picked up (Weick & Sutcliffe, 2001));

- Safety culture research (e.g., organizational-cultural preconditions for disaster);
- Concepts such as the normalization of deviance (Vaughan, 1996), procedural drift (Snook, 2000), and drift into failure (Dekker, 2011), which all refer to disaster incubation in one way or another. Turner is implicated in all research that focuses on social shifts in what is considered (or even noticed as) unsafe in a group or organization;
- Control-theoretic notions about erosion and loss of control (Leveson, 2012): the kind that Turner talked about in sociological, managerial, and administrative terms.

Failures of foresight was the preliminary title of *Man-Made Disasters* before it was published. One way or another, this central notion has helped inspire the ideas above. Others, like Janis' 1972 groupthink—the psychological drive for social consensus in the face of risky decisions—coincided with the appearance of Turner's work.

7.1.2 Disasters Do not Come Out of the Blue

Disasters do not come out of the blue, says man-made disaster theory. They are preceded by sometimes lengthy periods of gradually increasing risk. Yet this risk, or the buildup of risk, goes unnoticed or unrecognized. Turner referred to this as the incubation period. During this period, he suggested, "latent errors and events accumulate which are culturally taken for granted or go unnoticed because of a collective failure of organizational intelligence" (Pidgeon & O'Leary, 2000, p. 15). The accumulation of these events can produce a gradual drift toward failure. Intervening in, and preventing incubation has become a practical and theoretical preoccupation. This might seem to make such sense today, and such a large slice of what safety science tries to do. Why was Turner's work largely left unnoticed and forgotten when it first came out? Today there is a critical mass of scholars and a huge and still growing body of literature, as Diane Vaughan writes in the foreword to the second edition, "in a world only too aware of the social harm that results from organizational-technical system failures" (p. xiii).

But not back in 1978. Writing in retrospect, Turner himself admitted having the sense that the original book had disappeared without a trace. The world in which he wrote it was concerned with disasters, for sure. But it was concerned with their aftermath, not their incubation. The major risk in the minds of sociologists, political scientists, and many others was that of a nuclear holocaust: a potential fight to the death between the West and the rest. In North America particularly, the study of disasters was inextricably related to the nuclear threat, always reverting to civil protection as both solution and policy imperative. What could be done in the wake of a nuclear disaster and how should societies prepare for that? What were the chances of survival, the responsibilities and the possibilities? Disaster research was postapocalyptic: drawn to what was dire, desperate, severely abnormal, and extreme. Its focus was on the onset and aftermath of catastrophe, not on the preconditions. Indeed, as Turner (1978, p. 33) put it, "almost all studies take the onset of disaster as the starting point for their research, treating prior events in a cursory manner, if at all." What, then, "are the conditions which tend to be associated with the pre-disaster situation"?

The pre-disaster situation, Turner would go on to discover, is actually not spectacular at all. It isn't even really interesting for any deviant or strange human or social or organizational phenomena. If the aftermath of catastrophe was about the severely abnormal, Turner's incubation of catastrophe was about the normal, the banal even. Turner found that disasters brew among precisely the sorts of typical interhuman processes that play out in any organizational setting. These are the same processes that are largely responsible for mostly letting things tick along just fine. Turner noted that people were prone to discount, neglect, or not take into account relevant information. He identified usual organizational phenomena such as information not being fully appreciated, not correctly assembled, or the discounting of information that conflicted with prior understandings of risk. As later authors summed it up: "despite the best intentions of all involved, the objective of safely operating technological systems could be subverted by some very familiar and 'normal' processes of organizational life" (Pidgeon & O'Leary, 2000, p. 16).

Turner saw the making of a disaster in such very human and very normal organizational terms. He was the first to pay serious, scholarly attention to the combination of normal, everyday technical, and social processes, and to the institutional and administrative arrangements that, in combination, could systematically produce man-made disasters.

7.2   THE INCUBATION PERIOD

In 1966, a portion of a coal mine tip (unusable material) near Aberfan, South Wales, slid down into the village and engulfed its school. It killed 144 people, including 116 children (Turner, 1978). Investigating the period leading up to the disaster, Barry Turner found that it was characterized by events that went unnoticed or were disregarded. Why was this? These events, small as they might have seemed at the time (high reliability theory would later call these "weak signals," see the next chapter), were at odds with taken-for-granted beliefs about hazards, and with norms for their control and avoidance. Risk calculations or quantifications about tip stability were not available or relevant to the organization, so that is not where Turner went looking for answers. Instead, he zeroed in on the organization's managerial and administrative processes. In safety science, it was a first. From here on, an organization's or even industry's management and administration was going to be a promising target. How was it possible that they believed one thing about risk, while something else altogether was going on in the world around them? Barry Turner became preoccupied by this discrepancy between a buildup of risk on the one hand, and the continued belief that it was under control on the other. For Turner, the space that gradually opened up between beliefs and growing risk was filled with human action or inaction. It was about perceptions, assessments, decisions, and actions. It was about what people did or did not do, about what they said or shared. A disaster, then

> ... rarely develops instantaneously. Instead, there is an accumulation over a period of time of a number of events which are at odds with the picture of the world and its hazards represented by existing norms and beliefs. Within this 'incubation period' a chain

of discrepant events, or several chains of discrepant events, develop and accumulate unnoticed.

(p. 72)

Thus, for Turner, accidents and disasters were primarily sociological phenomena. He shifted the focus from engineering and structures to social processes. He also suggested that failure does not just happen when an engineering calculation has been exceeded. It needs to be understood over time, and in needs to be understood in terms of people and groups and organizations—many of which would paradoxically have been tasked with preventing or containing the disaster in one way or another. In the wake of the Fukushima nuclear disaster in Japan, Downer writes about the lack of readiness for it:

> This blindness was evinced by the degree to which Japan seemed underprepared for the disaster. At the plant itself, for instance, procedures and guidelines were woefully insufficient for a meltdown, forcing operators to respond on an almost entirely ad hoc basis. When the lights went out they had to borrow flashlights from nearby homes to study the plant's gauges. On-site dosimeters maxed out at levels that were far below those that could be expected from a catastrophe, unable to display readings any higher. And, as the plant's former safety manager would later testify, emergency plans' ... had no mention of using sea-water to cool the core, an oversight that caused unnecessary, and perhaps critical, delays. The bureaucratic infrastructure beyond the plant evinced similar shortcomings. Official announcements were often ill-considered and characterized by denial, secrecy, and refusal to accept outside help. Important medications were scarce, with byzantine rules hindering their effective distribution). Evacuations were poorly managed, with vital information being withheld in ways that exposed many families to unnecessary danger. We should be wary, moreover, of seeing such failures of foresight as a specifically Japanese problem. Very few observers suggest that other nations are substantially better prepared for such a contingency. ... All societies routinely make choices about nuclear power that seem myopic when considered in relation to a potential disaster.
>
> **Downer (2013, pp. 2–3)**

7.2.1 STAGES OF INCUBATION

So what exactly does the incubation period look like? And how and why is it possible for a disaster to brew unnoticed or unchecked during such a period? Remember that man-made disaster theory saw disasters as fundamentally sociological. This enabled the theory to trace existing cultural assumptions and how they fit or diverge from reality—and build a stage model for how this divergence happened. What mattered to Turner was that in the majority of the accidents and disasters he had studied, basically no one saw it coming. He made an effort to point out that it was not as if the disaster was 'out there,' waiting to happen and visible if only people had tried a little harder to look for it. Given how people and their responsibilities were configured in institutional roles and organizational relationships, it would have been sheer impossible for anyone to see the disaster coming. His stages reflected the patterns that he saw in all the disasters. As you will see, the incubation of the disaster was technically only one of the stages, rather than consisting of multiple stages itself (Table 7.1).

TABLE 7.1

Six Stages Associated with the Incubation of Disaster According to Man-Made Disaster Theory (see Turner & Pidgeon, 1997, p. 72)

Stage I	Notionally normal starting points	a. Initial culturally accepted beliefs about the world and its hazards; b. Associated precautionary norms set out in laws, codes of practice, mores, and folkways.
Stage II	Incubation period	The accumulation of an unnoticed set of events which are at odds with the accepted beliefs about hazards and the norms for their avoidance.
Stage III	Precipitating event	Brings attention to itself and transforms general perceptions of Stage II.
Stage IV	Onset	The immediate consequences of the collapse of cultural precautions become apparent.
Stage V	Rescue and salvage—first-stage adjustment	The immediate post-collapse situation is recognized in ad hoc adjustments which permit the work of rescue and salvage to be started.
Stage VI	Full cultural readjustment	An inquiry or assessment is carried out and beliefs and precautionary norms are adjusted to fit the newly gained understanding of the world.

In Stage I(a), the set of culturally held beliefs about the world and its hazards are sufficiently accurate to enable individuals and groups to survive successfully. What people do and believe about risk is noted down in codes of practice, represented by mores and enacted in folkways (Stage I(b)). This does not mean that accidents do not happen. But they do not rock people's world; they do not change beliefs or assumptions. Indeed, unfortunate events during this stage might typically be attributed to a violation of the current codes of practice, mores, or standard ways of controlling risk. They do not represent a fundamental surprise that proves people's beliefs to be wrong. They would not trigger any deeper reflection about the nature and control of risk.

NOAH WEBSTER AND A FUNDAMENTAL SURPRISE

Woods, Dekker, Cook, Johannesen, and Sarter (2010, pp. 230–232) relate a story about the famous bibliographer Noah Webster (known for his dictionaries) that illustrates the difference between fundamental and local surprise quite nicely.

One day, Webster arrived home unexpectedly to find his wife in the arms of his servant. 'You surprised me,' said his wife. 'And you have astonished me,' responded Webster. Webster's choice of words captured the difference that Turner wants us to understand: A local surprise, as he suggests, can be explained away on the basis of a violation of current practices or rules, but it leaves beliefs and assumptions intact. A fundamental surprise, in contrast, necessitates the revision of assumptions and beliefs about how the world works. A fundamental surprise is an astonishment. Webster's belief about his relationship with his wife was suddenly proven false.

Mrs. Webster who, although surprised by the incident, still could maintain her image of herself, her environment, her husband, and the relationship between them.

This distinction between surprise and astonishment also shows up in one's ability to define in advance the issues for which one must be alert. Surprises relate to a specific event, but they can be accommodated in current models of how the world works. In contrast, Mr. Webster's shocking incident revealed something much larger, something fundamental that showed his model of the world to be wrong.

Another distinction, relevant to Turner's ideas, concerns the value of information. Mrs. Webster lacked only one piece of information which, had she had it in advance, would have allowed preventing her surprise: the information that her husband would return early that day. No revision of beliefs would have been necessary to get or recognize such information.

No single piece of information could have prevented Mr. Webster's astonishment. As Turner suggests as well, the critical incident is preceded by precursors from which an outside observer might have deduced the state of the couple's relations. But if you are invested in that relationship, then you may be (again, as Turner proposes) be very reluctant to imagine the worst—which is precisely why you may not be prepared for the fundamental surprise.

Stage II is characterized by a steadily growing discrepancy between what is happening in the world on the one hand and how it is thought to operate on the other. This is the actual incubation period. Based on his retrospective analyses, Turner attributed the growing gap between world and belief to a range of problems in human communication and interpretation, such as:

- erroneous assumptions and misunderstandings;
- rigidities of human belief and perception;
- disregard of complaints or warning signals from outsiders;
- a reluctance to imagine worst outcomes;
- decoy phenomena that suck up organizational resources but that do not represent the real growing threat out there.

Something else that Turner found in his analysis was warnings of impending disaster that *were* recognized and accurately assessed, but that failed to trigger an appropriate or adequate response. There are many institutional reasons for this. One could be the perceived credibility of the source (e.g., a whistleblower), who identified the

warning event. Another could be the lack of resources or political will to actually do
something about the warning; another, leaving the industry with nothing but 'fantasy
documents' that do not really help prevent or manage a disaster (Clarke & Perrow,
1996). For example, disaster planning in the nuclear power industry:

> ...is routinely insincere and insufficient. Clarke and Perrow (1996), for instance,
> describe some of the evacuation planning undertaken for Shoreham Nuclear Power
> Station on Long Island, but argue that the assumptions of these plans were so unrealistic
> that they are more properly understood not as earnest contingency preparation, but as
> elaborate public performances—bureaucratic *Kabuki*. When organizations do attempt
> to think more earnestly about nuclear disaster, meanwhile, the plans they produce
> routinely lack any institutional authority. The US Nuclear Regulatory Commission
> (NRC), for instance, developed 'Severe Accident Management Guidelines' (SAMGs)
> for directing reactor operations in the event of 'unanticipated accident sequences'
> (i.e. events like Fukushima), but training in these guidelines was voluntary. At the time
> of the crisis the NRC did not require that operators demonstrate any knowledge of the
> SAMGs or their application, and a recent audit of US plants has found the guidelines
> to have been largely neglected.
>
> **Downer (2013, p. 3)**

The incubation period starts when the first discrepant event happens but goes unnoticed or is misinterpreted. The period ends with the fundamental surprise. This is
an event that:

- Is evident and undeniable: it cannot go unnoticed;
- Unveils the latent structure of the prior events during the incubation period.
  It is the event that 'connects the dots.' People will now look differently upon
  the occurrences during the incubation period and wonder how they missed
  what it pointed to;
- Shows people's prior beliefs and assumptions about the world to be wrong.

In Turner's model, this constitutes Stage III, the stage that triggers what he called
'cultural collapse.' Stage IV never got much of Turner's interest, because it was the
stage that many of his contemporaries were worried about: the immediate aftermath of a catastrophe and how people and organizations respond to it. Stage V was
more interesting to Turner again, as that is where interesting sociology takes place.
What do people and organizations do to either salvage or overhaul their definitions
of reality? In many of the cases he studied, Turner found that participants could have
significant difficulty during these stages, for psychological, political, or emotional
reasons. In fact, some might be tempted to once again convert their shock and fundamental surprise into a merely local surprise, so as to rescue their previously held
assumptions about the world. An accident that reveals fundamental vulnerabilities
in a particular system design, for example, may get attributed to a procedural violation or human error. Such an attribution requires neither an expensive redesign,
nor a revision of the assumption that the system was basically safe. Stage VI offers
yet another opportunity for the fundamental surprise to be treated as such, said

Turner. During this stage a more thorough and leisurely investigation can take place. Existing ideas about hazards and the means of controlling them must be reviewed and revised. Investigations should show not only the true nature of the events during the incubation period, but lay out what adjustments need to be made to make sure 'it never happens again.'

Though Turner was in sparse company, some other researchers also noted the relevance of social and psychological processes ahead of large-scale disasters. For example, Stech applied the man-made disasters idea to the failure of Israeli intelligence to foresee the Yom Kippur attack by Syria and Egypt on October 6, 1973, even though all necessary data was available across the intelligence apparatus (Stech, 1979). The situation faced by intelligence organizations, or course, is fertile ground for disaster incubation in Turner's sense:

• Complexity, dynamics, ambiguity, and noise;
• Interorganizational competition and institutional barriers;
• An incoherent environment in which adversaries may deliberately create false traces to wrong-foot the other.

In such a setting, people need to accurately estimate the intention of others. And then they need to match it with one's own capacity to meet the threat emanating from it. It is a setting ripe for all kinds of sociological and psychological processes that would make achieving these aims difficult. Erroneous assumptions (both about adversaries and one's own capacities) and the minimization of emergent danger are the characteristics of man-made disaster theory that Stech found especially relevant to military intelligence failure. Interestingly, Stech then turned man-made disaster concepts on their head to explain military intelligence successes, including the British defeat of German U-boat operations in the North Atlantic in the World War II. Responsible for such success, he explained, was that people entertained multiple hypotheses and that they employed evidence to falsify those hypotheses and prove assumptions wrong— rather than relying on it to confirm an interpretation or a course of action that had already been chosen.

7.2.2 Failures of Foresight

How do people inside of organizations miss the incubation of disasters, even if it is part of their jobs to foresee them? From his analysis of past accidents, Turner concluded that the incubation of disaster is possible because a chain of discrepant events, or several chains of discrepant events, develop and accumulate unnoticed or are misunderstood. This happens for one of two reasons:

> For discrepant events to build up this way, it is clear that they must all fall into one of two categories: either the events are not known to anyone, or they are known but not fully understood by all concerned, so that their full implications are not understood in the way that they will be after the disaster.

Turner and Pidgeon (1997, p. 72)

He offered the following reasons for events to go either unnoticed altogether, or get misinterpreted (p. 73):

- Events go unnoticed or are misunderstood because erroneous assumptions are made and maintained by the people responsible for controlling risk. These erroneous assumptions may have arisen because of rigidities of belief and perception, particularly because of institutional inflexibility. They may also arise because of decoy problems that draw attention away from the discrepant events that people should be noticing instead. Lastly, erroneous assumptions can persist when nonexperts from outside the organization are dismissed as uninformed, irrelevant, or even alarmist.
- Discrepant events can also go unnoticed or misunderstood because of problems in the handling of information in complex situations. Excessive information can mean that crucial messages get lost in a mass of noise or that those handling those messages were busy with other matters.
- Danger can get belittled because of the human reluctance to fear the worst, or the inability to imagine worst outcomes. This means that warnings of approaching danger can pass unnoticed. As Perrow would say a few years later: "warning of an incomprehensible and unimaginable event cannot be seen, because it cannot be believed" (1984, p. 23).
- Violations of rules and regulations can become normal in those situations where the formal guidance is no longer fully up to date, or not seen as relevant or applicable to how work is done. This, too, can hide warnings of impending disaster.
- 'Strangers on sites' was yet another problem identified by Turner. In his study of disasters that involved the public (e.g., a building fire), he discovered that rules and instructions did not typically extend to those 'strangers on sites'—a wide and amorphous group—and could easily contribute to the toll of a disaster. Disasters can easily get worse when this vulnerability is not noticed and dealt with in planning beforehand.

A decade before Turner, an important insight came from Roberta Wohlstetter's 1962 *Pearl Harbor: Warning and Decision*. Pearl Harbor had been a surprise, and there was the belief that it could have been avoided if all information had been available at the time. Wohlstetter showed that that was only true in part, or in a more complex way. "All" information about the impending 1941 Japanese attack, she showed, *was* available. There was sufficient data in the U.S. intelligence apparatus to both foresee and perhaps forestall the attack. The failure was not insufficient information, but misunderstanding that information. Or, more to the point: it was about not putting the data together; not connecting the dots. Intelligence-gathering mechanisms, Lanir concluded, unavoidably accumulate not only useful information, or signals, but also noise, or irrelevant information, which is, of course, really hard to figure out in advance. But it gets even more difficult if information is distributed in ways that make it harder to see the pattern of signals in the noise. "Broadly speaking," said Lanir (1986, p. 10), "intelligence's record shows that it gets details right, but understands big issues poorly" (Lanir, 1986). Turner reflected on that as follows:

There is an additional aspect of this concern with communication in the pre-disaster period which was thrown up by some of the cases examined, and which has not so far been discussed. The analysis ... makes the rather over-simple assumption that information is either available or not available, that knowledge of immediate hazards, or factors leading to them, is either present or absent. This of course is not true, for it is often the case that the information which could prevent a disaster is available to someone, but that that individual may not realize the significance of the knowledge that he has, or he may not be able to pass this information on in time, either because he does not know precisely where it is needed, or because the constraints of habit, lack of authority or lack of resources stop him from passing on his knowledge.

(1978, pp. 105–106)

The limits of organizational intelligence about impending disaster, according to this view, consists of obstacles to transfer, communication, and transmission, but also lies in the inability to put the big picture together. Turner urged those studying the origins of disaster to not just pay attention to the aggregate amount of information, but also to:

- The distribution of this information;
- The structures and communication networks within which it is located;
- The nature and extent of the boundaries which impede the flow of this information.

Turner's insights were prescient. In 2003, the Columbia Accident Investigation Board (CAIB) observed that:

Management decisions made during Columbia's final flight reflect missed opportunities, blocked or ineffective communications channels, flawed analysis, and ineffective leadership. Perhaps most striking is the fact that management ... displayed no interest in understanding a problem and its implications. Managers failed to avail themselves of the wide range of expertise and opinion necessary to achieve the best answer to the debris strike question—'Was this a safety-of-flight concern?' ... In fact, their management techniques unknowingly imposed barriers that kept at bay both engineering concerns and dissenting views, and ultimately helped create 'blind spots' that prevented them from seeing the danger the foam strike posed.

CAIB (2003, p. 170)

7.2.3 THE CREATION OF LOCAL RATIONALITY

The question, says Turner, and it is largely a social-scientific one, is "what stops people from acquiring and using appropriate advance warning information, so that large-scale accidents and disasters are prevented?" (1978, p. 195). He found four groups of information needed to prevent disasters:

1. That which is completely unknown;
2. That which is known but not fully appreciated;

3. That which is known by someone, but is not brought together with other information at an appropriate time when its significance can be realized and its message acted upon;
4. That which was available to be known, but which could not be appreciated because there was no place for it within prevailing models of understanding. As Weick would later say: "Seeing what one believes and not seeing that for which one has no beliefs are central to sensemaking. Warnings of the unbelievable go unheeded" (1995, p. 87).

How do insiders make those numerous little and larger trade-offs about the value of nature of information that together contribute to incubation? It is easy to say, in hindsight, that certain data or patterns of data constituted a 'signal' even if people at the time dismissed it as noise. Gephart, in his reflections on man-made disaster theory not long after the appearance of Turner's book, observed that:

> the communication problems and unheeded warnings conceived by Turner as central to the pre-onset stages are actually only seen in retrospect. They are present in all disasters and in a variety of non-disasters as well. Considerable noise blends with potential warning signals to mask the warnings; they are distinguishable from normal signals and false warnings only after the event.
>
> **Gephart (1984, p. 211)**

In the second edition of the book, Pidgeon and Turner respond that this kind of critique can be directed at almost any post-hoc analysis of disasters. Indeed, judgmental language, essentially blaming people for not seeing the something coming that seems so obvious now, is extremely common. Hopkins, for example, has used phrases such as "poor decision-making" or a "failure to learn" (Hopkins, 2010). And high reliability theory (Weick & Sutcliffe, 2007) talks about it in the following (equally retrospective and judgmental) terms:

> ...failure means that there was a lapse in detection. Someone somewhere didn't anticipate what and how things could go wrong. Something was not caught as soon as it could have been caught.
>
> *(p. 93)*

Both the hindsight and outcome biases in how we look at other people's decisions in retrospect have been studied extensively (Fischhoff, 1975). The biases are pervasive. It remains a challenge to illuminate why, to key decision makers, "warning signs" seemed normal or unremarkable at the time. The easy case is one in which people in organizations with an ability to influence events deliberately overlook warning signs; when "crisis-prone organizations may even exert considerable effort to block warning signals" (Rosenthal & Kouzmin, 1996, p. 122). A small example of that, at least in some sense, is below.

MECHANIC DISCIPLINED FOR REPORTING CRACKS IN 737

An airline has agreed to remove the disciplinary action from the mechanic's file and to pay him $35,000 in legal fees. The lawsuit was filed under the whistleblower protections of the so-called AIR-21 statute (the Wendell H. Ford Aviation Investment and Reform Act for the 21st century.) The statute provides an appeal process for airline workers who are fired or otherwise disciplined for reporting safety information. The settlement was reached after a January 8, 2015 Department of Labor Administrative Judge dismissed the airline's motion for summary judgment and granted in part the mechanic's motion for summary judgment.

The judge's decision summarizes the allegations as follows: "On the evening of July 2, 2014, the [mechanic] was assigned to perform a [maintenance] check on a Boeing 737-700 aircraft. This maintenance check is part of the airline's Maintenance Procedural Manual (MPM). This check requires a mechanic to follow a task card which details the tasks to be accomplished." The task card requires the mechanic to "walkaround" the aircraft to visually inspect the fuselage. During his inspection, the [mechanic] discovered two cracks on the aircraft's fuselage and documented them. Discovery of these cracks resulted in the aircraft being removed from service to be repaired."

Thereafter, the mechanic was called into a meeting with his supervisors to "discuss the issue of working outside the scope of his assigned task." He was then issued a "Letter of Instruction" advising the mechanic that he had acted outside the scope of work in the task card and warning him that further violations could result in further disciplinary actions. The mechanic alleged in his whistleblower complaint that the letter from his airline "was calculated to, or had the effect of, intimidating [him] and dissuading him and other [mechanics] from reporting the discovery of cracks, abnormalities or defects out of fear of being disciplined."

The airline responded to the mechanic's allegations claiming that the mechanic went outside the scope of his duties when he observed the cracks and reported them. The airline further claimed that its Letter of Instruction was issued because the mechanic worked "outside the scope of his task" and not because he reported a safety problem. It further claimed that the letter was not a disciplinary action and the mechanic was not entitled to whistleblower protection.

Fortunately, the administrative judge sided with the mechanic in dismissing the airline's claims and finding that the mechanic engaged in activities protected by AIR-21 and that the airline was aware of it. Although no final decision was reached on the merits of the mechanic's case, the settlement followed close on the heels of the judge's decision (Goglia, 2015).

Turner was sensitive to the problem of hindsight in explaining incubation after the fact. He recognized that it is so easy to show what was wrong in the organization and the way it gathered and communicated intelligence. How could we go beyond the notion that things *go* wrong because things *are* wrong (deeper inside of the organization or higher up)? Because according to man-made disaster theory, things go wrong even though they were not wrong (or not seen as such) at the time at all.

Sometimes, Turner found, individuals or organizations are unaware of their areas of ignorance. They operate with implicit ideas or theories of their world, their environment, which do not get tested or challenged until a disaster is triggered. Only then might they discover that those models or theories are false or outdated. What they do, or conclude, however, makes sense within those models or theories. That, Turner quickly discovered, has everything to do with people's bounded, or local, rationality. Invoking Herb Simon, he reminded his readers:

> ...that the capacity of the human mind for formulating and solving complex problems is very small compared with the size of those problems whose solution is required for objectively rational behavior in the real world... or even for a reasonable approximation of such rationality.

> **Simon (1957, p. 198)**

As you will recall from Chapter 5, globally (as opposed to local) rational would mean that people's knowledge and decisions can be understood with reference to some norm, some optimum. Decisions or knowledge may be optimal when the decision maker has perfect, exhaustive access to all relevant information, takes enough time to consider it all, and applies clearly defined goals and preferences to make the final choice. In such cases, errors are explained by reference to deviations from this 'rational' norm, this ideal. If the decision turns out wrong it may be because the decision maker did not take enough time to consider all information, or that s/he did not generate an exhaustive set of choice alternatives to pick from. Errors, according to this logic, are deviant. They are departures from a standard. Errors are irrational in the sense that they require a motivational (as opposed to cognitive) component in their explanation. If people did not take enough time to consider all information, it is because they could not be bothered too. They did not try hard enough, and they should try harder next time, perhaps with the help of some training or procedural guidance. Many safety investigations are of course still rife with such globally rationalist assumptions, as illustrated above. And as you have seen, even Turner had a hard time getting away from normative and judgmental language.

Simon pointed out how humans could not or should not even behave like perfectly rational decision makers. While economists clung to the normative assumptions of decision-making (decision makers have perfect and exhaustive access to information for their decisions, as well as clearly defined preferences and goals about what they want to achieve), psychology, with the help of artificial intelligence, posited that there is no such thing as perfect rationality (i.e., full knowledge of all relevant information, possible outcomes, relevant goals), because there is not a single cognitive system in the world (neither human nor machine) that has sufficient computational capacity to deal with that all. Rationality is bounded. From around the time of Turner's writing, psychology increasingly charted people's imperfect, or bounded, or local rationality. Here are some of its key findings:

- The gathering of information and making of decisions is governed by people's local understanding, by their focus of attention, by their goals and knowledge, rather than some global ideal.
- Human decision-making is embedded in, and systematically connected to, the situation in which it takes place: it can be understood (i.e., makes sense) with reference to that situational context, not by reference to some universal standard, or to knowledge of a disaster that has not yet happened.
- Human actions and assessments can be described meaningfully only in reference to the localized setting in which they are made (without knowledge of outcome). Human actions and assessments can be understood by intimately linking them to details of the context that produced and accompanied them.
- If a decision is locally rational, it makes sense from the point of view of the decision maker—which is what matters if we want to learn about the underlying reasons for what from the outside looks like a miscommunication or bad decision.

Turner's sociological angle on studying disasters has helped legitimate the study of people's information environment as one way of understanding why what they do made sense to them at the time.

7.2.4 STUDYING THE 'INFORMATION ENVIRONMENT'

How is it, then, that locally rational decisions about certain information can incrementally move a system to the edge of disaster? As Lanir alluded to (good at the details, but poor at the big issues), a critical aspect of this dynamic is that people in decision-making roles on the inside of a socio-technical system miss or underestimate the global side effects of their locally rational decisions. These decisions are sound when set against local judgment criteria; given the time and budget pressures and short-term incentives that shape behavior. Given the knowledge, goals, and attentional focus of the decision makers and the nature of the data available to them at the time, it made sense. It is in these normal, day-to-day processes, where we can find the seeds of organizational failure and success. And it is these processes we must turn to in order to find leverage for making further progress on safety. As Rasmussen and Svedung (2000) put it:

> To plan for a proactive risk management strategy, we have to understand the mechanisms generating the actual behavior of decision-makers at all levels ... an approach to proactive risk management involves the following analyses:
>
> - a study of normal activities of the actors who are preparing the landscape of accidents during their normal work, together with an analysis of the work features that shape their decision making behavior

- a study of the present information environment of these actors and the information flow structure, analyzed from a control theoretic point of view. (p. 14)

Reconstructing or studying the "information environment" in which actual decisions are shaped and in which local rationality is constructed, can help you get a better idea of the processes of organizational sensemaking. These processes lie at the root of organizational learning and adaptation, and thereby at the source of drift into failure.

THE 'INFORMATION ENVIRONMENT' AND DRIFTING INTO FAILURE

The two Space Shuttle accidents (Challenger in 1996 and Columbia in 2002) are highly instructive here, if anything because the CAIB, as well as later analyses of the Challenger disaster (e.g., Vaughan, 1996) represent significant (and, to date, rather unique) departures from the typical structuralist probes into such accidents. These analyses take normal organizational processes toward drift seriously, applying and even extending a language that helps us capture something essential about the continuous creation of local rationality by organizational decision makers.

> One critical feature of the information environment in which NASA engineers made decisions about safety and risk was "bullets". Richard Feynman, who participated in the original Rogers Presidential Commission investigating the Challenger disaster, already fulminated against them and the way they collapsed engineering judgments into crack statements: "Then we learned about 'bullets'—little black circles in front of phrases that were supposed to summarize things. There was one after another of these little goddamn bullets in our briefing books and on the slides"
>
> **Feynman (1988, p. 127)**

Eerily, "bullets" appeared again as an outcropping in the 2003 Columbia accident investigation. With the proliferation of commercial software for making "bulletized" presentations since Challenger, bullets proliferated as well. This too may have been the result of locally rational (though largely unreflective) trade-offs to increase efficiency: Bulletized presentations collapse data and conclusions and are dealt with more quickly than technical papers. But bullets filled up the information environment of NASA engineers and managers at the cost of other data and representations. They dominated technical discourse and, to an extent, dictated decision-making, determining what would be considered as sufficient information for the issue at hand. Bulletized presentations were central in creating local rationality and in nudging that rationality ever further away from the actual risk brewing just below.

Edward Tufte analyzed one Columbia slide in particular, from a presentation given to NASA by a contractor in February 2003 (CAIB, 2003). The aim

of the slide was to help NASA consider the potential damage to heat tiles created by ice debris that had fallen from the main fuel tank. (Damaged heat tiles
triggered the destruction of Columbia on the way back into the earth's atmosphere, see Figure 2.5.) The slide was used by the Debris Assessment Team in
their presentation to the Mission Evaluation Room. It was entitled "Review of
Test Data Indicates Conservatism for Tile Penetration", suggesting, in other
words, that the damage done to the wing was not so bad (CAIB, 2003, p. 191).
But actually, the title did not refer to predicted tile damage at all. Rather, it
pointed to the choice of test models used to predict the damage. A more appropriate title, according to Tufte, would have been "Review of test data indicates
irrelevance of two models". The reason was that the piece of ice debris that
struck the Columbia was estimated to be 640 times larger than the data used
to calibrate the model on which engineers based their damage assessments.
(Later analysis showed that the debris object was actually 400 times larger.)
So the calibration models were not of much use: they hugely underestimated
the actual impact of the debris.

The slide went on to say that "significant energy" would be required to
have debris from the main tank penetrate the (supposedly harder) tile coating
of the shuttle wing, yet that test results showed that this was possible at sufficient mass and velocity, and that, once the tiles were penetrated, significant
damage would be caused. As Tufte observed, the vaguely quantitative word
"significant" or "significantly" was used five times on one slide, but its meaning ranged all the way from the ability to see it using those irrelevant calibration tests, through a difference of 640-fold, to damage so great that everybody
onboard would die. The same word, the same token on a slide, repeated five
times, carried five profoundly (yes, significantly) different meanings, yet
none of those were really made explicit because of the condensed format of
the slide. Similarly, damage to the protective heat tiles was obscured behind
one little word. The word was 'it.' And it was part of a sentence that read "Test
results show that it is possible at sufficient mass and velocity" (CAIB, 2003,
p. 191). The slide weakened important material, and the life-threatening nature
of the data on it was lost behind bullets and abbreviated statements.

A decade and a half before, Feynman (1988) had discovered a similarly
ambiguous slide about Challenger. In his case, the bullets had declared that
the eroding seal in the field joints was "most critical" for flight safety, yet that
"analysis of existing data indicates that it is safe to continue flying the existing
design" (p. 137). The accident proved that it was not. The solid rocket boosters
(SRBs or solid rocket motors (SRMs)) that help propel the space shuttle out
of the earth's atmosphere are segmented. This makes ground transportation
easier and has some other advantages. A problem that was discovered early
in the shuttle's operation, however, was that the solid rockets did not always
properly seal at these segments, and that hot gases could leak through the
rubber O-rings in the seal, called blow-by. This eventually led to the explosion of Challenger in 1986. The pre-accident slide picked out by Feynman had

declared that while the lack of a secondary seal in a joint (of the solid rocket motor) was "most critical," it was still "safe to continue flying." At the same time, efforts needed to be "accelerated" to eliminate SRM seal erosion (1988, p. 137). During Columbia as well as Challenger, slides were not just used to support technical and operational decisions that led up to the accidents. Even during both post-accident investigations, slides with bulletized presentations were offered as substitutes for technical analysis and data, causing the CAIB (2003), similar to Feynman years before, to grumble that: "The Board views the endemic use of PowerPoint briefing slides instead of technical papers as an illustration of the problematic methods of technical communication at NASA" (p. 191).

The overuse of bullets and slides illustrates the problem of information environments and how studying them, can help us understand something about the creation of local rationality in organizational decision-making. NASA's bulletization shows how organizational decision makers are configured in an "epistemic niche" (Hoven, 2001). That which decision makers can know is generated by other people, and gets distorted during transmission through a reductionist, abbreviationist medium. The narrowness and incompleteness of the niche in which decision makers find themselves can come across as disquieting to retrospective observers, including people inside and outside the organization. It was after the Columbia accident that the Mission Management Team "admitted that the analysis used to continue flying was, in a word, 'lousy.' This admission—that the rationale to fly was rubber-stamped—is, to say the least, unsettling." (CAIB, 2003, p. 190). "Unsettling" it may be, and probably is—in hindsight. But from the inside, people in organizations do not spend a professional life making "unsettling" decisions. Rather, they do mostly normal work. Again, how can a manager see a "lousy" process to evaluate flight safety as normal, as not something that is worthy reporting or repairing? How could this process be normal? The CAIB (2003) itself found clues to answers in pressures of scarcity and competition:

> The Flight Readiness process is supposed to be shielded from outside influence, and is viewed as both rigorous and systematic. Yet the Shuttle Program is inevitably influenced by external factors, including, in the case of STS-107, schedule demands. Collectively, such factors shape how the Program establishes mission schedules and sets budget priorities, which affects safety oversight, workforce levels, facility maintenance, and contractor workloads. Ultimately, external expectations and pressures impact even data collection, trend analysis, information development, and the reporting and disposition of anomalies. These realities contradict NASA's optimistic belief that pre-flight reviews provide true safeguards against unacceptable hazards

> *(2003, p. 191)*

Perhaps there is no such thing as "rigorous and systematic" decision-making based on technical expertise alone. Expectations and pressures, budget priorities and mission schedules, contractor workloads, and workforce levels all impact technical decision-making. All these factors determine and constrain what will be seen as possible and rational courses of action at the time. This dresses up the epistemic niche in which decision makers find themselves in hues and patterns quite a bit more varied than dry technical data alone. But suppose that some decision makers would see through all these dressings on the inside of their epistemic niche, and alert others to it. Tales of such whistleblowers exist. Even if the imperfection of an epistemic niche (the information environment) would be seen and acknowledged from the inside at the time that still does not mean that it warrants change or improvement. The niche, and the way in which people are configured in it, answers to other concerns and pressures that are active in the organization—efficiency and speed of briefings and decision-making processes, for example. The impact of this imperfect information, even if acknowledged, is underestimated because seeing the side effects, or the connections to real risk, quickly glides outside the computational capability of organizational decision makers and mechanisms at the time.

In 2003, David Woods was invited to give testimony to U.S. congress on the Columbia Space Shuttle Accident (Woods, 2003). The investigation, he argued, showed how NASA failed to balance safety risks with intense production pressure. As a result, the accident matched a classic pattern—a drift toward failure as defenses eroded in the face of production pressure. When this pattern was combined with a fragmented problem-solving process that was missing cross-checks and unable to see the big picture, the result was an organization that could no longer see its own blind spots about risks. Further, NASA was unable to revise its assessment of the risks it faced and the effectiveness of its countermeasures against those risks as new evidence accumulated. What made NASA's safety/production trade-offs so insidious, Woods stated, was that evidence of risks became invisible to people working hard to produce under pressure. Safety margins erode over time.

As an organizational or socio-technical accident, Woods argued how Columbia showed the need for organizations to monitor their own practices and decision processes. The aim is to detect when they are beginning to drift toward safety boundaries. The critical role for the safety group within the organization would be to monitor the organization itself—to measure organizational risk—the risk that the organization is operating nearer to safety boundaries than it realizes. The information environments surrounding decision makers play a key role. Studying information environments, how they are created, sustained, and rationalized, and in turn how they help support and rationalize complex and risky decisions, is one route to understanding organizational sensemaking. More will be said on these processes of sensemaking elsewhere in this book. It is a way of making what sociologists call the macro-micro connection. How is it that global pressures of production and scarcity find their way into local decision niches, and how is it that they exercise their often invisible but powerful influence on what people think and prefer; what people then and there see as rational or unremarkable? Although the intention was that NASA's flight safety evaluations be shielded from those external pressures, these pressures nonetheless seeped into even the collection of data, analysis of trends, and reporting of anomalies. The information environments thus created for decision makers were continuously and insidiously tainted by pressures of production and scarcity (and in which organization are they not?) pre-rationally influencing the way people saw the world. Yet even this "lousy" process was considered "normal"—normal or inevitable enough, in any case, to not warrant the expense of energy and political capital on trying to change it. The incubation of disaster can be the result.

One area for intervention, then, is the information that people in an organization use for decision-making (Rasmussen & Svedung, 2000). Studying and influencing information environments, how they are created, sustained, and rationalized, and in turn how they help support and rationalize complex and risky decisions, can help illuminate the small incremental steps that mark an incubation period. This, in a sense, is where risk is constructed.

Managing the information environment, of course, cannot be done through a priori decisions about what is important: that simply displaces the problem. Rather, high reliability theorists and others (Janis, 1982) recommend decision makers to remain what Weick calls complexly sensitized: to situate decision-making in an information environment full of inputs from different angles—and from below (Weick, 1993). Westrum called it an organizational culture that has requisite imagination, a play on Ashby's cybernetic notion of requisite variety (Ashby, 1956). With requisite imagination,

> the organization is able to make use of information, observations or ideas wherever they exist within the system, without regard for the location or status of the person or group having such information, observations or ideas.
>
> **Westrum (1993, p. 402)**

This can encourage decision makers to defer to expertise and take minority opinion seriously (Weick & Sutcliffe, 2007). Yet potentially meaningful signals can remain weak or few in information-intense environments. And as the CAIB reported, this can once again encourage a tendency to oversimplify, categorize, bulletize, and thus exclude (CAIB, 2003).

In addition, even with attention to information environments, decisions inside of them are pre-rationally influenced by forces such as production pressures, budget priorities and schedules, contractor workloads, employee qualifications, and workforce levels. This codetermines and constrains what is possible and rational for decision makers at the time. Although the intention was, for instance, that NASA's flight safety evaluations be shielded from external pressures (essentially turning it into a closed system, as per the high reliability recommendation), these pressures nonetheless affected data collection, trend analysis, and anomaly reporting (Feldman, 2004; Vaughan, 1996). Indeed, the idea of a closed system is probably illusory: the boundaries between the world and any system are not only arbitrary, but the outside is 'folded into' the system of interest at many points of individual contact (Cilliers, 1998; Dekker, Cilliers, & Hofmeyr, 2011).

7.2.5 DATA OVERLOAD

What if there is too much data? Data overload is a common problem for decision makers in all kinds of organizations (Woods, Patterson, & Roth, 2002). And it has not got any easier with the vast proliferation of electronically available and mediated data over the past decades. These are among the questions that decision makers might well be pursuing:

- Which issue should be addressed first?
- What are the postconditions of these issues for the remainder of operations?
- Is there any trend?
- Are there noteworthy events and changes in the process right now?
- Will any of this get worse?

One way of looking at data overload is as a workload problem. The way to reduce workload is to either reduce the work, or to increase the time available. This may not be practical, of course. And by how much should we reduce work or increase time, if those options are even available? This assumes that there are maximum processing rates, which in turn treats data as something objective, something fixed and inert. But people are not passive recipients of observed data; they are active participants in the intertwined processes of observation, action, and sensemaking. People employ all kinds of strategies to help manage data, and to impose meaning on it. For example, they redistribute cognitive work (to other people, to artifacts in the world); they re-represent problems themselves so that solutions or countermeasures become more obvious.

Another way of looking at data overload is as a clutter problem—there is simply too much for people to cope with. The solution to data overload as a clutter problem is to remove stuff or collapse it into bigger categories or increasingly meaningless words (like the use of 'significant' in the example from the section above). Seeing data overload as clutter, however, is completely insensitive of context. What seems clutter in one situation may be highly valuable, or even crucial in another situation.

Seeing data overload as a workload or clutter problem is based on certain assumptions about how human decision-making works. Questions about maximum human data-processing rates are misguided because this maximum, if there is one at all, is highly dependent on many factors, including people's experience, goals, history, and directed attention. As alluded to earlier in the book, Clutter and workload characterizations treat data as a unitary input phenomenon, but people are not interested in data, they are interested in meaning. And what is meaningful in one situation may not be meaningful in the next. De-clutter functions are context insensitive, as are workload-reduction measures. What is interesting, or meaningful, depends on context. How can someone intent on preventing disaster incubation know what the interesting, meaningful, or relevant pieces of data will be in a particular context? This takes a deep understanding of the organization, the risks it faces, and work as it is done inside of it.

7.2.6 GROUPTHINK

Around the same time that Barry Turner was formulating his thoughts on man-made disaster, Irving Janis, on the other side of the Atlantic Ocean, was trying to understand how groups of smart people could generate glaring policy failures. This, too, would become an important contribution and enrichment to safety science by importing social concepts, in this case from psychology and group dynamics, into the previously engineering-dominated the field. Like Turner, Janis relied on forensic case material for his data. His focus, however, was not on entire organizations or institutions, but on the small group dynamics that gradually help congeal a really bad decision as the most logical, legitimate, and desirable one. His poster child was the 1961 Bay of Pigs invasion, in which a CIA-sponsored paramilitary group of some 1,400 Cuban exiles launched a botched attack on the south coast of Cuba. It lasted 3 days, until the invaders surrendered and mostly ended up in Cuban jails. But other cases animated his theorizing as well, including the escalation of the Vietnam War and Watergate. He also used contrast cases in which policy makers produced good outcomes, such as the ending of the Cuban missile crisis of 1962 and the making of the Marshall plan, which was intended to help Europe recover from World War II (and to prevent recurrence of such massive armed conflict there).

Groupthink, at its heart, is a theory about how cohesive teams develop informal norms and intragroup social dynamics that aim—as if driven by a hidden, non-explicit, unwitting agenda—to preserve harmony and team relations.[1] This can lead to the identification and solidifying of a single policy option as the one that everyone agrees on, to the expense of any other (and often better) options. Key is the increasing lack or impotence of alternative conflicting viewpoints while policy teams deliberate and settle onto decisions. Echoes of this 1970s observation can now be found throughout the literature with an affinity to safety—from high reliability theory to resilience engineering. The symptoms of groupthink are below. They are cast in the slightly normative and judgmental fashion that is common to this type of study, but which has probably helped us identify important phenomena that called for more investigation and greater understanding.

---

[1] Others before Janis, or contemporaneous to him, were also interested in group dynamics in general, and this phenomenon of cohesiveness and 'group think' specifically, and studied it within fields of Social Psychology and Political Science, including Kurt Lewin, Philip Tetlock, Stanley Schachter, Hans Morgenthau, Arnold Wolfers, Phil Zimbardo, and Raymond Aron.

- *Incomplete survey of alternatives.* It is difficult to say, of course, what a 'complete' survey would be. But Janis did observe a striking paucity of alternatives generated in homogeneous teams (even though certain policy-making situations actually offer very few alternative options).
- *Incomplete survey of objectives.* Discussing policy options often took precedence over agreeing on the goals these policy decisions were meant to achieve. And even if that was present initially in discussions, cohesive teams could quickly start taking knowledge of and agreement about their objectives for granted.
- *Failure to examine the risks of the preferred choice.* This is of course easy to say in hindsight (when the risks of a failed approach are all too apparent), but Janis often did not even see conscious, let alone systematic efforts to try to identify or enumerate the risks of the preferred choice.
- *Failure to reappraise initially rejected alternatives.* Once a policy choice was, or seemed to become, the preferred one, these cohesive teams had no way to go back to square one. Their social dynamics would have in fact discouraged any member of the team from initiating such a return.
- *Poor information search.* Once a policy decision had (and often even before it had) become obvious as the preferred one, teams paradoxically stopped investing in finding out everything about it. Whether this was an unconscious decision to avoid coming up with contradictory data, or a logical choice to invest effort in planning the execution of the decision now that it seemed all but taken, crucial aspects of the situation were often left unexamined.
- *Selective bias in processing information at hand.* Whatever information seemed to support the preferred policy direction was generally deemed just, moral, and reliable.
- *Failure to work out a contingency plan.* In a sign of confidence that the decision (to be) taken was the right one, and would not fail precisely because it was the right one, cohesive teams often expended no energy on working out a plan for if it were to run into trouble or fail after all.

The subtle intragroup pressures through which these symptoms played out and became visible would seem to be highly dysfunctional. But of course, cohesiveness does have things going for it. Research during the 1950s and 1960s had shown that:

Other things being equal, as cohesiveness increases there is an increase in a group's capacity to retain members and in the degree of participation by members in group activities. The greater a group's cohesiveness, the more power it has to bring about conformity to its norms and to gain acceptance of its goals and assignment to tasks and roles. Finally, highly cohesive groups provide a source of security for members which serves to reduce anxiety and to heighten self-esteem.

Dorwin Cartwright (1968), as quoted in Janis (1982, p. 4)

And indeed, diversity—in socioeconomic background, race, and gender—was not generally regarded as a necessary virtue during that time. The interacting social dynamics that Janis gleaned from his casework were as follows:

Overestimation of the group's power and morality
- An illusion of invulnerability, shared by most or all the members, which creates excessive optimism and encourages taking risks.
- An unquestioned belief in the group's inherent morality, inclining the members to ignore the ethical or moral consequences of their decisions.

Closed-mindedness
- Collective efforts to rationalize in order to discount warnings or other information that might lead the members to reconsider their assumptions before they recommit themselves to their past policy decisions.
- Stereotyped views of out-group people as too evil to warrant genuine attempts toward engagement, or as too weak or stupid to counter whatever risky attempts are made to defeat their purposes.

Pressures toward uniformity
- Self-censorship of deviations from the apparent group consensus, reflecting each member's inclination to minimize to himself the importance of his doubts and counterarguments.
- A shared illusion of unanimity concerning judgments conforming to the majority view (partly resulting from self-censorship of deviations, augmented by the false assumption that silence means consent).
- Direct pressure on any member who expresses strong arguments against any of the group's stereotypes, illusions, or commitments, making clear that this type of dissent is contrary to what is expected of all loyal members.
- The emergence of self-appointed mindguards—members who protect the group from adverse information that might shatter their shared complacency about the effectiveness and morality of their decisions.

On the basis of his and others' research, which allowed him to formulate initiatives to counter the kinds of dynamics generated by groupthink, Janis suggested that teams and leaders do—among other things—this to prevent it from happening:

- The leader of a policy-forming group should assign the role of critical evaluator to each member, encouraging the group to give high priority to airing objections and doubts. This practice needs to be reinforced by the leader's acceptance of criticism of his or her own judgments in order to discourage the members from soft-pedaling their disagreements.
- The organization should routinely follow the administrative practice of setting up several independent policy-planning and evaluation groups to work on the same question, each carrying out its deliberations under a different leader.

- Each member of the policy-making group should discuss periodically the group's deliberations with trusted associates in his or her unit of the organization and report back their reactions.
- One or more outside experts or qualified colleagues within the organization who are not core members of the policy-making group should be invited to each meeting on a staggered basis and should be encouraged to challenge the views of the core members.
- At every meeting devoted to evaluating policy alternatives, at least one member should be assigned the role of devil's advocate.

The contribution from Janis to safety science is probably fairly obvious, even though his work was focused on policy makers in political settings. The sorts of countermeasures he suggested might well have been considered as useful in a number of accidents and disasters that followed his work.

7.2.7 Addressing the Barriers: Safety Imagination

But even if we invest in such countermeasures as suggested by Janis, as Leveson (2012) put it, organization members do their best to meet local goals, constraints, and conditions. In the busy daily flow and complexity of activities they may be unaware of any potentially dangerous side effects of their decisions, almost independent of whether they got there with unwitting aims to retain harmony or not. It is, again, only with the benefit of hindsight or omniscient oversight (which is utopian) that the effects as observed by Janis can be linked to actual risk. Jensen (1996) describes it as such:

> We should not expect the experts to intervene, nor should we believe that they always know what they are doing. Often they have no idea, having been blinded to the situation in which they are involved. These days, it is not unusual for engineers and scientists working within systems to be so specialized that they have long given up trying to understand the system as a whole, with all its technical, political, financial, and social aspects.
>
> *(p. 368)*

Being a member of a system, then, can make systems thinking all but impossible. Perrow (1984) made this argument very persuasively, and not just for the system's insiders. An increase in system complexity diminishes the system's transparency: Diverse elements interact in a greater variety of ways that are difficult to foresee, detect, or even comprehend. Influences from outside the technical knowledge base (those "political, financial, and social aspects" of Jensen, 1996, p. 368) exert a subtle but powerful pressure on the decisions and trade-offs that people make, and constrain what is seen as a rational decision or course of action at the time (Vaughan, 1996). How can experts and other decision makers inside organizational systems make sense of the available indicators of system safety performance?

Making sure that experts and other decision makers are well informed is in itself an empty pursuit. What well informed really means in a complex organizational setting is infinitely negotiable, and clear criteria for what constitutes enough information are impossible to obtain. As a result, the effect of beliefs and premises on decision-making and the creation of rationality can be considerable. Others after Turner would find the same thing, for example Weick (1995) and Vaughan (1996). Can we address problem of decision makers not seeing something because they have no paradigm or belief system for it? In subsequent work on man-made disaster theory, Pidgeon and O'Leary (2000, p. 23) scoured the literature for 'best practices' related to what they called "safety imagination." What they found was these guidelines for fostering safety imagination:

- Attempt to fear the worst
- Use good meeting management techniques to elicit varied viewpoints
- Play the 'what if' game with potential hazards
- Allow no worst-case situation to go unmentioned
- Suspend assumptions about how the safety task was completed in the past
- Approaching the edge of a safety issue, a tolerance of ambiguity is required, as newly emerging safety issues will never be clear
- Force participants to visualize 'near miss' situations that can develop into accidents.

The point of such practices of safety imagination, Pidgeon and O'Leary argued, is to counter some of the information difficulties and rigidities of thinking that Turner found across his disaster incubation periods. This includes (2000, p. 23):

- Extending the scope of potential scenarios that are considered relevant to the risk issue at hand (elicit varied viewpoints, play the 'what if' game, visualize near-misses becoming accidents).
- Countering complacency and the view that it would not happen to us (fear the worst, consider the worst-case scenarios).
- Forcing the recognition that during an incubation period the most dangerous ill-structured hazards are by definition shrouded in ambiguity and uncertainty (tolerate ambiguity).
- Attempting to step temporarily beyond, or even suspend, institutionally or culturally defined assumptions about what the likely hazard and its consequences will be (suspend assumptions about how the safety task was completed in the past).

What this ultimately requires, they pointed out, is a culture of safety that has at least these aspects (p. 18):

1. Senior management commitment to safety;
2. Shared care and concern for hazards and a solicitude over their impacts on people;
3. Realistic and flexible norms and rules about hazards;
4. Continual reflection on practice through monitoring, analysis, and feedback systems (organizational learning).

These later developments in man-made disaster theory suggest that it takes skepticism and disbelief in the sources of an organization's own success. That is very hard, of course. As long as things are going well, then they may continue to go well—if you just keep doing the same. But that is not so, says man-made disaster theory. Doing more of the same, and doing it even better, may well set you up for that disaster:

> A key part of the organizational etiology of disaster incubation is the way in which the 'negentropic' (or order-producing) tendencies of social systems contribute to the generation of extreme hazard from relatively safe situations, through the structured amplification of the consequences of earlier errors. That is, unintended consequences of errors are not propagated in purely random fashion, but may emerge as *anti-tasks* which make non-random use of large-scale organized systems of production. For example, consider the recent serious outbreaks of E-coli. Food poisoning in Scotland: here the consequences of the original contamination of cooked meat in one location were greatly amplified as the products were then distributed, unknowingly contaminated, to many people via the normal food distribution system.
>
> **Pidgeon and O'Leary (2000, p. 17)**

That which is responsible for the system's success, in other words, is also responsible for its collapse. And if it has been responsible for creating a lot of success (in terms of efficiency, productivity, competitiveness), then it will exacerbate the failure as well. The failure will piggyback nonrandomly on everything that people have created in their system, their organization, to guarantee successful outcomes. The counterintuitive conclusion that man-made disasters leaves us with is that humans make disasters *because* they create success.

7.3 MODELS OF DRIFT AND DISASTER INCUBATION AFTER TURNER

Sociological research as well as human factors work and research on safety has begun to sketch the contours of answers to the why of drift (Dekker, 2011; Leveson, 2012; Rasmussen & Svedung, 2000; Vaughan, 1996). Rasmussen and Svedung, looking back on accidents from the 1970s through the 1990s, concluded that

> …reports from accidents such as Bhopal, Flixborough, Zeebrugge and Chernobyl demonstrate that they have not been caused by a coincidence of independent failures and human errors. They were the effect of a systematic migration of organizational behavior toward accident under the influence of pressure toward cost-effectiveness in an aggressive, competitive environment.
>
> **Rasmussen and Svedung (2000, p. 14)**

Rasmussen illustrated the competing priorities and constraints that affect sociotechnical systems in a well-known figure (Figure 7.1). He suggested that there is an operating point inside an operating envelope. The envelope is bounded not just by the risk of failure, but also by economic collapse and work-overload. Indeed, organizations have to be safe. People's workload must be doable. But enough needs to get done to remain economically viable. All three of these objectives compete and conflict in various ways. For example:

- If the system reduces output too much to reduce workload, it will fail economically;
- If the system adds to many safety devices and barriers, it will make it difficult to remain competitive and get work done;
- If the system workload increases too far, the burden on workers and equipment will be too great;
- If the system moves in the direction of increasing risk, accidents are more likely to occur.

Together, pressures to become 'cheaper' (creating a greater margin to economic failure) and 'faster' (renegotiating what counts as acceptable workload under pressure to be more efficient) can push the operating point closer to the safety boundary. The drift of the operating point within the envelope can be a slow process. Each step is usually small and can go unnoticed, and no significant problems are noticed until it might be too late. Behavior that is acquired and honed in practice, and that is seen to achieve the goals important at that time, becomes legitimized through unremarkable repetition (Snook, 2000).

The boundaries themselves can move too:

- Changes in technology or other productivity changes can increase or decrease the workload that people might be able to handle, for example. Even supposed safety requirements can change the workload boundary. New procedures to carry out a task, for instance, and the requirement to complete elaborate checklists with signatures for each step can take significant time away from the tools and the actual task, thus pushing the workload boundary inward.
- The acceptable (safe) performance boundary can also move over time, for example, in response to risk events, regulatory pressure, political or media exposure of risk (all this tends to move the safety boundary inward). The introduction of new technologies can either increase or decrease the safety margin and thus move the boundary outward or inward.
- The economic boundary itself can move as well. Issuing new shares, for instance, may generate more capital that can be used up to create more space for the operating point (this would move the boundary outward). Economic times may also be tough, for example, as a result of sanctions or the imposition of tariffs, or because of cyclical economic movements (this would move the boundary inward, making it more likely that economic failure gets triggered by pressure from the other constraints).

What other theories have been proposed to look more deeply into this 'systematic migration?' First, we look at the contributions from Vaughan, Snook, and Dekker.

Then, we explore the common features of drift and disaster incubation that can be found across these theories.

7.3.1 Normalization of Deviance

Diane Vaughan developed her theory of normalization of deviance while studying the Space Shuttle Challenger Launch Decision in the late 1980s and early 1990s. The shuttle design (as you have seen earlier in this chapter) carried within it a potentially dangerous design feature: the field joints in the SRBs that could rotate, fail to seal (particularly in lower ambient temperatures) and let hot gases escape uncontrolled. The group that was assessing the joints on the SRBs conducted analyses after each flight. The purpose was to find the limits and capabilities of joint performance. It was this evidence (of blow-by and other damage to the joints and seals) that became the subject of normalization of deviance. Vaughan discovered that normalization of deviance followed a recurring pattern:

1. It starts with an assumption that risk is under control because of a redundantly engineered system. In this case, there was not one, but there were two O-rings to help seal the joint.
2. Then there is a signal of potential danger. Test and real flights showed that the O-rings were not behaving exactly as designed or assumed. The rotation between the two parts of the joint, and the difficulty of the O-rings to seal the gap in time, was not anticipated, and challenged the original ideas about risk.
3. An official action acknowledging elevated risk would then be taken. It could be assigning a team to study the problem, or the evidence, more closely. This of course required extra resources to be made available.
4. A review of the evidence was often what such actions amounted to. Through deeply technical discussions, against the background of an unruly and not entirely known technology, the group would come to the conclusion that the risk was, or could be kept, under control after all.
5. This would be followed by an official action indicating the normalization of deviance. The evidence just examined would be accepted as normal system behavior and enter the minds of group members as the new basis to work and evaluate any future deviance.

Each time, evidence initially interpreted as a deviation from expected performance was reinterpreted as within the bounds of acceptable risk. Success with launches, even those that created damage to the O-rings, meant that engineers recurrently observed the problem with no consequence. Flying with the flaw became normal and acceptable. The acceptance of this risk led to the Challenger breaking apart on January 28, 1986. Vaughan was keen to counter the idea that the Challenger launch decision was, at its core, a story of amoral, calculating people who pushed production schedules over safety concerns. That is of course an easy and convenient story to tell, and superficially, the Challenger launch decision offers enough support for it.

But just beneath this, Vaughan found normal people, experts even, under normal pressurized conditions of an engineering organization wrestling to interpret the behavior of unproven, unruly (and probably operationally unready) technology.

A critical ingredient of this is the apparent insensitivity to mounting evidence that, from the position of retrospective outsider, could have shown how bad the judgments and decisions actually are. This is how it looks from the position of retrospective outsider: the retrospective outsider sees a failure of foresight. From the inside, however, the abnormal is pretty normal, and making trade-offs in the direction of greater efficiency is nothing unusual. In making these trade-offs, however, there is a feedback imbalance. Information on whether a decision is cost-effective or efficient can be relatively easy to get. An early arrival time is measurable and has immediate, tangible benefits. How much is or was borrowed from safety in order to achieve that goal, however, is much more difficult to quantify and compare. If it was followed by a safe landing, apparently it must have been a safe decision.

Starbuck and Milliken, in their own analysis of the Challenger Space Shuttle accident, called this process "finetuning until something breaks." As a system is taken into use, it learns, and as it learns, it adapts:

> Experience generates information that enables people to fine-tune their work: fine-tuning compensates for discovered problems and dangers, removes redundancy, eliminates unnecessary expense, and expands capacities. Experience often enables people to operate a sociotechnical system for much lower cost or to obtain much greater output than the initial design assumed.
>
> **Starbuck and Milliken (1988, p. 333)**

Starbuck and Milliken highlighted how an organization can learn to "safely" borrow from safety while achieving gains in other areas. Counterintuitively, fine-tuning until something breaks, or drifting into failure has, at its heart, a form of organizational learning (learning the wrong thing, as Vaughan (1999) would later say). Each consecutive empirical success seems to confirm that the way you currently work is just fine, and that borrowing just a little bit more from safety in order to achieve other goals will probably not hurt. The system can operate equally safely, yet more efficiently.

As Weick (1993) pointed out, however, safety in those cases may not at all be the result of the decisions that were or were not made. Rather, it may be an underlying stochastic variation that hinges on a host of other factors, many not easily within the control of those who are tasked or believed to be in control. Empirical success, in other words, is not proof of safety. Past success does not guarantee future safety. Borrowing more and more from safety may go well for a while, but you never know when you are going to hit. This moved (Langewiesche, 1998) to say that Murphy's law is wrong: everything that can go wrong usually goes right, and then we draw the wrong conclusion.

7.3.1.1 Continued Belief in Safe Operations

Research on so-called high reliability organizations (HROs, see next chapter) stretches across decades and diverse high-hazard complex domains (aviation, nuclear power, utility grid management, Navy). It has tried to dig deeper into organizations' beliefs in their own infallibility. The HRO paradigm suggests that accidents are incubated when the organization's belief in continued safe operations is left to grow and solidify (Rochlin, 1999). High reliability theory has concluded (and counseled) that the past is not a good basis for a belief in future safety, and tapping into only a limited number of channels of information will render this belief narrow and unchallenged (LaPorte & Consolini, 1991; Weick, 1987). This can happen because of overconfidence in past results, suppressing of minority viewpoints, and the giving of priority to acute performance expectations or production pressures (Dekker & Woods, 2009). To guard against these drift-inducing impulses, HRO theorists suggest we stay curious, open-minded, complexly sensitized, inviting of doubt, and ambivalent toward the past (Weick, 1993). Thus, practitioners in HROs are described as skeptical, wary, and suspicious of quiet periods. Success, after all, or the absence of symptoms of danger (Starbuck & Milliken, 1988):

> ...breeds confidence and fantasy. When an organization succeeds, its managers usually attribute success to themselves or at least to their organization, rather than to luck. The organization's members grow more confident of their own abilities, of their manager's skills, and of their organization's existing programs and procedures. They trust the procedures to keep them apprised of developing problems, in the belief that these procedures focus on the most important events and ignore the least significant ones.

> *(pp. 329–330)*

Weick and colleagues echoed this two decades later (Weick & Sutcliffe, 2007):

> Success narrows perceptions, changes attitudes, reinforces a single way of doing business, breeds overconfidence in the adequacy of current practices, and reduces the acceptance of opposing points of view.

> *(p. 52)*

It is this complexity of possible interpretations of events that allows organizations to anticipate and detect what might go wrong. An important part is to understand the gap between how work is imagined and how work is done in practice (Dekker, 2003; Hollnagel, 2014), which requires leadership involvement. It includes managerial and supervisory visibility at the sharp end and interest in what goes on there beyond whether it complies with pre-understood notions of protocol and procedure (Dahl & Olsen, 2013). Of course, even deference to expertise, a diversity of viewpoints (including dissenting ones) and a sensitivity to operations does not protect a system from failure (Hayes, 2012). To some extent, most theories (including man-made disaster theory) are pessimistic about people's ability to pick up on the shifting of norms and erosion of margins—a conclusion Diane Vaughan drew after her extensive analysis of the Space Shuttle Challenger launch decision (Vaughan, 1996).

7.3.1.2 Goal Interactions and Normalization of Deviance

A key ingredient suspected in any incubation period is the organization's preoccupation with production and efficiency (Turner & Pidgeon, 1997; Vaughan, 1996;

Woods, 2003). Pressures to achieve production goals can be felt acutely and the effect of operational or managerial decisions on the ability to achieve them can often be measured directly. But there is a feedback imbalance: the extent to which these decisions create pressure on safety margins (while obscuring chronic safety concerns) is not typically easy to see or quantify (Woods, 2006). Vaughan (1996) has traced in detail how pressures of production find their way into local decision settings and exercise an invisible, powerful influence on what practitioners see as rational at the time. Sociology refers to this as the *macro-micro connection*—which links macro-level forces operating on the entire organization, and the micro-level cognitions and decisions of individual people within. However, this link is far from straightforward, and cannot just provide a roadmap or action list in any prescriptivist, managerial sense. A suggested starting point is to trace the organization's diversity of goals and how they might conflict (Dörner, 1989; Vaughan, 1999; Woods et al., 2010), creating basic incompatibilities in what its members need to achieve. As Dörner observed (Dörner, 1989), "Contradictory goals are the rule, not the exception, in complex situations" (p. 65). Some organizations pass goal conflicts on to individual practitioners quite openly, but many are never made explicit (Dekker, 2005); left to emerge from multiple irreconcilable expectations from different levels and sources or from both subtle and tacit pressures and from management or customer reactions to past trade-offs (Woods et al., 2010).

Incompatible goals emerge from the organization and its interaction with its environment. The managing of these conflicts is typically transferred to local operating units (the sharp-end), such as control rooms, patient wards, airline cockpits. The conflicts are negotiated and resolved in the form of countless daily decisions and trade-offs. These are decisions and trade-offs made by individual operators or crews vis-à-vis operational demands: external pressure becomes internalized: the macro becomes micro where global tension between efficiency and safety seeps into local decisions and trade-offs by individual people or groups—typically into what they see as normal, professional action (Dekker, 2011; Woods & Cook, 2002). When success is achieved, this either stays unnoticed, or practitioners might be celebrated and rewarded. When "failure" is the result, then these same assumptions and statements are then compared to the implicit or explicit guidance and any gaps are offered as a causal explanation for system collapse.

Some might consider these trade-offs between production and protection to be amoral calculations by managers, engineers, or operators (Goldman & Lewis, 2009; Hopkins, 2010; Woolfson & Beck, 2004), but cost and efficiency are taken-for-granted goals in most professions committed to problem-solving under constraints (Petroski, 1985; Wynne, 1988). "Satisficing" irreconcilable constraints and demands (which can mean doing more with less), can be part of a professional culture, code-termining what organization members will see as rational. Vaughan, again, tried to develop a more nuanced and socially patterned vision of the incubation period (Vaughan, 1996). The "normalization of deviance" describes a process whereby a group's construction of risk can persist even in the face of continued (and worsening) signals of potential danger. This can go on until something goes wrong, which (as Turner would have predicted) reveals the gap between the presence of risk and how it was believed to be under control (cf. Starbuck & Milliken, 1988). Small departures from an earlier established norm are often not worth remarking or reporting on. Such incrementalism contributes to normalization (Dekker, 2011). It allows normalization and rationalizes it (Starbuck & Milliken, 1988):

> Experience generates information that enables people to fine-tune their work: fine-tuning compensates for discovered problems and dangers, removes redundancy, eliminates unnecessary expense, and expands capacities. Experience often enables people to operate a socio-technical system for much lower cost or to obtain much greater output than the initial design assumed.
>
> *(p. 333)*

Actions that are interpreted as "bad decisions" after an adverse event (like using unfamiliar or nonstandard equipment) are, at the same time, actions that seemed reasonable—or people would not have taken them (Dekker, 2002; Vicente, 1999). This literature does not see wrongdoing, but rather tries to understand how people can see their actions as being right. Success can entail risk, of course: things almost always go right, repeatedly (Weick, Sutcliffe, & Obstfeld, 1999). As alluded to earlier, success, or the absence of symptoms of danger, can help engender confidence in future results (Clarke & Perrow, 1996; Starbuck & Milliken, 1988). Contributing to this is what Vaughan called "structural secrecy," the mix of bureaucracy and specialized knowledge that leads to people in one department or division lacking the expertise to understand the implications of the work in another (or even the work of their own specialists). This echoes Turner's views on the formalized information exchanges (e.g., meetings, PowerPoint presentations, memos) can worsen the problem: such efforts to communicate can ironically result in people knowing less because of the way in which knowledge is organized and summarized, and, most importantly, left out (Vaughan, 1996).

7.3.2 PRACTICAL DRIFT

Snook's ideas build on a case from 1994. Two U.S. Air Force F-15 fighters accidentally shot down two U.S. Army Black Hawk Helicopters over Northern Iraq, killing all 26 peacekeepers onboard. In response to this disaster, the complete array of military and civilian investigative and judicial procedures ran their course. After almost 2 years of investigation with virtually unlimited resources, no culprit emerged, no bad guy showed up, no smoking gun was found. Snook attempts to make sense of the tragedy that seemed to make no sense at all: a tragedy without a cause. In his work, Snook develops individual, group, organizational, and cross-level accounts of the accident. Snook proposed a dynamic, cross-level mechanism, or theory that he called 'practical drift.' This is the slow, steady uncoupling of practice from written procedure in individual units over time, which is no problem—until a stochastic event brings them together in a tightly coupled situation with sudden interdependencies (e.g., helicopters, airborne controllers, and fighter jets). Then mismatches in practices and assumptions suddenly show up and can produce lethal effects. Interestingly, practical drift happens because (or perhaps in spite of) everyone behaving just the way their colleagues would expect them to behave.

As Snook liked to say (see the next chapter), the shoot down was a normal accident in a highly reliable organization.

The possibility of 'procedural drift' begins with a premise by Weick (1987). The problem is not that systems are tightly coupled per se, but that there is a dynamic patterning of loose and tight couplings. Practical drift develops along stages, which Snook (2000) identified as follows:

1. *Designed stage*: An organization set up to deal with a particular problem is typically rather loosely coupled when it is designed. Bonds among sub-units are relatively loose, so that there is not a lot of interdependence. But in Snook's case, the operational plan for the organization delivering aid in Northern Iraq was *overdesigned*, written precisely to prevent or manage the highly rare occurrences of units being in each other's way. The overspecified rules were written by planners who would never have to actually work in an organization that applied the rules.

2. *Engineered stage*: When this organization went 'live' as Operation Provide Comfort in the Middle East in 1991, it took with it the rules designed to handle the worst-case scenario of a tightly coupled situation where the actions of any one subunit were assumed to have an immediate and substantial impact on the operations of others. For the first months, even though units operated independently, in loosely coupled situations, work as it had been imagined was largely the way work was done. But gradually, the accumulation of practical demands of getting the work done, goal conflicts, resource constraints, and learning about possible local efficiencies created pressure for change. Practical drift is then set in motion. The rules written for worst-case scenarios were bound to become seen as overbearing, overspecified, and overcontrolling. Pragmatic demands of everyday work, and success in carrying it out in ways that had less connection to the designed organization and its rules, meant that there was a growing gap between global design rules and local adaptations. Different units each did this in their own ways, as only they could decide what made most sense to them. The predictability of each unit for others declined as a result. But this did not matter as long as there were no tight couplings or interdependencies. "A loosely coupled system may be a good system for localized adaptation. If all of the elements in a large system are loosely coupled to one another, then any one element can adjust to and modify a local unique contingency without affecting the whole system. These local adaptations can be swift, relatively economical, and substantial" (Snook, 2000, p. 195).

3. *Applied stage*: The net effect of practical drift, says Snook, is a decline in global rationality. Global rationality for a coupled system is not simply the sum of all local rationalities. On the contrary. The tightly logical and synchronized rationale that governed system design is no longer present, and the 'engineered' world has slipped out of view and people's memories as well. In an applied world, units have organized their work, and their norms, around locally pragmatic responses to the intimate and immediate demands they face. Over time, the globally engineered, standardized

system is replaced by a series of locally adaptive subunit logics, each justifying their own interpretations and adaptations—if they even feel they have to justify them. Imperceptibly, each unit has followed its own unique path in angling away from the originally designed baseline.

4. *Failed stage*: In what Snook calls 'a rare stochastic fit' (or perhaps 'bad luck'), the total system can suddenly become tightly coupled and highly interdependent. Subunits are now forced to act on the basis of assumptions about the behavior of others, and likely the only reliable assumptions they have is that the others will behave according to the original global design rules—even if they themselves do not. Because of the very nature of the local adaptations that have occurred over time and in intimate contact with the context and setting of that particular unit, it would be almost impossible for all units to know how all other units have adapted. So they turn to a frozen image of the old, designed organization instead. But that is not what others are using at all. As Snook concludes: "During situations of tight coupling, it is this disconnect, this delta, this gap between the locally emergent procedures actually being followed in various subgroups from those that engaged actors assume are dictating action, that constitutes a general set of conditions that increases the likelihood of disastrous coordination failures" (2000, p. 199).

In response to the accidental shoot down, Snook saw how "higher headquarters in Europe dispatched a sweeping set of rules in documents several inches thick to 'absolutely guarantee' that whatever caused this tragedy would never happen again" (2000, p. 201). It is a common, but not typically satisfactory reaction. Introducing more procedures and rules does not necessarily avoid the next incident, nor do exhortations to follow rules more carefully necessarily increase compliance or enhance safety. In fact, it is likely that more tightly scripted organizations will invite even more practical drift. And in the end, of course, a mismatch between procedures and practice is not unique to accident sequences, (as you will recall from Chapter 2). The accident, and the sudden shift to tight coupling and interdependency that precedes it, merely makes highly *visible* what would otherwise remain imperceptible. The lesson, for Snook, is that practical drift is very difficult, or impossible (and perhaps undesirable) to prevent. And the sudden contractions into tight coupling are virtually impossible to predict. Given this, his prescription is to note the following:

- A complex high-hazard organization that cannot afford to learn from trial and error has the tendency to overdesign and overcontrol.
- A long enough period of loosely coupled operations is sufficient to generate substantial gaps between globally designed rules and locally rational subgroup practices. The more overdesign and overcontrol there has been from the outset, the more likely this practical drift is likely to be triggered and pick up pace.
- There is always a reasonable chance that isolated subgroups will come into sudden contact with each other, and base their actions on assumptions that the others are still behaving according to the global rules.

When it comes to control, Snook concludes, less is sometimes more. But is there nothing organizations can do, other than trying not to overcontrol and hope for the best?

7.3.3 Drift into Failure

Both sociology and complexity science offer possible insights into the dynamics, relationships, habituation, adaptive capacity, and complexity of the systems we want to understand and control. The theory of drift into failure (Dekker, 2011) is organized around inspirations from those ideas. It suggests that the following five concepts together may characterize drift:

1. Scarcity and competition
2. Decrementalism, or small steps
3. Sensitive dependence on initial conditions
4. Unruly technology
5. Contribution of the protective structure

Any organization operates and must try to survive in an environment that has real constraints, such as the amount of capital available, the number of customers reachable, and the qualifications of available employees. There are also hard constraints on how fast things can be built, developed, and driven. And, most importantly, there are other organizations that try to do the same thing.

Recall how Jens Rasmussen suggested that work in complex systems is bounded by three types of constraints (Rasmussen, 1997) (see Figure 7.1). There is an economic boundary, beyond which the system cannot sustain itself financially. Then there is

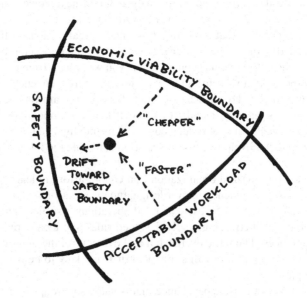

FIGURE 7.1 Rasmussen's depiction of the various pressures and constraints acting on a system's operating point. (Interpretation and drawing by author.)

a workload boundary, beyond which people or technologies can no longer perform the tasks they are supposed to. And there is a safety boundary, beyond which the system will functionally fail. For Rasmussen, these three constraints, or boundaries, hem in the operation of any safety-critical system from three sides. There is no way out, there is only some room for maneuvering inside of the space delineated by these three boundaries. Managerial and economic pressure for efficiency will push the system's operations closer to the workload and safety boundaries. Indeed, the likely result of increasing competitive pressure on a system, and of resource scarcity, will be a systematic migration toward workload and safety boundaries. Noncommercial enterprises experience resource scarcity and Rasmussen's three constraints too.

Adapting to resource pressure, approvals can be delayed or rushed, surveillance can get reduced. Yet doing business under pressures of resource scarcity is normal: Scarcity and competition are part and parcel even of doing inspection work. Few regulators, for instance, will claim that they have adequate time and personnel resources to carry out their mandates. Yet the fact that resource pressure is normal does not mean that it has no consequences. Trade-offs are made. The pressure expresses itself in the common organizational, political wrangles over resources, managerial preferences for certain activities and investments over others, and in almost all engineering and operational trade-offs between strength and cost and between efficiency and diligence. A driver for drift hides somewhere in this conflict, in this tension between operating safely and operating at all, and between building safely and building at all. It can help explain the steady disengagement of practice from earlier established norms or design constraints (Leveson, 2012). This disengagement can eventually become drift into failure.

One key issue is a feedback imbalance. Information on whether a decision is cost-effective or efficient can be relatively easy to get. How much is or was borrowed from safety in order to achieve that goal, however, can be much more difficult to quantify and compare. If the outcome was desirable, then it may even be invisible altogether, and encourage just another small step next time around (as Vaughan also demonstrated). Such fine-tuning constitutes incremental experimentation in essentially uncontrolled settings. On the inside, incremental 'nonconformity' is an adaptive response to scarce resources and production pressures. Departures from the norm, as Vaughan showed, can become the norm. Seen from the inside of people's own work, deviations become compliant behavior. They are compliant with the emerging, local ways to accommodate multiple goals important to the organization (maximizing capacity utilization but doing so safely; meeting technical or clinical requirements, but also deadlines). The erosion of safety constraints may at some point have consequences, however. As Langewiesche said: Murphy's Law is wrong. Everything that can go wrong usually goes right, and then we draw the wrong conclusion (Langewiesche, 1998). Decrementalism, or small changes, can eventually lead to big events.

This, in complexity and systems thinking, is related to a sensitive dependency on initial conditions (also known as the butterfly effect). Whether technology takes an unruly trajectory toward failure may depend on seemingly innocuous features or infinitesimally small differences in how it got started. Relocating an organizational process from the safety group to the maintenance group, for example, and calling the problem dealt with a maintenance problem rather than a safety problem, might seem innocuous enough. But it was one of those small, banal changes (iced-up foam striking the wing on lift-off) that might have ended up in a huge event when Space Shuttle Columbia burned up on reentry in 2003 (CAIB, 2003). Understanding how a system may sensitively depend on initial conditions is something that certification processes could perhaps be counted on to do. But certification does not typically take lifetime wear of parts or other complex system interactions into account. Certification processes do not really know how to take socio-technical adaptation of new equipment, and the consequent potential for drift into failure, into account when looking at nascent technologies. Systemic adaptation or wear is not, or hardly, a criterion in certification decisions, nor is there a requirement to put in place an organization to prevent or cover for anticipated wear rates or pragmatic adaptation, or fine-tuning. As a certification engineer for one regulator testified, "wear is not considered as a mode of failure for either a system safety analysis or for structural considerations" (NTSB, 2002, p. 24). Because how do you take wear into account? How can you even predict with any accuracy how much wear will occur?

The problem of certifying a system as safe to use can become even more complicated if the system to be certified is socio-technical and thereby even less calculable. What does wear mean when the system is socio-technical rather than consisting of pieces of hardware? In both cases, safety certification should be a lifetime effort, not a still assessment of decomposed system status at the dawn of a nascent technology. Safety certification should be sensitive to the coevolution of technology and its use and its adaptation. Using the growing knowledge base on technology and organizational failure, safety certification could aim for a better understanding of the ecology in which technology is released—the pressures, resource constraints, uncertainties, emerging uses, fine-tuning, and indeed lifetime wear.

So what should the standards be? The introduction of a new piece of technology is followed by negotiation, by discovery, by the creation of new relationships and rationalities. "Technical systems turn into models for themselves," said Weingart (1991, p. 8), "the observation of their functioning, and especially their malfunctioning, on a real scale is required as a basis for further technical development." Rules and standards do not exist as unequivocal markers against incoming operational data (and if they do, they are quickly proven useless or out-of-date). Rather, rules and standards are the constantly updated products of the processes of conciliation, of give and take, and of the detection and rationalization of new data. As Brian Wynne said:

> Beneath a public image of rule-following behavior and the associated belief that accidents are due to deviation from those clear rules, experts are operating with far greater levels of ambiguity, needing to make expert judgments in less than clearly structured situations. The key point is that their judgments are not normally of a kind—how do we design, operate and maintain the system according to 'the' rules? Practices do not follow rules, rather, rules follow evolving practices.
>
> **Wynne (1988, p. 153)**

Nor is there a one-way and unproblematic relationship between the original rules or requirements and subsequent operational data. Even if the data, in one reading, may prove the original requirements or rules wrong, this does not mean that complex systems, under the various normal pressures of operating economically, reject the requirements and rules and come up with new ones. Instead, the meaning of the data can get rene-gotiated. People may want to wait for more data. The data can be denied. The data can be said to belong to a category that has nothing to do with safety but everything with normal operational variance (as happened in the Columbia Space Shuttle).

Safety certification, then, is not just about seeing whether components meet crite-ria, even if that is what it often practically boils down to. Safety certification is about anticipating the future. It is an envisioning problem (Dekker, Mooij, & Woods, 2002). Safety certification is about bridging the gap between a piece of new technology in the hand now, and its adapted, coevolved, grimy, greased-down wear and use further down the line. But we are not very good at anticipating the future. Certification prac-tices and techniques oriented toward assessing the standard of current components do not translate well into understanding total system behavior in the future. Making claims about the future, then, often hangs on things other than proving the worthi-ness of individual parts.

Unruly technology can make this extra difficult. Unruly means that something is disorderly and not amenable to discipline or control. It is a better word than uncer-tain. Very little to none of the technology we put into our complex systems is believed to be uncertain. Uncertain means unknown. And the whole point of engineering and validation and verification is to calculate our way out of such uncertainty, to not have such unknowns. We run the numbers, do the equations, and build the computer sim-ulations, so that we know under what load something will break and what the mean time between failures will be.

But despite all these efforts, there will always be unknowns, even if the technology is considered or judged or believed to be certain. The term "unruly technology" was introduced by Brian Wynne in 1988 to capture the gap between our image of tidiness and control over technology through design, certification, regulation, procedures, and maintenance on the one hand and the messy, not-so-governable interior of that tech-nology as it behaves when released into a field of practice. Technology and safety is about how things behave in context, not on the drawing board. Universal proclama-tions or assurances about reliability figures are mute when ideas or designs are put into working solutions in this or that situation. A crucial skill involves finding a practical balance between universality of safety assumptions and their contextualization.

This is a balance (and a possible gap) that not only operates between those on the outside (the public or consumers of the technology) and insiders (engineers, manag-ers, regulators), but applies even to insiders themselves: to practitioners very close to the technology. If the operational system is not itself following the rules by which it was predicted or supposed to operate, insiders can also reconcile such data with their beliefs—not by changing beliefs, but by looking differently at the data. This works particularly when:

- It is a relatively common problem, among a mass of other relatively common problems;
- There are routine operational ways of compensating for it (providing more redundancy); and
- Alternative approaches would severely disrupt the economic or operational viability of the system.

The piling up of mine tailings (as in Turner's original example), even against the background of a technology that might behave differently in practice than what was predicted, is one example. As Turner suggested, insiders can keep their beliefs intact, they can retain their image of tidiness and controllability. But the technology may remain unruly, no matter what.

The operation of unruly technology is of course surrounded by structures meant to keep it safe. The protection offered by these structures takes a lot of forms. In addition to the organization necessary to actually run the technology, there are regulatory arrangements, quality review boards, nominated post holders, safety departments, and a lot more. These deal with all kinds of information through meetings, committees, international or cross-industry collaborations, data monitoring technology, incident reporting systems, and a host of informal and formal contacts. All of which produces vast arrays of rules, routines, procedures, guidance material, prescriptions, expert opinions, engineering assessments, and managerial decisions. And the people who populate this system change jobs and might even move from industry to regulator and back again. This is one of the features that makes complex systems complex: their boundaries are fuzzy. It is almost impossible to say where the system that makes the maintenance rules begins or ends.

There is, in other words, a large web of relationships in which the operation of risky technology is suspended—a web, moreover, that has no clear boundaries, no obvious end or beginning. The whole structure that is designed (and has evolved) to keep the technology safe, can make the functioning and malfunctioning of that technology more opaque. The safer and better protected a system becomes, the less visible risk becomes too (Amalberti, 2013). The meaning of signals about the technology (e.g., that it is not behaving according to original manufacturer specifications, or that people are not abiding by the latest procedure) get constructed, negotiated, and transacted through the web of relationships that is strung throughout this structure. The 'weak signals' that are left over trigger only weak organizational responses, if any at all (Weick & Sutcliffe, 2007). In this paradox, it is possible that the protective structure itself contributes to the construction and treatment of weak signals, and by extension to drift—in ways that are inadvertent, unforeseen, and hard to detect.

7.3.4 SIMILARITIES AND OVERLAP IN DRIFT MODELS

Though different in background, pedigree, and much substantive detail, these works converge on important commonalities about the drift into failure:

- The first echoes Turner's important observation that accidents, and the drift that precedes them, involve normal people doing normal work in normal organizations—not miscreants engaging in spectacular acts or immoral deviance. We can call this the banality-of-accidents thesis: the potential for having an accident grows as a normal by-product of doing normal business under normal pressures of resource scarcity and competition.
- Second, most theories from Turner onward have at their heart a conflictual model: Organizations that involve safety-critical work are essentially trying to reconcile irreconcilable goals (e.g., staying safe and staying in

business). The major engine of drift hides somewhere in this conflict, in this tension between operating safely and operating at all, between building safely and building at all. This tension provides the energy behind the slow, steady disengagement of practice from earlier established norms or design constraints.

- Third, drifting into failure is incremental. Accidents do not happen suddenly, nor are they preceded by monumentally bad decisions or bizarrely huge steps away from the ruling norm. The organizational decisions that are seen as "bad decisions" after the accident (even though they seemed like perfectly good ideas at the time) are seldom big, risky, order-of-magnitude steps. Rather, there is a succession of decisions, a long and steady progression of small, incremental steps that unwittingly take an organization toward disaster. Each step away from the original norm that meets with empirical success (and no obvious sacrifice of safety) is used as the next basis from which to depart just that little bit more again. It is this incrementalism that makes distinguishing the abnormal from the normal so difficult. If the difference between what "should be done" (or what was done successfully yesterday) and what is done successfully today is minute, then this slight departure from an earlier established norm is not worth remarking or reporting on—particularly not if it generates productive gain. Incrementalism is about continued normalization: it allows normalization and rationalizes it.

An interesting difference between Vaughan's normalization of deviance and Snook's procedural drift is that in procedural drift the process is largely confined to the operational end of the organization. It is usually unrecognized by higher echelons. In normalization of deviance, on the other hand, there is an incremental change in the acceptance of risk which is incorporated upwards, and brought into the organization's norms and rules over time. As this increase in the acceptance of risk ascends through the layers of the organization, its origins and rationale from the technical and engineering experts at the sharp end becomes less transparent to those at the higher levels. The progressively increased acceptance of risk thus becomes formally incorporated in the documented policies of the organization, unlike procedural drift, which essentially remains undocumented. Normalization of deviance can be argued to be rule-based, whereas procedural drift happens generally outside the rules structure. Instead, it emerges as locally based workarounds or learning. This modifies the rules in the interests of apparent local efficiency, or fills operational gaps where the rules and procedures do not reach.

7.3.5 Drift into Failure and Incident Reporting

Can incident reporting not reveal a drift into failure? This would seem to be a natural role of incident reporting, but it is not so easy. The normalization that accompanies drift into failure challenges the ability of insiders to define incidents. What is an incident? Many of the things that would, after the fact, be constructed as incidents and worthy of reporting, are the normal, everyday workarounds, frustrations, and improvisations needed to get the job done. They are not reportworthy. They do not qualify as incidents. Even if the organization has a reporting culture, a learning culture, and a just culture so that people would feel secure in sending in their reports without fear of retribution, such 'incidents' would not turn up in the system. As Amalberti explained:

> accidents are different in nature from those occurring in safe systems: in this case accidents usually occur in the absence of any serious breakdown or even of any serious error. They result from a combination of factors, none of which can alone cause an accident, or even a serious incident; therefore these combinations remain difficult to detect and to recover using traditional safety analysis logic. For the same reason, reporting becomes less relevant in predicting major disasters.
>
> **Amalberti (2001, p. 112)**

Even if we were to direct greater analytic force onto our incident-reporting databases, this may still not yield predictive value for accidents in very safe systems, simply because the data is not there. The databases do not contain, in any visible format, the ingredients of the residual accidents that may still happen. Learning from incidents to predict and prevent such accidents may well be impossible. Incidents are about independent failures and errors, noticed and noticeable by people on the inside. But these independent errors and failures no longer make an appearance in the accidents that happen. The etiology of accidents in very safe systems, then, may well be fundamentally different from that of incidents, hidden instead in the residual risks of doing normal business under normal pressures of scarcity and competition. This once again raises the question whether predictive power can be derived by looking at negative events at all any more (recall this from the previous chapter). Perhaps the ability for an already safe system to resist harmful influences derives from the presence of its positive capacities, not (just) from its brushes with the supposed dress rehearsals that take the form of incidents.

7.4 MAN-MADE DISASTER THEORY AND SOCIETAL EMANCIPATION

Man-made disaster theory was born during 'The Great Social Upheaval' of the 1960s and 1970s. On the one hand, this may have hampered the uptake of man-made disaster theory among scholars and practitioners in Europe and elsewhere. Sociology and political science were deeply absorbed in more pressing social problems and challenges. These not only fundamentally overhauled their objects of study (e.g., the role of women, the nature of hierarchy, and sources of authority and expertise), but called for an overhaul of their very fields of study and the institutions and universities at which their research and teaching was carried out.

But in another way, this great social upheaval would have provided a fertile soil for something like man-made disaster theory. In the West, a generation had found a voice to rebel against 'men in gray flannel suits,' to express their concern at how power was concentrated in unassailable social (often patriarchal) structures, at the environmental destruction they wrought, the materialism they sponsored, their resource depletion, minority repression, geopolitical adventures and follies. Their parents may have taken the existence, legitimacy, and functioning of institutions of government and corporations for granted. But the generation coming of age in the 1960s and 1970s no longer did so: suspicion about the motives, ethics, and efficacy of administrations, organizations, and institutions bloomed like never before. These institutions had produced or would soon produce accidents (e.g., Tenerife, Three Mile Island, Chernobyl, Space Shuttle Challenger, Herald of Free Enterprise, Clapham Junction, Piper Alpha) that showed how flawed and limited their control of highly risky systems might actually be. Not only that, these same institutions often shirked 'accountability.' In many cases, they upset the public with their callousness at, or denial of, the social, environmental, and human cost of the failure.

Ultimately, critics looked at modernity itself as the culprit. Recall from earlier chapters how modernity followed on the back of the Enlightenment and the Scientific and Industrial Revolutions. These instilled in the West a fundamental belief in unstoppable progress. Things would only get better. Technology and bureaucracy would ensure this. Rational, bureaucratic, and technocratic control of systems of production was the way to do it. The clock, the office, public schooling, the production line, Taylorism—these all helped to rationally organize society, to make it controllable, predictable, and more efficient. And everyone would benefit: remember from Taylor's time how everyone could in principle buy a Model T Ford. The deep disillusionment of the 1960s and 1970s put such modernism itself on trial. It was deemed to be flawed to the core.

Bureaucracies, which had grown up along with the Industrial Revolution through the 19th century, were an example of modernist control. Supposedly neutral to societal status, run on a set of fixed and formalized rules, bureaucracies were thought to be the equalizing, democratizing answer to the whims of earlier systems of governance (the King, the Church, the Aristocracy). They were also seen as an answer to the needs of a rapidly modernizing and developing West. With the move to mass production, a growing number of things needed to be quantified, measured, kept track of, recorded, stored, and controlled: personnel files, production figures, financial results, stock levels, coal bills, and indeed accident and incident rates (remember Chapter 3). Bureaucracy could do this like nothing else.

A fully rational structure, which allowed no room for emotion, humanity, status, feelings, or dreams, could supposedly produce only rational results. At the close of the 19th century, Weber, of course, had already warned that bureaucracies would, on occasion, do the opposite. The very goal of rationalizing everything, he indicated, would itself produce irrational results. The accidents of the 1970s and 1980s put this on full display. Bureaucratically organizing the gathering of data about and the making of decisions on, safety-critical systems showed lots of flaws and holes. But these accidents were not evidence of a temporary malfunctioning of otherwise well-maintained bureaucratic control. Instead, they were produced by bureaucratic control itself. The launch of the Space Shuttle Challenger in 1986, for example, could be traced to the irrational effects of rational bureaucracy: fixed and formalized rules for gathering launch data and reporting results up the line, a specialization of labor that kept expertise away from decision makers, hierarchical structures that did not allow speaking up across the invisible lines of status, rank, and paycheck.

At a much larger level, the modernist idea of one objective reality had nestled in the so-called 'grand narratives' or 'meta-narratives.' These were the overarching ideologies through which administrations, institutions, and entire societies rallied people, locking them into 'one right way' of thinking about the world and their role in it. That no single ideology or way of thinking about the world could inoculate us against the risks of socio-technical failure became clear in the 1970s. Accidents seemed blind to political ideology, or to the persuasion of the economic rules under which the system had been run. As you will see in Chapter 10, the assumption that governance failures produced by capitalism (leading to the 1979 Three Mile Island nuclear power incident) would not happen under communism came to grief with the 1986 accident at Chernobyl. Grand narratives of capitalism versus communism seemed irrelevant: things went wrong anyway. The very notion of 'safety culture,' as you will see, was driven by this—it was a relatively neutral label under which the International Atomic Energy Agency could diplomatically critique the administrative and regulatory apparatus under both Soviet and capitalist rule.

It was the 1970s suspicion of modernity by a new generation that coalesced into theory. If the organizations and administrations and institutions of modernity were to be mistrusted with the governance of complex socio-technical systems, then what exactly was it that they did not do well? Where did they go wrong? What was it about the workings of bureaucracies, and the people inside them that produced such failures of foresight? Man-made disaster theory could help locate the failings of modern administrative structures. In a sense, of course, these pinpointed what people at the time did not see or do. It may have yet fallen short of helping us understand why it made sense to those people at the time. To be fair, man-made disaster theory was really *out there*. It was a pioneer, riding well out front of other thinking at the time. No wonder that it was not always well equipped to name its discoveries appropriately. It was too early, and thus lacked access to a vocabulary that helped it understand and clarify the psychological and social life of organizational decision makers—their 'sensemaking.' That vocabulary would only be supplied later.

STUDY QUESTIONS

1. Turner's man-made disaster theory introduced social science to the field of safety, and vice versa. How is the central notion of his theory indeed a sociological concern, rather than an engineering one?
2. What are the stages of disaster incubation? Which stage, do you think, allows the most opportunities for discovery and successful intervention?
3. What is the relationship between the notion of 'fundamental surprise' and Turner's stages of disaster incubation?
4. What does a 'failure of foresight' mean? Do you think it is actually possible to recognize a failure of foresight at any other time than in hindsight?
5. What do the creation of local rationality and decision makers' information environment and data have to do with failures of foresight?
6. Can you recognize the possibility for groupthink in man-made disaster theory? If so, how or where?

7. What are Turner's own prescriptions, and those of his successors (e.g., in high reliability theory) for addressing failures of foresight?
8. What are the commonalities and differences between Vaughan's normalization of deviance and Snook's practical drift?
9. What is drift into failure? Is it possible to use event/incident reporting to detect and stop it?

REFERENCES AND FURTHER READING

Amalberti, R. (2001). The paradoxes of almost totally safe transportation systems. *Safety Science, 37*(2–3), 109–126.

Amalberti, R. (2013). *Navigating safety: Necessary compromises and trade-offs—theory and practice*. Heidelberg: Springer.

Ashby, W. R. (1956). *An introduction to cybernetics*. New York, NY: J. Wiley.

CAIB. (2003). *Report volume 1, August 2003*. Washington, DC: Columbia Accident Investigation Board.

Cilliers, P. (1998). *Complexity and postmodernism: Understanding complex systems*. London, UK: Routledge.

Clarke, L., & Perrow, C. (1996). Prosaic organizational failure. *American Behavioral Scientist, 39*(8), 1040–1057.

Dahl, O., & Olsen, E. (2013). Safety compliance on offshore platforms: A multi-sample survey on the role of perceived leadership involvement and work climate. *Safety Science, 54*(1), 17–26.

Dekker, S. W. A. (2002). Reconstructing the human contribution to accidents: The new view of human error and performance. *Journal of Safety Research, 33*(3), 371–385.

Dekker, S. W. A. (2003). Failure to adapt or adaptations that fail: Contrasting models on procedures and safety. *Applied Ergonomics, 34*(3), 233–238.

Dekker, S. W. A. (2005). *Ten questions about human error: A new view of human factors and system safety*. Mahwah, NJ: Lawrence Erlbaum Associates.

Dekker, S. W. A. (2011). *Drift into failure: From hunting broken components to understanding complex systems*. Farnham, UK: Ashgate Publishing Co.

Dekker, S. W. A., Cilliers, P., & Hofmeyr, J. (2011). The complexity of failure: Implications of complexity theory for safety investigations. *Safety Science, 49*(6), 939–945.

Dekker, S. W. A., Mooij, M., & Woods, D. D. (2002). Envisioned practice, enhanced performance: The riddle of future (ATM) systems. *International Journal of Applied Aviation Studies, 2*(1), 23–32.

Dekker, S. W. A., & Woods, D. D. (2009). The high reliability organization perspective. In E. Salas (Ed.), *Human factors in aviation* (pp. 123–146). New York, NY: Wiley.

Dörner, D. (1989). *The logic of failure: Recognizing and avoiding error in complex situations*. Cambridge, MA: Perseus Books.

Downer, J. (2013). Disowning Fukushima: Managing the credibility of nuclear reliability assessment in the wake of disaster. *Regulation & Governance, 7*(4), 1–25.

Feldman, S. P. (2004). The culture of objectivity: Quantification, uncertainty, and the evaluation of risk at NASA. *Human Relations, 57*(6), 691–718.

Feynman, R. P. (1988). *"What do you care what other people think?": Further adventures of a curious character*. New York, NY: Norton.

Fischhoff, B. (1975). Hindsight≠foresight: The effect of outcome knowledge on judgment under uncertainty. *Journal of Experimental Psychology: Human Perception and Performance, 1*(3), 288–299.

Gephart, R. P. (1984). Making sense of organizationally based environmental disasters. *Journal of Management, 10*, 205–225.

Goglia, J. (2015). *Southwest airlines settles whistleblower suit by mechanic disciplined for reporting cracks in 737*. Forbes.

Goldman, L., & Lewis, J. (2009). Corporate manslaughter legislation. *Occupational Health, 61*(2), 12–14.

Hayes, J. (2012). Operator competence and capacity: Lessons from the Montara blowout. *Safety Science, 50*(3), 563–574.

Hollnagel, E. (2014). *Safety I and safety II: The past and future of safety management*. Farnham, UK: Ashgate Publishing Co.

Hopkins, A. (2010). *Failure to learn: The BP Texas City refinery disaster*. Sydney: CCH Australia Limited.

Hoven, M. J. V. (2001). *Moral responsibility and information and communication technology*. Rotterdam, NL: Erasmus University Center for Philosophy of ICT.

Janis, I. L. (1982). *Groupthink*, (2nd ed.). Chicago, IL: Houghton Mifflin.

Jensen, C. (1996). *No downlink: A dramatic narrative about the Challenger accident and our time* (1st ed.). New York, NY: Farrar, Straus, Giroux.

Langewiesche, W. (1998). *Inside the sky: A meditation on flight* (1st ed.). New York, NY: Pantheon Books.

Lanir, Z. (1986). *Fundamental surprise*. Eugene, OR: Decision Research.

LaPorte, T. R., & Consolini, P. M. (1991). Working in practice but not in theory: Theoretical challenges of "High-Reliability Organizations". *Journal of Public Administration Research and Theory, 1*(1), 19–48.

Leveson, N. G. (2012). *Engineering a safer world: Systems thinking applied to safety*. Cambridge, MA: MIT Press.

NTSB. (2002). *Loss of control and impact with Pacific Ocean, Alaska Airlines Flight 261 McDonnell Douglas MD-83, N963AS, about 2.7 miles north of Anacapa Island, California, January 31, 2000 (AAR-02/01)*. Washington, DC: National Transportation Safety Board.

Perrow, C. (1984). *Normal accidents: Living with high-risk technologies*. New York, NY: Basic Books.

Petroski, H. (1985). *To engineer is human: The role of failure in successful design* (1st ed.). New York, NY: St. Martin's Press.

Pidgeon, N. F., & O'Leary, M. (2000). Man-made disasters: Why technology and organizations (sometimes) fail. *Safety Science, 34*(1–3), 15–30.

Rasmussen, J. (1997). Risk management in a dynamic society: A modelling problem. *Safety Science, 27*(2–3), 183–213.

Rasmussen, J., & Svedung, I. (2000). *Proactive risk management in a dynamic society*. Karlstad, Sweden: Swedish Rescue Services Agency.

Reason, J. T. (1990). *Human error*. New York, NY: Cambridge University Press.

Rochlin, G. I. (1999). Safe operation as a social construct. *Ergonomics, 42*(11), 1549–1560.

Rosenthal, U., & Kouzmin, A. (1996). Crisis management and institutional resilience: An editorial comment. *Journal of Contingencies and Crisis Management, 4*(3), 119–124.

Simon, H. A. (1957). *Models of man: Social and rational; mathematical essays on rational human behavior in a social setting*. New York, NY: Wiley.

Snook, S. A. (2000). *Friendly fire: The accidental shootdown of US black hawks over northern Iraq*. Princeton, NJ: Princeton University Press.

Starbuck, W. H., & Milliken, F. J. (1988). Challenger: Fine-tuning the odds until something breaks. *The Journal of Management Studies, 25*(4), 319–341.

Stech, F. J. (1979). *Political and military intention estimation* (N00014-78-0727). Bethesda, MD: Defense Technical Information Center.

Turner, B. A. (1978). *Man-made disasters*. London, UK: Wykeham Publications.

Turner, B. A., & Pidgeon, N. F. (1997). *Man-made disasters* (2nd ed.). Oxford: Butterworth Heinemann.

Vaughan, D. (1996). *The challenger launch decision: Risky technology, culture, and deviance at NASA.* Chicago, IL: University of Chicago Press.

Vaughan, D. (1999). The dark side of organizations: Mistake, misconduct, and disaster. *Annual Review of Sociology, 25*(1), 271–305.

Vicente, K. J. (1999). *Cognitive work analysis: Toward safe, productive, and healthy computer-based work.* Mahwah, NJ: Lawrence Erlbaum Associates.

Weick, K. E. (1987). Organizational culture as a source of high reliability. *California Management Review, 29*(2), 112–128.

Weick, K. E. (1993). The collapse of sensemaking in organizations: The Mann Gulch disaster. *Administrative Science Quarterly, 38*(4), 628–652.

Weick, K. E. (1995). *Sensemaking in organizations.* Thousand Oaks: Sage Publications.

Weick, K. E., & Sutcliffe, K. M. (2001). *Managing the unexpected: Assuring high performance in an age of complexity* (1st ed.). San Francisco, CA: Jossey-Bass.

Weick, K. E., & Sutcliffe, K. M. (2007). *Managing the unexpected: Resilient performance in an age of uncertainty* (2nd ed.). San Francisco, CA: Jossey-Bass.

Weick, K. E., Sutcliffe, K. M., & Obstfeld, D. (1999). Organizing for high reliability: Processes of collective mindfulness. *Research in Organizational Behavior, 21*, 81–124.

Weingart, P. (1991). Large technical systems, real life experiments, and the legitimation trap of technology assessment: The contribution of science and technology to constituting risk perception. In T. R. LaPorte (Ed.), *Social responses to large technial systems: Control or anticipation* (pp. 8–9). Amsterdam: Kluwer.

Westrum, R. (1993). Cultures with requisite imagination. In J. A. Wise, V. D. Hopkin, & P. Stager (Eds.), *Verification and validation of complex systems: Human factors issues* (pp. 401–416). Berlin: Springer-Verlag.

Woods, D. D. (2003). *Creating foresight: How resilience engineering can transform NASA's approach to risky decision making.* Washington, D. C.: US Senate Testimony for the Committee on Commerce, Science and Transportation, John McCain, chair.

Woods, D. D. (2006). How to design a safety organization: Test case for resilience engineering. In E. Hollnagel, D. D. Woods, & N. G. Leveson (Eds.), *Resilience engineering: Concepts and precepts* (pp. 296–306). Aldershot, UK: Ashgate Publishing Co.

Woods, D. D., & Cook, R. I. (2002). Nine steps to move forward from error. *Cognition, Technology & Work, 4*(2), 137–144.

Woods, D. D., Dekker, S. W. A., Cook, R. I., Johannesen, L. J., & Sarter, N. B. (2010). *Behind human error.* Aldershot, UK: Ashgate Publishing Co.

Woods, D. D., Patterson, E. S., & Roth, E. M. (2002). Can we ever escape from data overload? A cognitive systems diagnosis. *Cognition, Technology & Work, 4*(1), 22–36.

Woolfson, C., & Beck, M. (Eds.). (2004). *Corporate responsibility failures in the oil industry.* Amityville, NY: Baywood Publishing Company, Inc.

Wynne, B. (1988). Unruly technology: Practical rules, impractical discourses and public understanding. *Social Studies of Science, 18*(1), 147–167.

8 The 1980s and Onward
Normal Accidents and High Reliability Organizations

Verena Schochlow and Sidney Dekker

KEY POINTS

- Accidents and disasters gained social visibility and political attention in the 1970s and 1980s and put the limits of safety engineering and risk management on display. Questions arose: Was there a limit to the complexities we could handle? Were there things that we perhaps should not do, or build at all?
- Normal Accident Theory (NAT) suggests that some accidents are 'normal'—and thus in a sense predictable—because they can be traced to the interactive complexity and coupling of the systems we design, build, and operate. Interactive complexity and tight coupling built into the very structure of these systems *will* generate certain accidents, NAT says, independent of how much risk management we do.
- Yet there are interactively complex and tightly coupled systems that do not generate accidents, or that have not yet. So are there characteristics of 'high reliability' that can somehow be distilled from the things that these systems are doing? This is what high reliability organizations (HROs), also known as high reliability theory (HRT) suggests.
- A comparison of the two approaches can serve as an introduction to the debate that was triggered by the publication in the early 1990s of the book *The Limits of Safety*. Both theoretical schools have had considerable influence on what has happened in safety science since.

8.1 NORMAL ACCIDENT THEORY

Why are certain accidents so difficult to foresee and prevent? Perrow, a sociology professor at Yale, concluded that their very potential is baked into the way we design, build, and operate the systems that produce them. Perrow wanted to understand why it is so difficult to apply the proven techniques of system safety and reliability to accidents that were not just component failures. He figured that component failure accidents are relatively easy to predict. Specific characteristics of interactive complexity and tight coupling, however, make systems vulnerable to unforeseen escalations and interactions. This can have consequences that can barely be pre-dicted, understood, or contained:

> Minor events are typically component failure accidents. They are caused by a failure of one or two components in a system, and they do not involve any unexpected inter-actions. The potential for component failure accidents can to a considerable extent be identified through standard risk analysis methods. For instance, in a Failure Mode and Effect Analysis (FMECA), the analyst considers one system component at a time, and identifies the possible failure modes. This analysis should capture a fair share of com-ponent failure accidents triggered by hardware failure, provided the analyst is able to cover each component, failure mode and relevant system state.
>
> In contrast to component failure accidents, system accidents involve the unanticipated interaction of several latent and active failures in a complex system. Such accidents are difficult or impossible to anticipate. This is partly because of the combinatorial problem—the number of theoretically possible combinations of three or four component failures is far larger than the number of possible component failures. Moreover, some systems have properties that make it difficult or impossible to predict how failures may interact.
>
> **Rosness et al. (2010, p. 47)**

Complex interactive systems, such as nuclear power plants, can inevitably produce accidents. The one that alerted Perrow (as well as large numbers of other researchers and engineers around the world) was Three Mile Island (TMI). The accident that happened there might have fallen in the original 'incredible' category. But once it had happened, was it so incredible that it *could* happen at all?

> The answer lies in an institutionally deep-rooted confidence that contingency planning is unnecessary for nuclear disasters. For, as Beck puts it, our risk assessment bureaucra-cies have found ways to deny systemic hazards. Policymakers treat the risks of nuclear meltdowns as if they are objectively calculable in way that the risks of terrorism are not, and entirely preventable in a way that floods are not. Thus, they defer to expert assurances that nuclear accidents, unlike floods and acts of terror, will not happen, and so need not be prepared for.
>
> **Downer (2013, p. 3)**

Going back to the structural characteristics (even though he was not an engineer), Perrow suggested that the interactive complexity and tight coupling of a system like nuclear power generation were setting it up for a 'normal accident': an accident that was the 'normal' consequence of how it was designed, built, and operated. His sense was that systems that rank highly on both dimensions (i.e., systems with complex interactions and tight coupling, such as nuclear power), will inevitably produce 'normal' accidents. Even though rare, such accidents were, to Perrow, an inevitable and therefore 'normal,' integral characteristic of those systems.

THE TMI NUCLEAR ACCIDENT

The TMI nuclear power plant, located 10 km south-east of the city Harrisburg in Pennsylvania, consisted of two reactor units. Unit 1 was put into operation in 1974, unit 2 in 1978. Simplified, the nuclear reactors were supposed to produce energy in the following way: the centerpiece of each unit is the nuclear core. Nuclear fission creates heat that is transferred to surrounding water, the primary cooling system. The water is under high pressure so that it does not boil. The heat of this first water cycle is then transferred to a second one, which cools the water in the first water cycle so that it can flow back to the core, ready to absorb heat again. The heating of the water in the physically separated second cycle, the secondary cooling system, produces steam. With the help of steam generators electricity is produced and transferred to the power grid. Whereas the water in the primary system is radioactive, the water in the secondary system is not. Thus, the steam from the secondary system is ejected through the cooling tower.

On March 28, 1979, the accident happened in unit 2. First, a feedwater pump stopped working in the cooling system. However, the reason for that remained uncertain. As described above, the cooling system in TMI unit 2 consisted of a primary and secondary system. The failing pumps were supposed to feed water into the secondary cooling system. With the pumps stopped, no water was supplied to secondary parts of the steam generators, which led to a lack of steam. When the amount of steam dropped, the safety system automatically shut down the steam turbine and electric generator. This happened within 2 s after the pumps failed.

As the flow of the feedwater stopped in the secondary system, the temperature of the water increased. This in turn led to an increase in the temperature of the primary system, causing the water to expand and the water level to rise. Pressure in the system increased. The pressurizer tank is part of the primary system and its function is to maintain the pressure. By keeping a high pressure, it prevents water from boiling in the primary system even though it gets hot. In response to the increasing pressure, the fission process in the core was shut down automatically and the pilot-operated-relieve-valve (PORV) opened in the pressurizer tank. This happened all within 8 s after the initial failure of the pump.

The core gave off a lot less heat with the fission process shutdown, but nevertheless continued to give off residual heat for a long time. That heat would need to be removed to keep the core from overheating as that would damage the vessel and cause radioactive materials to escape. As the fission process had stopped and the PORV was open, the pressure in the primary system dropped. The dropped pressure should have closed the PORV valve at about 13 s after the pump failure. However, there was a mechanical failure with the PORV and it got stuck open. The operators were not informed about this, as the PORV light on their control panel responded to a power signal of the PORV, not the actual status of the PORV. Through this open valve the water of the primary, radioactive system leaked out.

At the same time, three emergency feedwater pumps had automatically turned on to replace the failed pumps in the secondary system. However, the water could not reach the steam generators as a valve was closed on the emergency feedwater lines. These valves were always supposed to be open, except during special tests. It has never been found how or why these valves were closed at that time. The operators noticed that the emergency pumps started. However, they did not know that the valves for feedwater were closed. There were two lights on the extensive control panel that indicated that the emergency feedwater valves were closed, but a maintenance tag covered one; the other one did not catch the attention of the operators.

Water was now leaking out of the primary system through the PORV. The secondary system was losing water as it boiled away and it did not receive new water. Meanwhile, the control room had turned into a hectic environment with many alarms going off and the operators trying to make sense of what was going on.

The open PORV and steam leaving the steam generators caused the water temperature in the primary system to drop. The water level dropped and the operators turned on pumps to add water to the primary system. As water was being pumped in faster than it leaked out through the PORV water levels began to rise again in the primary system and pressurizer, but the pressure remained low. The water levels further increased when the water in the primary system was heating up and expanding again. This was due to the primary system no longer losing heat in the steam generators, as the secondary system had by now boiled dry with the emergency water lines blocked.

Two minutes after the initial pump failure, the automation turned on two high pressure injection (HPI) pumps to add water to the primary system. With this system operating, the pressure was dropping, the water levels were rising and the temperature became constant. Still, the operators were left with the question how the water level could be high, but the pressure low, as they did not know the PORV was open. The operators saw the water from the HPI as a risk, as the extra water that was being provided could make the system go 'solid.' A system goes solid when the entire reactor is filled with water. This could cause damage and the operators were taught to avoid that. The operators saw no indication that the system had too little water and so they decided to reduce the flow of the HPI 2 min after it was turned on and 4 min after the initial pump failure.

As heat rose and pressure dropped, the water in the primary system started to boil and 5 min after the pump failure bubbles and steam appeared. Consequently, water was pushed away from the core to the pressurizer tank, increasing the water levels further. The operators saw this as an indication that there was plenty of water and began to drain water out of the primary system. However, water was leaving from around the core in the form of steam as well and the core was on its way of being uncovered. Uncovering of the core is very dangerous. It causes the core to overheat, as water is needed to transfer the heat away. During this the vessel of the core can melt and radiation can be released.

At 8 min after the pump failure, the operators discovered that no emergency feedwater was reaching the steam generators through the secondary system. The operators opened the blocked valves and water rushed in the secondary system toward the steam generators. How the valves could have been closed and whether this had something to do with the problem was not understood at that time.

The water that was leaking from the open PORV had filled up a drain tank and was now dripping into the sump at the bottom of the reactor. At 11 min an alarm went off that water was high in the sump. Four minutes later a disc on this drain tank burst, leaking water on to both the floor and into the sump. The water in the sump was being pumped away to the auxiliary building. At 39 min during the accident, the operators decided to stop the sump pump. They did not know where the water was coming from and thought it might be radioactive.

At 20 min, instruments measured more radiation than normal. At the same time, the temperature and pressure in the entire building rose quickly because of the escaping steam from the PORV. The operators turned on the cooling equipment and fans of the building to counteract it. They did not realize that it was coming from the open PORV in the primary cooling, as they believed the PORV was closed.

After an hour, four feedwater pumps in the primary system began vibrating heavily. This was because the water had boiled into steam. The operators were worried the violent shaking would damage the pumps or the piping. This would mean a loss of circulation of the water in the primary system. About 14 min later they decided to shut down two of the pumps, and 27 min later they decided to turn off the remaining pumps as well. This means that there was a loss of circulation and cooling around the core. After 2 h, radiation alarms went off in the containment building. As water left through the PORV and little water was being added, the top of the core had been uncovered. The fuel rods had been damaged and radioactive gases had escaped. In this process, hydrogen had formed and seeped into the containment building.

By now many of the TMI officials had gathered, technicians and experts were contacted over the phone to assist in solving the problem. In the group assessment of the situation, the blocked valve between the pressurizer and PORV was checked and closed. This blocked the path to the open PORV, stopping the leak and increasing pressure again. The damage to the core, however, continued, as the core was still partially uncovered. High levels of radiation were found over the entire building.

Unknown to the operators, at about 3 h into the accident approximately two-thirds of the core was uncovered. The operators turned on one pump for the primary system, but shut it down again after 19 min because it was shaking violently. Around this time the situation was declared a site emergency. Additional tasks were given to people to monitor radiation on and off site and authorities were informed.

Fifteen minutes later the radiation levels in the auxiliary building became too high and the personnel were evacuated. Alarms were sounded 5 min later indicating high levels of radiation in the containment building as well. The operators turned the HPI back on for 18 min. Alarms kept going and the station manager declared a general emergency, as it was now assumed that there might be danger for the public around the TMI plant. Again more authorities were informed about the situation.

At 4 h into the accident, based on the radiation levels he was reading, the station manager became certain that there was damage to the fuel rods. How the fuel rods had been damaged was uncertain to him.

At 4.5 h, the HPI was turned on at high rate. It would take another 2 h before the core was covered again. At 5.15 h, it was found that radioactive particles had escaped the plant. At this time, the control room tried to create natural circulation in the primary system. Natural circulation is a flow without the usage of pumps. The idea was to heat water in the core to create movement to the steam generators. This plan, however, failed as there were steam gas bubbles blocking the pipes and the primary system was not filled with enough water anymore.

The core had been covered again at 6 h and 30 min. However, as the pressure was still high the operators tried to depressurize the primary system. In this process, water of the primary system was lost and the core became uncovered again. Attempts to depressurize the system continued for 3.5 h. At about 10 h there was a thud noise in the control room. It was unclear what this was and interpreted as something minor in the control room, like the slamming of a ventilation damper. However, in hindsight of the accident it was found that this was an explosion of the hydrogen that had escaped.

The following days many experts tried to figure out what had happened and what risks existed for the public. Experts disagreed about the state of the plant and the risk of another hydrogen explosion. During these days many conflicting messages were given to the public. About 140,000 people left the area around the plant soon after the accident, of whom most returned in the following weeks. Later, it became clear that in the accident radioactive materials had been released into the surrounding environment although there are different studies and opinions on the effects of the release. As the core in unit 2 had been damaged, the unit could not be put in operation again. However, unit 1 is still in operation today.

8.1.1 Linear versus Complex Interactions

A system's complexity, Perrow suggested, is related to its size and the number of subunits in it. It is *interactively* complex when it exhibits:

- unfamiliar sequences;
- unplanned or unanticipated propagations;
- invisible or not immediately comprehensible interactions;
- vulnerability to common mode errors (this means that different parts or subsystems fail at or around the same time, even though they have functionally nothing to do with each other. Their failure goes back to a 'common mode' or shared feature hidden more deeply in the system somewhere).

The 'unplanned' and 'unfamiliar' nature of some of these interactive complexity problems is of course a function not just of the system, but also of the operators' knowledge and understanding of it. This has always been one of the points of critique of Perrow's classification: neither interactive complexity nor tight coupling can be separated out as characteristics of purely the (engineered) system alone. In addition, the two dimensions (complexity and coupling) may not be so separate or independent either. Tight spacing of equipment, for instance, can give rise to common mode failures and thus to interactive complexity. But it also shows up in tight coupling (as you'll see below), since it affects the ability to steer away from a rapid escalation of problems. As Kates put it:

> The absence of clear criteria for measuring complexity and coupling make his many examples seem anecdotal, inconsistent, and subjective, limiting its usage in the growing effort at hazard taxonomy as a way of ordering and simplifying the hazard domain.

> *(1986, pp. 121–122)*

Perrow contrasts interactively complex systems against what he called linear systems (see Table 8.1). Linear systems have simple production sequences. They features clearly separable phases of production and spatial segregation. If a component failure occurs, it can be quickly tracked down, isolated, and removed without much of a disturbance to the rest of the system. Typically, there is a good understanding of the production process and direct information is available. An example of a linear system is serial production, such as manufacturing. Complex systems, on the other hand, are characterized by proximity, interconnected subsystems, and multiple interacting components. Removal of a failed component is difficult, since it is likely to be intricately linked to other components and processes. Operators tend to be specialized in their roles and knowledge, making it less likely that they are able to detect or correctly diagnose an interaction before it cascades through the system and leads to an accident.

TABLE 8.1

The Difference between Complex and Linear Systems According to Perrow (1984)

Complex Systems	Linear Systems
Tight spacing of equipment	Equipment spread out
Proximate production processes	Segregated production processes
Common mode connections with parts that are not in the production sequence	Order of sequence changeable
Personnel specialization limits operators awareness of interdependencies and understanding of complex interactions	Less personnel specialization
Many control parameters with unanticipated interactions	Parameters are few and direct
Indirect information on the process and anomalies (inference is necessary)	Direct observation and information is available

COFFEE MAKER CAUSES AIRLINER DEPRESSURIZATION

There is nothing inherently difficult about putting a coffee maker aboard a large airplane, yet (it is) a simple part in a complexly interactive system. On a flight during a very cold winter, passengers had been told there would be no coffee because the drinking water was frozen. Then the flight engineer noticed he could not control cabin pressure. It was later found that frozen drinking water had cracked the water tank, heat from ducts to the tail section then melted the ice in the tank, and because of the crack in the tank, and the pressure in it, the newly melted water near the heat source sprayed out. It landed on an outflow valve in the pressurizing system, which allows excess cabin pressure to vent to the outside. The water, which had gone from water to ice and then back to water, now turned to ice again because the outside of the air valve is in contact with −50° air outside the plane at 31,000 feet. Ice on the valve built up pressure in the valve and caused it to leak, and the leak made it difficult to maintain proper cabin pressure (Perrow, 1984, p. 135).

8.1.2 LOOSE VERSUS TIGHT COUPLING

Coupling is Perrow's second structural characteristic. It affects a system's ability to prevent, or recover from, an escalation of problems.

> Tightly coupled systems are characterised by the absence of "natural" buffers. A change in one component will lead to a rapid and strong change in related components. This implies that disturbances propagate rapidly throughout the system, and there is little opportunity for containing disturbances through improvisation. Tight couplings are sometimes accepted as the price for increased efficiency. For instance, Just-in-time production allows companies to cut inventory costs but makes them more vulnerable if a link in the production chain breaks down. In other cases, tight couplings may be the consequence of restrictions on space and weight. For instance, the technical systems have to be packed more tightly on an offshore platform than on a refinery, and this may make it more challenging to keep fires and explosions from propagating or escalating.
>
> **Rosness et al. (2010, p. 48)**

Tightly coupled systems feature:

- Time-dependent processes;
- Invariant production order;
- Little slack;
- Safety devices, buffers, and redundancies are limited to those that were designed in beforehand.

Time dependence means that delays, extensions, and storage are all impossible. The pace of activities is driven in large part by the environment or by the nature of the process itself (e.g., a chemical reaction). This has consequences for the cognitive activities of those operating the system. They need to stay mentally ahead of the process, and any fault diagnosis or trouble-shooting needs to go hand with maintaining process integrity. You cannot just park the system for a moment while you figure out what is wrong with it (Woods & Patterson, 2001).

Invariant production order means that there is only one way to produce an outcome. Each step needs to be taken in sequence. In loosely coupled systems, in contrast, the outcome can be produced in a number of ways, and parts can be added later (and even from elsewhere). An example of a loosely coupled system that Perrow uses is university education: courses can be added later, and only in some cases is the order invariant (e.g., in a sequence of calculus courses). Courses can also be brought in from outside the university, by requesting credit for prior learning elsewhere.

The lack of slack means that quantities used for the production process need to be highly precise. Wasted supplies can overload or disrupt the process (such as the presence of too much catalyst in a chemical reaction, as you might recall from Chapter 6). In a tightly coupled system, one kind of resource cannot be substituted for another, and failed equipment typically entails a shutdown, as substitution with other equipment is not possible. Because of the time dependency of the process, invariant order and lack of slack, tightly coupled systems offer little or no opportunity to improvise or jerry-rig a solution once things have started going wrong. Loosely coupled systems, in contrast, do allow for unplanned responses to individual failures.

In tightly coupled systems (see Table 8.2) "processes happen very fast and can't be turned off, the failed parts cannot be isolated from other parts, or there is no other way to keep the production going safely. Then recovery from the initial disturbance is not possible; it will spread quickly and irretrievably for at least some time" (Perrow, 1999, pp. 4–5).

Perrow created a grid that set one dimension (complexity) against the other (coupling), so as to get a visual impression of the potential to create 'normal' accidents (see Figure 8.1).

TABLE 8.2
The Difference between Tightly and Loosely Coupled Systems (Perrow, 1984)

Tight Coupling	Loose Coupling
Delays in processing not possible	Processing delays possible
Invariant sequences	Order of sequences changeable
Only one method to achieve the goal	Alternative methods available
Little slack possible in supplies, equipment, personnel	Slack in resources available
Buffers and redundancies have to be designed in deliberately	Buffers and redundancies can be found or created during operations
Substitutions of supplies, equipment, and personnel are impossible or limited	Substitutions are readily available

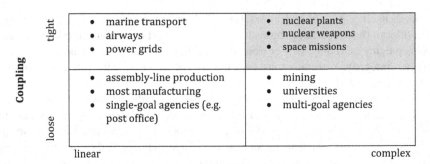

FIGURE 8.1 Perrow's (1984) interaction/coupling grid.

Through his analysis, Perrow became preoccupied with systems that feature both complex interactions and tight coupling. These of course include nuclear plants and space missions (marked in the gray quadrant in Figure 8.1). In such interactively complex and tightly coupled systems, failures can interact in multiple unexpected ways, which human operators can neither anticipate nor comprehend during the escalation of events (Woods, 1988). That is what also distinguishes a component failure accident from a normal or system accident. In a component-failure accident, the problem happens in components that are separable or recognizable from the rest, even if they are connected in some (anticipated) sequence. Normal or system accidents, in contrast, are characterized by multiple unexpected interactions.

In complexly interactive and tightly coupled systems, the 'size' of the failure is also not proportional to its outcome (which is of course consistent with complexity theory: small changes can have huge consequences). Not only catastrophic failures can lead to catastrophic outcomes, but minor ones can too. This is due to system characteristics. Perrow (1999) pointed at a shirt caught on the handle of a circuit breaker, which caused a 4-day shutdown; or a dropped light bulb, which led to a rapid and dangerous cooling of the core in nuclear power plant. Consistent with the principles of human factors (as you will recall from Chapter 5), Perrow refused to see the human operator as the cause of these sorts of accidents. Accidents are caused by the system as a whole; not by its components. As he put it, when a human operator is "confronted by unexpected and usually mysterious interactions among failures, saying that he or she should have zigged instead of zagged is possible only after the fact" (1999, p. 9).

8.1.3 THE PARADOX OF CENTRALIZED DECENTRALIZATION

Tight coupling, said Perrow, requires centralization. That is, operators need to be able to rely on preplanned or pre-engineered anomaly responses because of the rapidity of escalation. There is no time to figure things out 'on the fly' and improvise a response. Centralization in this context means that operators are told what to do through a response that all of them have had to learn; a response that was pushed out from the center to the operational frontline(s). The problem, though, is that complex interactions require decentralized responses. Specialized, localized responses may well be necessary to deal with complex situations. Rosness and colleagues sum up the dilemma:

1. With a complex system, you should try to reduce the degree of interactive complexity;
2. With a tightly coupled system, you should seek ways to loosen the couplings;
3. If you have to live with a high degree of interactive complexity, you should build a decentralized organization;
4. If you have to live with tight couplings, you should centralize your organization;
5. If your system has catastrophic potential, and you are not able to apply any of the above strategies, then you should discard your system (Rosness et al., 2010, p. 49).

This is the paradox: having to decentralize centralized responses. Procedures and relying on human operators' discretion to apply them well or appropriately, is the only way to do this (Dekker, 2003; Woods & Shattuck, 2000). And even then, Perrow (1999) argued that:

complex but loosely coupled systems are best decentralized; linear and tightly coupled systems are best centralized; linear and loosely coupled systems can be either; but complex and tightly coupled systems can be neither—the requirements for handling failures in these systems are contradictory.

(p. 331)

Loosely coupled systems with linear interactions have to deal neither with unforeseeable interactions among failures nor with limited time to reflect and check on failures. In that way, either centralization or decentralization can work for these systems. In loosely coupled systems with complex interactions, operators who experience failures first need to have the freedom to use their resources in order to diagnose and recover from them. But loosely coupled systems do not require immediate action. So even if the failure is unexpected, people have time to reflect on them and improvise or experiment their way to a solution. To do so, they also need the authority to make changes—which of course means decentralization. In systems with linear interactions and tight coupling, failures are visible and interact in expected ways. The process is well understood. This enables people to use programmed responses provided by some centralized authority in the case of a failure. Due to tight coupling, these must be applied immediately and without question. The dilemma occurs in tightly coupled systems with interactive complexity. They demand centralization to handle tight coupling but decentralization to handle interactive complexity. A mixture of these two might be difficult but possible for some systems, though Perrow notably excluded his love-to-hate system at the upper end of both scales: nuclear power. His recommendations are not very surprising then:

1. "The first would be systems that are hopeless and should be abandoned because the inevitable risks outweigh any reasonable benefits" (Perrow, 1999, p. 304). In this category he put nuclear weapons and nuclear power;
2. "...and others, which we cannot abandon because we have built much of our society around them, should be modified" (p. 4). In this category he put chemical plants and airliners, for example. He suggested to either restrict certain industries or to reduce interactive complexity and tight coupling by better organization and so-called 'technological fixes.' As a successful example he mentioned air traffic control (designed in a relatively linear and therefore less complex way).

THE FUKUSHIMA DAIICHI ACCIDENT

The Fukushima Daiichi plant (Fukushima I) consists of six units of which three were in operation at the time of the accident. Nevertheless, two more units held nuclear material. The plant is located at the eastern Japanese coast in an active seismic rim. During the afternoon of the March 11, 2011, the plant was affected by a severe earthquake with a magnitude of 9.0 on the Richter scale. The epicenter was located about 160 km east of the plant in the Pacific Ocean. The earthquake directly led to an emergency shutdown of the active units 1–3. Similar to the TMI Accident, the cores needed to be cooled even after a shutdown due to the residual heat to prevent them from melting and releasing radioactivity. But the earthquake also caused a 15 m (50 feet) high Tsunami wave that flooded and destroyed the plant's emergency generators. Cooling of the cores was now only partially possible with the help of car batteries and other sources of electricity. Eventually, the cooling of the cores was not sufficient and a meltdown of the cores in units 1–3 ensued. Radioactivity was released into the environment and hydrogen built up, causing several explosions and further damage to the plant during the following days. In total, nearly 500,000 people in the surrounding region had to be evacuated. Fire fighters still tried to cool the plant by pumping water in the units, but due to climbing levels of radioactivity most personnel were evacuated on March 15th. It has been estimated that the long-term human fatalities including the increased cancer mortality will reach over 10,000.

This accident shows some of the characteristics discussed in Perrow's NAT. First, of course, the accident took place in a nuclear power plant—a tightly coupled and interactively complex system vulnerable to 'normal' accidents. Component failures almost always play a role and can trigger a normal accident. Perrow (1999) gave a similar example describing component failure accidents:

> If a valve fails there is good chance that the pump will overheat and fail, and if this happens, the boiler is quite likely to overheat because the coolant is insufficient. Designers know this, and so do operators—though they may not be able to prevent it, or intervene in the series of events
>
> *(p. 71)*

In Fukushima, the failure of the emergency generators to provide cooling of the cores after the shutdown of the reactor, could be considered a component failure of sorts. Of course, the shutdown of a reactor is technically not regarded as a failure: this is a planned and designed reaction of the plant to seismic activity. Workers were aware of the possibility of earthquakes, and also of possible Tsunami waves. What was not anticipated was the magnitude: the earthquake's intensity and the height of the wave were beyond the design limit of the plant. It was not anticipated that the wave would flood the emergency backup generators and thereby cut off the electricity. The sequence of events, however, was actually anticipated and comprehensible:

- external power was lost due to a short-outs caused by the earthquake;
- water cut off the emergency backup generators providing electricity;
- due to the lack of electricity the emergency cooling system stopped working;
- the cores overheated due to the residual heat;
- the cores started melting, releasing radioactivity and hydrogen;
- the escaping hydrogen exploded, causing further damage and release of radioactivity.

Concentrating systems, in this case combining six reactor units in one plant, increased complexity and coupling. As an example Downer (2013) mentioned that the hydrogen explosion in one reactor created a "radioactive hot zone" around the others, which hindered workers in their efforts to prevent further damage. After the flooding of the emergency generators, workers in Fukushima soon became aware of the problem and tried to prevent the series of events above. This can be seen in their interventions such as using car batteries to provide electricity for the emergency cooling system and later trying to cool the reactors with seawater. Nevertheless, the workers were not able to stop the chain of events with the anticipated outcome, the meltdown of the cores. Tight couplings and complex interactions existed, but not enough to amount to a 'normal' accident. Yet Fukushima showed that accidents in certain systems are inevitable regardless of the efforts undertaken to prevent or mitigate them. For Perrow, Fukushima was yet another example of the catastrophic potential of nuclear power, and he renewed his calls to try to abandon it (Perrow, 2011, 2013).

NAT was provocative and intriguing enough to encourage considerable further research (Rosness, Guttormsen, Steiro, Tinmannsvik, & Herrera, 2004):

> Perrow drew attention to the possibility that some technologies may force us to adopt organisational structures and practices that are incompatible with central values in western democracies. In order to attain the degree of centralisation that is required to handle some tightly coupled systems, we may be forced to create work environments characterised by harsh discipline and very little autonomy.
>
> Several objections have been raised against Normal Accident Theory:
>
> - The notions of interactive complexity and tight coupling are so vague that it is difficult or impossible to subject the theory to empirical tests;
> - It is difficult to derive a simple and effective prescription for assessing or monitoring major accident risk from Normal Accident theory, because it is difficult to measure or monitor such attributes as interactive complexity or decentralisation;
> - Analysis of recent major accidents suggests that most accidents result from other problems than a mismatch between complexity/coupling and degree of centralisation;
> - Some critics find the suggestion that some technologies should be discarded too pessimistic, too fatalistic, or politically unacceptable;
> - The assertion that an organisation cannot be centralised and decentralised at the same time sounds like a tautology. However, this assertion has been challenged by researchers that study so-called High Reliability Organisations.
>
> **Rosness et al. (2010, p. 52)**

As researchers got further into trying to apply NAT, they realized that the theory explains only a limited set of accidents, namely 'normal' accidents. Perrow himself admitted that TMI was "the only accident in my list that qualifies" (Perrow, 2011, p. 51). And it was not a full-blown accident at that time. Many, though, have seen continued value in viewing the concepts of tight coupling and interactive complexity as independent from one another. Rijpma (2003), relied on it to explain the rapid capsize of the Herald of Free Enterprise in 1987. Wolf (2002) also used the concept to describe how cost-cutting and the reduction of available resources leads to tight coupling and therefore increases the risk for accidents in petrochemical plants. Others see strong features of a normal accident in the Fukushima nuclear disaster in Japan. For example:

> A conspicuous example is the routine clustering of several nuclear reactors in a single facility, a choice that offers significant economic and political benefits but creates conditions where the failure of one unit can propagate to others (The Fukushima plant, for instance, was a cluster of six reactors—something that workers must have rued as the fallout from the first explosion created a deadly radioactive hot zone around its neighbors, ultimately thwarting efforts to contain the disaster).
>
> **Downer (2013, p. 3)**

The sense that the risk of a 'normal' accident is structurally linked to features of a system's design has been empowering. It can help redirect attention away from procedural or human factors interventions and toward safety-through-design.

8.2 HIGH RELIABILITY ORGANIZATIONS

High reliability organization (HRO) is not one theory described by one author. It is rather a set of common or affiliated principles that were developed over time and by different authors. While the researchers behind these ideas share a common way to look at highly complex and hazardous organizations, different authors have added ideas and perceptions, which has led to multiple HRO models or components of the theory. Of course that has been accompanied by slightly different methods, measures of safety, or assumptions about the human operator.

The roots of HRO go back to a group of researchers from the University of California, the so-called Berkeley group, among them Todd La Porte, Karlene Roberts, and Gene Rochlin. Karl Weick, a social and an organizational psychologist, also became involved in HRO research and added ideas. In 1995, Roberts started working with another group of safety scientists at the California State University in San Bernadino. This San Bernadino group in turn developed further HRO methods and applied HRO to industries where failures are business failures such as in the financial sector. Recent developments include the application of Resilience Engineering (RE) principles in HRO.

The following section describes the initial research of La Porte and his colleagues. Because the authors' views and theories developed and changed over time, the next section mainly refers to the initial publications as these ideas were subject to the debate between NAT and HRO that is explained in Section 8.3. The more recent approaches including the ideas of the San Bernadino group and their take on resilience engineering will be discussed in Section 8.4.

8.2.1 THE BEGINNINGS OF HRO: LA PORTE, ROBERTS, AND ROCHLIN

Todd La Porte, Karlene Roberts, and Gene Rochlin, as members of the original Berkeley group, started their research with a project "concerned with the design and management of hazardous organizations that achieve extremely high levels of reliable and safe operations" (Roberts, 1989, p. 111) that involved, for example, the U.S. Air Traffic Control system and U.S. Navy nuclear aircraft carriers (Rochlin, LaPorte, & Roberts, 1987). Roberts noted that earlier research mainly focused on damage control—assuming that an accident has happened or will happen. The group aimed to answer the question whether there were high-risk organizations operating nearly error-free. And if so, what were the characteristics of these organizations? What did they do well?

Roberts clustered organizations on two dimensions that were very different from Perrow's: technological risk and reliability. Organizations were labeled HROs when they repeatedly could have failed with catastrophic results—but did not (Roberts, 1990). This was measured not in terms of absolute safety but with the help of relative measures of performance, such as collisions related to number of movements

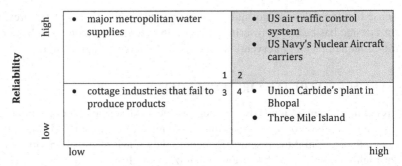

FIGURE 8.2   Roberts' (1989) reliability/technological risk chart.

or outages related to hours of service. The HROs studied were also characterized by complexity, tight coupling, and highly interconnected technologies (LaPorte & Consolini, 1991). Safety and reliability were the organizations' prioritized goals. Figure 8.2 provides examples for each category as identified by Roberts (1989):

NUCLEAR AIRCRAFT CARRIERS

At the beginning of their research, Roberts and her colleagues engaged in extensive field work on selected HROs. Their aim was to understand how such complex and highly interactive organizations achieved outstanding and unexpected safety records. How was that possible in practice, when theory predicted elsewise?

Therefore, they engaged in quarterly workshops with the selected organizations and also spent a lot of time within the organizations to familiarize themselves with the organization, its' "language," and tasks. Apart from observation, the researchers also collected archival and survey data. Among the first organizations intensely studied for a period of 3 years were two U.S. Navy Nuclear Aircraft carriers in peacetime operations.

Aircraft carriers are warships designed for the takeoff and landing of military aircraft. Their purpose is to provide a flexible base for military operations in areas where the military does not operate on own sites. Each studied carrier was manned with 6,000 persons and commanded by 9squadrons of aircraft. An airfield measuring about 1,100 feet in length and 250 feet in width was located on the upper deck. The ship was 24 stories high and carried ship fuel for 15 years.

Among the possible catastrophic accidents on an aircraft carrier, Roberts (1990) listed nuclear reactor accidents, aviation-related accidents, flooding, fire, grounding, collision, and fuel or weapon explosions.

Next to this catastrophic potential were the aircraft carriers' complexity and tight coupling: Different technologies from different stages of advancement were used and interacted with each other. The ship was operated on a 24-h schedule. On the flight deck up to 200 persons worked at the same time, loading, fueling, and maintaining aircraft starting and landing at intervals as short as 48 s. Aircraft were parked in very close proximity and needed to take-off and land in a very limited space.

One of the measures for reliability in this case was the so-called "crunch-rate." A crunch happens when two aircraft touch while being in motion on the flight deck, even if no damage is done. The crunch rate on the two air-craft carriers was one in 7,000 and one in 10,000 moves respectively, which to Roberts and her colleagues seemed an unexpectedly low number given the circumstances. They considered it a highly reliable operation (Roberts, 1989).

The following statement by an officer gives an impression of the work on an aircraft carrier and highlights the complexity as well as the potential hazards (Rochlin et al., 1987):

> So you want to understand an aircraft carrier? Well, just imagine that it's a busy day, and you shrink San Francisco Airport to only one short runway and one ramp and gate. Make planes take off and land at the same time, at half the pres-ent time interval, rock the runway from side to side, and require that everyone who leaves in the morning returns that same day. Make sure the equipment is so close to the edge of the envelope that it's fragile. Then turn off the radar to avoid detection, impose strict controls on radios, fuel the aircraft in place with their engines running, put an enemy in the air, and scatter live bombs and rockets around. Now wet the whole thing down with salt water and oil, and man it with 20-year-olds, half of whom have never seen an airplane close-up. Oh, and by the way, try not to kill anyone
>
> *(p. 77)*

How was it possible to have safe operations on a complex and tightly coupled carrier with catastrophic potential? In what way did the organization differ from other non-HROs?

The HRO researchers were fascinated and "awed" by the observed per-formance and the human and organizational effort and costs involved (La Porte & Rochlin, 1994). What they found were "layers and layers of activ-ities that allow changes in reaching goals" (Roberts, 1990, p. 167), flexible reactions to demands and circumstances. They observed intense training of personnel, redundancy for all processes and adaptive decision-making. Every person aboard seemed to experience a high sense of responsibility and every person had the possibility to intervene in safety-critical situations. Safety was of utmost priority. This was also expressed in the order "not to break the rules unless safety is at stake" (Roberts, 1990, p. 168). After the field study, the HRO group discussed and organized their findings in certain fields of research that are described in the following section.

The initial aim of HRO was not to give recommendations, but to explore HROs in regard to multiple areas of interest: "'Making things work better' was not the reason for our being there, nor the purpose of our work" (Rochlin, 1996, p. 56). Instead, the idea was to describe what makes HROs so reliable and to stimulate further research rather than to prescribe a set of rules. La Porte, Roberts, and Rochlin did not focus on human performance on an individual level; their interest was in human performance at a group or organizational level (LaPorte & Consolini, 1991). In contrast to other approaches, the HRO research did not take accidents as a dependent variable. Instead, their dependent variable was the nearly error-free operation. From that point of view, the researchers were not so much interested in what caused accidents, but in what contributed to high reliability and safety levels (Roberts, 1990).

The preliminary findings revealed some characteristics present in all or most of the studied HROs (La Porte, 1996; Roberts, 1989; Roberts, 1990):

- Organizational structure and rules: The organizational structures of HROs seemed to be marked by flexibility and redundancy. This was shown in parallel and overlapping processes, backups, cross-checks, and skills redundancy (e.g., through job rotation). Redundancy was seen as a measure to recover from errors and ensure reliability despite the shortcomings of individuals or single components.
- The structures were described in terms of "nested authority," which refers to the adaptive change from a hierarchic organization during routine operations to flatter decision patterns which deferred to experience and expertise in tightly coupled, demanding situations. This was interpreted as one way to handle the coexisting interdependence and complexity in HROs. Another way suggested by Roberts and Gargano (1989) was the existence of one unit with the sole purpose to cope with interdependencies aboard aircraft carriers.
- Operational decision-making and communication: The flexible adaptation of organizational structures also affected decision-making and communication patterns. The shift in structure went along with a shift in authority. If called for by circumstances, decision-making got decentralized and passed to the people with the information and experience at hand to the locus of the problem. This was also seen as one way to cope with the tension between quick and accurate decision-making, or between efficiency and reliability. Several channels of communication seemed to be used constantly to inform others, to maintain the "big picture" and to keep the team integrated. Communication across these channels served as a critical safety function (Rochlin, 1999).
- The culture of high reliability: This culture of safety is viewed by Rochlin (1999, p. 1549) as "dynamic, intersubjectively constructed belief in the possibility of continued operational safety, instantiated by experience with anticipation of events that could have led to serious errors, and complemented by the continuing expectation of future surprise." People in an HRO showed high levels of personal engagement and a "sense of mission" (La Porte, 1996). Compared to non-HROs the willingness to report errors and potential failures seemed to be untypically high in HROs. People reporting were not blamed but rather rewarded for their initiative.

- The adaptation to technology: Roberts (1989, p. 121) noted a "remarkable resilience" on the one hand and a resistance of decision makers to adopting new technologies on the aircraft carriers on the other hand. The resulting mix of old and new technologies ensured operations even without electricity and continuing practice of operator skills.

Based on the preliminary characteristics and especially the initial exploratory work on nuclear aircraft carriers, Roberts (1990) described the following strategies used by this HRO to counter high interactive complexity and tight coupling. Complexity was countered with:

- continuous training
- job design strategies to keep function separate
- main direct information sources

Tight coupling was countered with:

- redundancy
- hierarchical differentiation
- bargaining

Simultaneous complexity and tight coupling at the same time were countered with:

- redundancy
- accountability
- responsibility
- "culture" of reliability

Nevertheless, La Porte (1996) concluded that the characteristics of HROs found in the case studies were not simply implementable in other organizations which intended to become HROs. He stated that these characteristics might be necessary but not sufficient. This again shows the more descriptive character of the initial HRO project.

8.2.2   Weick and Sutcliffe's Concept of Mindfulness

Mindfulness was later added to the HRO lexicon as a unifying concept, denoting the "capability to induce a rich awareness of discriminatory detail and a capacity for action" (Weick, Sutcliffe, & Obstfeld, 1999, p. 82). Mindfulness putatively increases the capability of people in effective HROs to detect and handle unexpected events. This in turn leads to the reliability of the organization, hence its nearly error-free performance in a demanding social and political environment and despite the huge potential for errors in its complex processes and technologies. Weick regarded reliability as the ability to perform in the same way despite changing conditions for operation. Thus, the operators' activity is required to vary according to the circumstances whereas the cognitive processes making sense of the activity remain stable (Weick et al., 1999).

How can mindfulness be created? Weick suggested five cognitive processes of mindful organizing that seemed consistently present in effective HROs (see also Figure 8.3):

1. *Preoccupation with failure*: This process describes that in effective HROs people have an "apparent ongoing focus on failure" (Weick, Sutcliffe, & Obstfeld, 2008, p. 38) instead of success. In this sense, also near-failures, such as near misses, are seen as a danger to safety. Furthermore, mindfulness requires interpretative work of so-called weak signals. Weak signals were defined as signals that at the time of observation had no apparent clear or direct link to a potential danger, or as events that were judged to have happened in such rare, unlikely conditions that they would not happen again (Vaughan, 2002). Effective HROs tend to generalize failures or weak signals, instead of localizing and reducing safety problems. Success is not interpreted as demonstrating competence, as this would lead to inertia, to inattention, and to errors that go undetected. Instead people in effective HROs are even assuming that current successful operation makes future success less likely (Weick et al., 2008, p. 41).

2. *Reluctance to simplify* interpretations: Weick et al. (2008) identified simplification as a potential danger to HROs since it leads to a negligence of data, reduction of precautions, and an increased likelihood of undesired surprises. Effective HROs undertake effort to limit simplifications. That is done, for example, through processes of renewal and revision, job rotation, cross-checks, or selecting new employees with nontypical prior experience.

3. *Sensitivity to operations*: This process refers to having "the integrated big picture of operations in the moment" (Weick et al., 2008, p. 43). This "big picture" is difficult to achieve and maintain. It, therefore, requires active effort. Because of the individual's limited range and varying focus, the

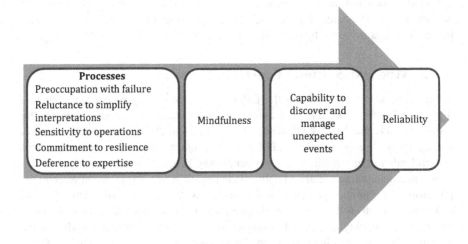

FIGURE 8.3 The relationship between mindfulness and reliability. (Drawing by authors.)

"big picture" is a shared accomplishment. Ways to achieve sensitivity to operations include collective story building, shared mental representations, and knowledge of system parameters.

4. *Commitment to resilience*: Weick et al. (2008, p. 46) pointed out that "people deal with surprises, not only by anticipation that weeds them out in advance, but also by resilience that responds to them as they occur." This cognitive process is preventative in that operators in effective HROs do not wait for errors to happen, but prepare for unexpected events beforehand. Accordingly, "they pay attention both to error-prevention and to error-containment" (Weick et al., 2008, p. 47). This is visible in the support of improvisation, the recombination of actions, and the parallel belief and doubt in past experiences in effective HROs.

5. *Deference to expertise*: Another feature of effective HROs is that organizations that are normally structured in a hierarchical way are able to subordinate hierarchical rank to expertise, if demanded. Depending on the situation, the organization allows an adaptive loosening of filters so that operators at the bottom of the hierarchical structure can rise to the top level of decision-making temporarily. This is also referred to as underspecification of structures.

Let's briefly pause at the last point there. Deference to operational or engineering expertise is generally deemed critical for maintaining safety—despite the vulnerability and limits of expertise (as you will recall from Chapter 2). HROs have been acclaimed for their sensitivity to operations and deference to expertise. They are attentive to their operational front-end, the sharp end. This is where the "real" work gets done, where workers are in direct contact with the organization's safety-critical processes. HROs push decision-making down and around, creating a recognizable "pattern of decisions 'migrating' to expertise" (Weick & Sutcliffe, 2007, p. 16). Expert practitioners might even need to be engaged for decisions that have, on the surface, little connection to safety-critical operations or system design. "Budgets," for example, "are often insensitive to operations" (Weick & Sutcliffe, 2007, p. 13), but budgetary decisions can in the long run very well have operational or safety consequences.

In hindsight, not deferring to expertise is often seen as a major safety shortcoming. Prior to the Texas City refinery explosion in 2005, for example, BP had eliminated several thousand U.S. jobs and outsourced refining technology work. Many experienced engineers left (J. A. Baker, 2007). And at NASA, the appointment of Sean O'Keefe (who had been Deputy Director of the White House Office of Management and Budget) to lead the organization was a clear signal from the new Bush administration that the focus should be on management and finances (CAIB, 2003). This affirmed and accelerated a trend that had been set years before. NASA had vastly reduced its in-house safety-related technical expertise in the 1990s. The Apollo-era research and development culture had once been prized for its deference to technical engineering expertise (Mindell, 2008; Murray & Cox, 1989). But this had become overridden by bureaucratic accountability—managing upwards with an allegiance to hierarchy, procedure, and following the chain of command (Feynman, 1988; Vaughan, 1996). Even under those conditions, expertise sometimes emerges spontaneously outside existing organizational structures, when knowledgeable people self-organize into ad-hoc networks to solve problems (Murray & Cox, 1989; Rochlin et al., 1987). But this can ultimately work effectively only in organizations that value expertise and experience more than rank and hierarchy, particularly when novel or unexpected situations arise (Rochlin, 1999; Schwenk & Cosier, 1980).

Subsequent investigations traced the Challenger and Columbia Space Shuttle accidents in 1986 and 2003 back to this shift away from expertise, and led to calls for organizations in general to take engineering and operational expertise more seriously. This appeal has become well-established in the literature on HROs and resilience (see Chapter 12). Contributing to the Columbia accident, the investigation report said that "managers failed to avail themselves of the wide range of expertise and opinion necessary." Their management techniques "kept at bay both engineering concerns and dissenting views, and ultimately helped create 'blind spots' that prevented them from seeing the danger the foam strike posed" (CAIB, 2003, p. 170). After the accident, NASA was told it needed "to restore deference to technical experts, empower engineers to get resources they need, and allow safety concerns to be freely aired" (p. 203).

Experts, however, are not immune to the very normal organizational processes that have become known under labels such as 'group think,' 'cognitive fixation,' or 'attentional blindness.' This in itself is not remarkable: experts are people too, and subject to the vicissitudes and vagaries of social dynamics and organizational life as much as other people. But what if others are urged to rely on the experts to know better? And who would be the experts to say that experts have it wrong? Note, of course, that deference to expertise is not the only general characteristic or prescription that comes out of HRO research. The original HRO studies conducted by La Porte, Roberts, and Rochlin, as well as Weick and Sutcliffe's later additions, acknowledge that certain highly complex and hazardous organizations achieve very high levels of safe operations. Both approaches focus on concepts and processes on the organizational and group level, such as communication, decision-making, and a "culture" of reliability or mindfulness. Nevertheless, there are some differences: First, the initial aim of study differed in that the original research was mainly descriptive. Weick and Sutcliffe, in contrast, developed "key practices" that could become generalizable and exportable. Where the initial research strictly focused on the distinctive features of three HROs—air traffic control, nuclear aircraft carriers, and power distribution—later HRO research functioned as a "window on a distinctive set of processes directed toward fostering effectiveness that can unfold in all organizations" (Weick et al., 1999, p. 83).

8.2.3 HRO AND THE CAPACITY FOR SAFE OPERATIONS

HRO was an early member, if not a founding member, of a family of theories that sees safety as the presence of capacity, rather than as the absence of accidents. In the case of HRO, the riddle to be solved was how organizations with interactively complex and tightly coupled operations, were able to reliably generate safe results. Given the case research conducted to answer the riddle, though, it has been difficult to extract the causal factors responsible for generating such reliability. Moreover, the descriptive approach also lacks the theoretical framework that explains the psychological or social mechanics of why and how the identified characteristics contribute to reliability (Lekka, 2011). Taking on faith, for instance, that deference to expertise through an underspecification (or temporary loosening) of structures can lead to high reliability, may miss relevant aspects of the nature of expertise (Chi, Glaser, & Farr, 1988; Farrington-Darby & Wilson, 2006), its limits (Dismukes, Berman, & Loukopoulos, 2007; Jensen, 1996), or its occasional by-products such as stale knowledge and inhibition of innovation (Dekker, 2014).

To collate the various ideas on HRO, and to bring out some of the contrasts with NAT, Table 8.3 offers an overview on the main ideas of the two approaches.

TABLE 8.3
Comparison of the Main Ideas of NATs and HROs

	NATs	HROs
Authors	Charles Perrow, Scott Sagan	Todd La Porte, Karlene Roberts, Gene Rochlin, Karl Weick, Kathleen Sutcliffe
Objective	Facilitate decisions on systems and technologies.	Describe the characteristics of organizations that work nearly error-free despite of high risk and complexity.
Focus of study	Characteristics of high risk technologies that make a system accident-prone: 1. Interactive complexity 2. Tight coupling 3. Catastrophic potential	HROs that are characterized by: 1. Complex, hazardous and interdependent technologies and processes 2. Huge potential for errors 3. High levels of safety and reliability
Level of analysis	Macro: System level.	Micro: Group/organizational level.
Research	Case studies.	Case studies.
Viewpoint	"Pessimistic": accidents in NAT systems are inevitable.	"Optimistic": HROs achieve unexpectedly high records of safety.
Safety	Absolute safety is not achievable in NAT systems.	Safety is not defined by absolute, but by relative measures.
Accidents	The causes of normal accidents lie in the inherent characteristics of the system. Minor failures can interact in multiple, unexpected ways and cause serious catastrophes.	The focus is not on what causes accidents as a dependent variable but on what contributes to accident prevention and nearly error-free operations.
Human operator	In NAT systems, the human operator is confronted with unexpected and interacting failures, which he can neither anticipate nor comprehend.	The human operator is fallible. However, in effective HROs the human operators, due to organizational strategies (e.g., redundancy, training, "mindfulness"), are capable to anticipate, detect, and cope with errors.

(Continued)

TABLE 8.3 (*Continued*)
Comparison of the Main Ideas of NATs and HROs

	NATs	HROs
Technology	Technology is a source of complexity. Even technological redundancies and safety devices can increase interactive complexity and tight coupling and lead to new accidents.	Technology is a source of complexity, but can be controlled in HROs given concepts such as redundancy, "mindfulness" or a "culture" of high reliability are established.
Organization	NAT systems require a centralized as well as a decentralized organization at the same time which is impossible.	In HROs, centralization and decentralization are balanced. In a flexible way authority can temporarily be shifted from management to front-end operators depending on the conditions.
Interventions & Recommendations	1. Abandon the hazardous technology; 2. If this is not possible, modify the system (e.g., reduce interactive complexity and tight coupling)	Measures that contribute to ongoing learning and help create a state of "mindfulness," such as: • Continuous training • Job rotation • Decentralized decision-making • Redundancy • Sharing of information and mental representations

Because of the plurality and continuing development within HRO, the table focuses on common principles as described in the sections above.

8.3   SAGAN AND "THE LIMITS OF SAFETY"

8.3.1   NAT AND HRO IN A HISTORICAL CASE

The TMI incident, and the concerns it generated about nuclear safety, encouraged both NAT and HRO. Concern with nuclear safety also drove Scott Sagan. As you might recall from Chapter 6, his *Limits of Safety* (1993) told of 'close calls' in the military branches that were tasked with handling nuclear weapons during the Cold War. Sagan relied on declassified government documents, interviews with people involved in the incidents, and congressional hearings. His aim was to "provide a clearer understanding of the origins of accidents and the causes of safety" (Sagan, 1993, p. 13). His analysis has since been turned into a veritable 'competitive test' between the two schools of thought with roots in organization theory: HRO and NAT. Sagan applied the assumptions and predictions of the two theories to his case studies, and tried to assess which one provided better insights into his data; which one could provide a better explanation of his findings.

To set up the theoretical framework of HRO, Sagan referred not only to the work of La Porte, Roberts, and Rochlin, but also to a study called *Averting Catastrophe:*

Strategies for Regulating Risky Technologies conducted by Marone and Woodhouse (1986) and Wildavsky's *Searching for Safety* (1988). Sagan (1993) concluded that all three approaches agree on four factors that explain high reliability in organizations (p. 17):

1. prioritization of safety and reliability as a goal;
2. high levels of redundancy in personnel and technical safety measures;
3. a 'high reliability culture' in decentralized and continually practiced operations;
4. sophisticated forms of organizational trial and error learning.

For NAT, Sagan referred mainly to Perrow's description of normal accidents. He highlights NATs' theoretical standpoint relative to the four factors of HRO's:

1. safety as one of the conflicting organizational goals;
2. redundancy can increase complexity and the likelihood of accidents;
3. an organizational structure requiring decentralization and centralization at the same time is not achievable;
4. learning is restricted (e.g., due to unclear causes of incidents and events, false reporting, or denial of responsibility).

As his case for comparison, Sagan used the alerting system called Defense Condition (DEFCON) used by the military (especially the Strategic Air Command (SAC)) to command and control the U.S. nuclear arsenal. DEFCON used five levels to code different stages of nuclear threat. These levels required different stages of readiness of the military. Level one indicated the highest threat and required maximum readiness. Sagan argued that the system, as well as the SAC operations linked to it, were highly complex and tightly coupled. He described several incidents involving the alerting system and the handling of nuclear weapons during the Cold War, in which HRO would predict safe operations whereas NAT would predict the potential for very dangerous outcomes. Sagan then set the predictions against the actual historical outcomes of these incidents.

On October 24, 1962, during the Cuban missile crisis, the alert level for the SAC was set on DEFCON 2, the highest level ever used (recall this from the previous chapter). This led to a massive increase in nuclear capability; it put bombers and nuclear weapons on high alert. The numbers surrounding the operation seemed to favor the HRO prediction of safe operations: SAC aircraft flew some 2,090 missions, totaling 47,000 flight hours and 20 million miles—without any crash or nuclear weapons incident. Compared to preceding and subsequent months, accident rates even *decreased* in the time of crisis. Furthermore, the importance of safety was emphasized by superiors; continuous training was conducted to minimize risk; and a high sense of mission could be observed in the crews, all of which was seen by Sagan as being in line with HRO.

Yet, Sagan stated, NAT would be skeptical about this history. It was, after all, mainly written by the organization itself. Sagan, therefore, analyzed declassified documents for the details of two telling incidents. First, the SAC was ready to accept a safety deficiency concerning a bomb that was not yet certified for its intended use, but was pressed into service due to time pressure in establishing readiness for DEFCON 2. The request to use the noncertified bomb was actually refused by superiors, though Sagan was able to demonstrate that safety was not the priority driving decisions in SAC. Second, the air routes used by aircraft during the operation were supposed to be safe and outside of Soviet territory to avoid the possible dangers of an incursion. Nevertheless, in August 1962 one aircraft committed a navigational error heading off-course toward Soviet-held airspace. Close to the boundary, the aircraft was picked up by ground control and commanded to immediately change its course. The situation was described by Sagan as very serious due to a possible encounter with a Soviet interceptor. A change of the route to prevent repetition of the incident took 2 months.

Of course, the two incidents offered only limited support for NAT. So Sagan went on a subsequent search for the "hidden history" (1993, p. 251) of the operation, to try to uncover the extent to which tight coupling and interactive complexity had given rise to incidents that could easily have turned into something more serious. Sagan found numerous incidents involving nuclear weapons as well as command and control systems—despite all efforts to prevent them. Among them were false alarms leading to potentially dangerous actions. One example is described by Sagan (1993) as follows:

> Around midnight on the night of October 25, an air force guard at the Duluth Sector Direction Center saw what appeared to be a saboteur climbing the base security fence. He immediately shot at the figure and set off a sabotage alarm... Immediately, throughout the area dozens of armed sabotage squads were sent into the night to patrol the base perimeters. At Valk Field in Wisconsin, however, the alarm was faulty, ordering an immediate launch of aircraft. Pilots of the nuclear-equipped F-106A interceptors rushed to their aircraft and started the engines. These pilots had been told that there would be no practice alert, because of the DEFCON 3 status, and they fully believed that a nuclear war had just started. Just before take-off, a car, flashing lights, raced from the command post of the tarmac, and an officer signaled the aircraft to stop. The ADC declassified history reports that further communications to Duluth had revealed that no nuclear attack was under way. The "Russian spetznaz saboteur" that caused the incident was, ironically, a bear.

(pp. 99–100)

Sagan found that redundancy, one main factor of HRO, seemed to add complexity in some cases, for example, in the Falling Leaves alarm system, and to cause incidents rather than to prevent them. He also noted that trial and error learning often did not occur.

Sagan admitted that he had started the book with the expectation to find HRO assumptions to provide the better insight in the "apparent success story" (Sagan, 1993, p. 251) of nuclear weapon handling. His findings and analysis led him to the opposite view and turned him into a supporter for NAT. He also viewed his findings as a reflection of the "inherent limits of organizational safety" (1993, p. 279) and questioned the widely assumed stabilizing effects of nuclear weapons in international politics. He also emphasized the influence of organizational, cultural, and economic factors on accidents (Rijpma, 2003). His recommendations are specifically targeted at the reduction of the risk of nuclear weapons accidents by reducing the characteristics interactive complexity and tight coupling:

- avoid collocation, e.g., nuclear warheads should not be close to missile testing facilities and nuclear material should be separated from detonation devices in peacetime;
- create time buffers, e.g., install timers or prolong the time to launch nuclear weapons;
- create opportunity to reverse launch, e.g., self-destruction of missiles in flight or ballistic missile defenses.

8.3.2   NAT AND HRO IN DEBATE

In 1994, the editors of the *Journal of Contingencies and Crisis Management* invited Perrow, Sagan, and La Porte to comment on the theoretical assumptions made in *The Limits of Safety* and on the subsequent question: Are accidents preventable? In a sequence of comments, the contributing authors addressed each other's statements, assumptions, and conclusions in something that went "beyond a strict review" of the book (La Porte, Perrow, Rochlin, & Sagan, 1994, p. 205). At the time, it was seen as an important debate with historical implications, and parts of it have resonated since. The main points of the debate are summarized here.

8.3.2.1   Competitive versus Complementary Approaches

Sagan contrasted NAT and HRO as "pessimistic" versus "optimistic" and compared them on seven basic principles. He was also the first to refer to the two schools of thought as "theories," a notion that was later adopted by Perrow in the label "Normal Accidents Theory" (1994), but seen as "labelled too grandly" by La Porte (1994, p. 207).

La Porte commented that the HRO group adopted a realistic rather than optimistic point of view: the possibility of failures is not denied and the immense difficulties, efforts, costs, and demands to achieve highly reliable operations are in fact highlighted. La Porte meant to be cautious with the word "optimistic" so as not to communicate a false perception of HRO research. Its initial aim was never to suggest strategies for "perfect" operations with hazardous technologies at minimum risk (La Porte, 1994). "Pitting complementary perspectives against each other as if they were in competition, misreads their conceptual relationships" (p. 211), he argued, and then referred to the comparison as "overdrawn and inaccurate" (p. 209).

La Porte indeed noted that NAT and HRO might be more complementary than competitive, mentioning, for example, that Perrow was part of the original advisory group for HRO. He stated that both approaches agree that most organizations have the characteristics of "natural, moderately inchoate social systems" in which failure-free operation over an extended period of time is unlikely. None of the approaches denied the possibility of failures and none tried to predict the frequency of accidents as a function of the organization's characteristics. What the approaches differ in, after La Porte and Rochlin (1994), are the phenomena they study: accidents versus safe operations, respectively. Perrow was more skeptical of the complementary nature of the two schools of thought: He appreciated the contrast pointed out by Sagan and, for example, stated that NAT doubted the HRO assumption of safety as the foremost goal of managers, customers, and operators alike (Perrow, 1994). The question of the inevitability of accidents was seen as another contradictory point by Perrow.

8.3.2.2 Are Accidents Preventable?

According to Sagan, HRO would argue that groups and organizations can be shown to possess the capacities to prevent accidents. NAT puts that into doubt, since even the best intentions can get defeated by the inherent complexity and coupling of the system that is managed. Perrow noted that HRO researchers pointed to organizations such as the Diablo Canyon Nuclear Power plant which operated nearly error-free for a long period of time as "evidence" for their position. These systems, Perrow argued, simply have not had the time to demonstrate their catastrophic potential: they would do so eventually. That is of course difficult evidence to either produce or reject. But even when one system did demonstrate its catastrophic potential, Vaughan wrote in her reflections on the Challenger Space Shuttle disaster:

> ...we should be extremely sensitive to the limitations of known remedies. While good management and organizational design may reduce accidents in certain systems, they can never prevent them ... The causal mechanisms in this case suggest that technical system failures may be more difficult to avoid than even the most pessimistic among us would have believed. The effect of unacknowledged and invisible social forces on information, interpretation, knowledge, and—ultimately—action, are very difficult to identify and to control.
>
> *(1996, p. 416)*

Sagan characterized the different viewpoints on the likelihood of adverse events as follows: "Perrow may look at a glass of safety and find it 1% empty; high reliability theorists may see the same glass of safety as 99% full" (1993, p. 48). But "since the risky systems being observed have not been around very long, settling the issue on the basis of outcomes—how many serious accidents—is impossible; the controversy is properly over processes and this means the models of organizations used" as Perrow said (1994, p. 217). This indeed is reflected in much of the debate (see below).

8.3.2.3 Tightly Coupled and Interactively Complex Systems

HRO researchers described the organizations they studied—Air Traffic Control and U.S. Navy nuclear aircraft carriers—as "tightly coupled, complex, and highly interdependent technologies" (LaPorte & Consolini, 1991, p. 22). But were they, according to Perrow? His understanding concerning tightly coupled systems with interactive complexity differed from that of the HRO researchers. Perrow decided that Air Traffic Control "is basically a linear system rather than a complexly interactive one." He judged flight operations on aircraft carriers to be loosely coupled, because landings or takeoffs could easily be stopped or postponed. The point may have been hard to prove or disprove. Perrow, after all, regarded interaction and coupling as stable dimensions of a system, connected to structural features, whereas HRO researchers were quick to identify and acknowledge system dynamics. As Weick et al. (2008) put it: "most organizations are not frozen into one of the four combinations that are possible in Perrow's 2×2 of loose/tight coupling and linear/complex interaction. Instead, whole organizations change character in response to changed demands" (p. 34).

8.3.2.4   Organizational Structure

For Perrow, tightly coupled systems require centralization, whereas interactively complex systems require decentralization. He stated that it was impossible for organizations to accommodate those contrary demands. LaPorte and Consolini (1991), however, described the structure of HROs in terms of "nested authority patterns" that allow a flexible switch between centralization and decentralization in different modes of operations. Hopkins (1999) even put the argument like this:

> HRO theorists claim that it is empirically possible to have both centralised and decentralised decision making. If so, there is no reason why complex, tightly coupled organisations could not in principle adopt such a decision-making structure. It follows that there is no theoretical reason why accidents in complex, tightly coupled systems should be inevitable.
>
> *(p. 98)*

Perrow viewed that argument as another example for the incompatibility of the two approaches.

8.3.2.5   Technology and Human Operators

One final difference between the two approaches is their view on technology and the human operator. Sagan suggested that NAT and HRO

> hold very different visions of the degree to which the basic forces of human 'agency'— the elements of culture, design, management and choice—can counter or compensate for these dangerous structural pressures within hazardous organizations.
>
> *(1993, p. 45)*

Perrow had concluded that neither the human operator nor any safety precaution is able to cope with the sorts of unexpected and interacting failures that are inevitably produced by tightly coupled and interactively complex systems. The operator is "confronted by unexpected and usually mysterious interactions" (Perrow, 1999, p. 9) that he or she can neither anticipate nor comprehend and therefore not handle adequately. But HRO researches recognized strategies in the studied HROs to counter the negative characteristics of highly hazardous and interactively complex organizations. As Weick et al. (2008) later described it:

> We differ from other analysts such as Perrow and Sagan because we do not treat technology as a given that dominates organizational life through its own imperatives... We see technology less as an intractable technological imperative and more as a controllable option if it is engaged by effortful, continuous collective mindfulness enacted by smart, trusting, trustworthy, self-respecting people willing and able to negotiate the differences among their diverse views under intense time pressure.
>
> *(p. 51)*

8.3.2.6   Outcome of the Debate

Perrow (1994) himself commented on the debate as follows: "If this exchange between Sagan, La Porte and myself does nothing to advance theory in this area, I will be both surprised and very disappointed" (p. 212). But what was the debate's outcome and how did other researchers respond? Perrow and La Porte themselves continued to disagree on the conclusion drawn by Sagan. Perrow (1994) claimed that Sagan found much more support for NAT. Hale and Hovden (1998) agreed that the outcome of the study conducted by Sagan was in favor of Perrow's approach. La Porte (1994), meanwhile, attributed the findings to Sagan's personal interest in showing that the military forces controlling nuclear weapons had perhaps been a (normal) accident waiting to happen.

Was it a fair debate at all? Hopkins (1999) commented that some of the NAT propositions tested by Sagan were not part of the actual theory. These could, therefore, not provide support for the theory, and only confused the issue. Rijpma (1997) reflected differently on the debate. He attributed part of the conflict between the authors to the different levels of theoretical aggregation. Whereas NAT differentiates certain types of accidents, HRO analyzes strategies that result in overall reliability. Rijpma was of the opinion that both theories have implications for organizational design. Instead of focusing on the debate itself, he called for a more balanced and integrated prescriptive approach. He suggested that both theories would profit from cross-fertilization.

Finally, the main underlying question—are accidents preventable—could not be answered throughout the debate. Nor could it be determined which of the schools of thought offered a 'better' point of view. Rijpma (1997) described the outcome of the debate as a tie with diametrically opposed answers to the core question, but also doubted the use of the debate for the advancement of the theories (2003). Leveson, Dulac, Marais, and Carroll (2009) saw that the debate "developed as a contest of concepts and illustrative examples" (p. 229). According to Lagadec (1997) the inconclusive outcome was not so bad, since the "value of this debate lies in the tension between two opposing points of view; it is important to leave the question open, without simplifying or caricaturing it" (p. 24). And although La Porte and Perrow disagreed on the validity of Sagan's conclusions, Perrow (1994) admitted: "I think we need both formulations and research from both perspectives" (p. 220). He saw an achievement of both schools of thought in involving organizational theorists in safety science and accidents theory—a continuation of the important tradition begun by Barry Turner (see the previous chapter).

8.4   FURTHER DEVELOPMENT

In 2003, Rijpma concluded that "each of these two perspectives on accidents and safety is alive and kicking" (p. 43). But in what way?

8.4.1   Further Development of NAT

In 1999, Perrow published an updated version of "Normal Accidents" (originally published in 1984), including an afterword referring to the debate and Sagan's theoretical assumptions. He adopted Sagan's label "Normal Accidents Theory" for his approach and acknowledged Sagan's more pointed discussion of the limits and dangers of redundancy compared to his own earlier notion that redundancy constituted a major safety device (Perrow, 1999, p. 368). There was a shift in Perrow's view, emphasizing the negative effects of redundancy. Redundancy added to complexity and created unanticipated interactions and couplings. Perrow also extended his approach, exploring the putative reliability-inducing factors identified by the HRO group. He looked at safety as prioritized goal, at continuous learning and training, and at adaptive empowerment of lower-level operators. He concluded that the system characteristics of complexity and coupling affect the degree to which these HRO-factors could be established, making a distinction between error-inducing and error-avoiding systems. It allowed him "an 'optimistic' element in what Sagan describes as a basically pessimistic theory" (Perrow, 1999, p. 372). He classified systems according to certain factors such as experience with the operating scale and critical phases, the degree of organizational control over members or the degree to which information on errors can be obtained and shared. Perrow unsurprisingly judged nuclear plants and nuclear defense as well as marine transport as error-inducing. Space programs and chemical plants were assessed as error-neutral whereas naval carriers and air transport were seen as error-avoiding. He maintained, however, that normal accidents were inevitable, and that the frequency of those accidents is driven largely by the system characteristics that make things safe or less safe (Perrow, 1994).

NAT was also tested in empirical studies and applied to multiple case studies in different fields of practice. Gephart (2004) studied NAT in research on ecosystems accidents and how people made sense of them in inquiries after the fact. Vaughan's 1996 analysis of the Challenger launch decision viewed NASA as an HRO, which was not able to prevent the accident from happening. She concluded her analysis more pessimistically: these sorts of accidents are incredibly difficult to foresee and prevent, despite our best efforts. Hopkins (2001) tried to apply NAT to a gas plant explosion in Longford, Australia, but found the theory of little use in explaining the accident, in part because of the underspecification and confusion of the system dimensions of complexity and coupling. Mezias (1994) even applied NAT to the U.S. financial crisis in the 1980s and produced some useful insights. He did find it possible to explain the crisis with the help of the system characteristic complexity and coupling. Another field of application was healthcare. Gaba (2000) analyzed the healthcare sector as a high-hazard industry in terms of HRO and NAT. He came to the conclusion that the sector had not yet achieved nearly failure-free and reliable performance as demanded by the public. In his opinion, the causes for medical errors were mainly rooted in the structural characteristics of the system, which needed to be changed to improve patient safety. Wolf (2001, 2002) aimed to test NAT empirically and conducted a study on petrochemical and gasoline refineries in the United States between 1993 and 1997. He even produced a computed complexity index, based on the number of potential states a system could be in. He operationalized tight coupling as a binary variable related to (financial) resource availability. Safety was operationalized through the number of hazardous substance releases and the total case incident rate (TCIR)—taking into account the number of recordable injuries and illnesses for each plant. Wolf found that more complex and tightly coupled refineries were associated with a higher frequency of hazardous substance releases. It seemed to support NAT premises. The TCIR, however, was not related to the complexity of the plants. Wolf also found two 'normal accidents' in the plants that he otherwise classified as HROs due to their safety statistics. He took the occurrence of 'normal accidents' in HROs as further support for NAT. Independent of the different authors' conclusions, the application of NAT to several (case) studies and practical fields demonstrates that Perrow's book "Normal Accidents" was and is of importance for Safety Science research and "stands as a classic in the field" (Rijpma, 2003, p. 38).

8.4.2 FURTHER DEVELOPMENT OF HRO

The application of HRO also accelerated after the debate, still often in comparison with NAT—whether deliberately or not (e.g., Gaba, 2000; Vaughan, 1996). Airport security (Frederickson & La Porte, 2002) and teamwork for achieving high reliability in healthcare as done by D. P. Baker, Day, and Salas (2006) are further examples. The HRO group itself got involved in further research too. Research on nuclear submarines (Bierly & Spender, 1995) was conducted, and HROs were compared to non-HROs (Mannarelli, Roberts, & Bea, 1996). Researchers started to search for common principles and strategies used by most (if not all) organizations to achieve high reliability (Rijpma, 2003). Examples for the principles identified are the processes for mindfulness as or the elements of an HRO as identified by Libuser (1994) who worked together with Roberts:

- Process auditing: e.g., testing equipment, checks to identify safety problems
- Appropriate reward system
- Avoiding quality degradation
- Risk perception: knowledge and mitigation of risks
- Command and control
- Migrating decision-making: most expertized person makes the decision regardless of the level in hierarchy
- Redundancy of people or hardware: unalike and complementary backups
- Senior managers with "the big picture"
- Formal rules and procedures
- Training

The identified strategies were then used in an attempt to 'build' an HRO tailored to the implementing organization. One example for this was the application of the principles to a pediatric-intensive care unit in a hospital in Vermont (Roberts, Madsen, Desai, & Van Stralen, 2005). Table 8.4 provides examples of the implementation of the principles.

After establishing those principles, mortality and consequential events rates decreased, while admission and transportation rates increased. This was interpreted as a sign for the successful functioning as HRO. Nevertheless, Roberts et al. (2005)

TABLE 8.4

Examples of the Implementation of HRO Principles in a Hospital ICU

Principle	Implementation (Examples)
Process auditing	• Any team member could question care at any time • Questions were used in terms of the education of all team members
Appropriate reward system	• Intrinsic reward system: demonstration of knowledge led to one's opinion more frequently sought
Avoiding quality degradation	• Quality (improvement reviews) with nationally accepted norms as reference
Risk perception	• Program of in-service lectures to increase awareness of covert physiological state of patients
Command and control	• Decisions were made by the best qualified team member, often the bedside caregiver • Measuring of signs with two methods

emphasize that HRO processes can easily fail if they are not given constant attention, time, effort, and energy. This was, for example, shown when a change in leadership in the unit led to a return to a hierarchical model of medical care followed by the degradation of the commitment to the principles above. Yet other organizations across different industries aimed to become HROs as well. Examples of attempts to achieve HRO status come from Lekka and Sugden (2011) in an oil refinery and by Xiao and Moss (2001) with teams working in trauma resuscitation. Bourrier (2011) commented on the establishment of 'HRO' as a label of excellence, sought after by organizations across industrial sectors. Other research groups started cooperating with the original HRO group, or conducted research influenced by the original group leading to new HRO models and approaches. Among them was the San Bernadino group that started working with Roberts in 1995. Researchers of the San Bernadino group added new perspectives to HRO, including a model of inductive reasoning, adaptive decision-making, collaboration with others, and the sharing of knowledge.

Another stream of HRO research focused on RE principles which can be seen as partly derived from the HRO work or even as "the latest action agenda of HRO" (Dekker & Woods, 2009, p. 135). Following the definition of resilience as "ability of systems to anticipate and adapt to the potential for surprise and failure" (Hollnagel, Woods, & Leveson, 2006, p. 4) RE is defined as "a paradigm for safety management that focuses on how to help people cope with complexity under pressure to achieve success" (Hollnagel et al., 2006, p. 6). Costella, Saurin, and de Macedo Guimarães (2009) saw parallels and overlaps between HRO and RE: both approaches are suitable for high-hazard organizations with a high level of complexity, interconnectedness, dynamics, and uncertainty. Therefore, RE has been studied in and applied to similar industries as HRO, such as aviation, petrochemical, or nuclear plants. Furthermore, both approaches share the premise that organizations are capable to build or engineer safety and reliability. Both see safety as something a system does rather than something a system has (Dekker & Woods, 2009; Hollnagel et al., 2006). Both approaches offer sets of principles enabling organizations to better recover from errors and quickly return to normal operations (Lekka, 2011). The following list describes some of the RE principles derived from a review (Wreathall, 2006) and their relation to the earlier described HRO principles, as well as insights from other research:

- *Fostering of a just and learning culture*: A 'just culture' as, for example, described by Dekker (2012) and a 'culture of high reliability' as described by Roberts (1989) share the focus on a supportive environment. There, in general, people experience a high 'sense of mission' toward safety. Errors are openly reported and seen as a possibility for learning and increasing safety rather than for blaming individuals.
- *Management commitment to safety*: HRO researchers emphasized how safety is the priority in effective HROs, not only for frontline staff and leadership, but also at top management levels. This is consistent with both the findings and recommendations of man-made disaster theory (see the previous chapter) (Pidgeon & O'Leary, 2000). In highly resilient organizations, top management is concerned with human performance, values it in word and deed, and provides continuous and extensive actions related to human performance.
- *Flexibility*: Both HRO and RE acknowledge the importance of the organization's flexible adaption to demanding situations and new, complex problems. Both emphasize the requirement of frontline practitioners, the people with the most expertise and first-hand knowledge of the situation, to be enabled to make safety-relevant decisions without having to wait for management instructions.
- *Awareness and preparedness*: Some of the processes to create mindfulness described by Weick such as the preoccupation with failure or the commitment to resilience adopt the notion of awareness and preparedness. People in effective HROs do not wait for accidents to happen, but have a certain degree of skepticism, and are always looking for possible safety gaps. They do not take past success as a prediction for future success and prepare for incidents through certain strategies such as training or redundancy. RE, too, focuses on being ahead of possible problems and on the active anticipation of future events.

Today, HRO represents an important research stream in Safety Science, one that continues to be developed. Dekker and Woods (2009) concluded that "high reliability organization research is, and will always be, a work in progress, as its language for accommodating the results, and the methodological persuasions for finding and arguing for them, evolves all the time" (p. 139). In fact, both approaches influenced and contributed to today's views on safety. Some of the characteristics of HROs as identified by La Porte, Roberts, and Rochlin as well as ideas on complex tightly coupled systems as described by Perrow are still important concepts in current approaches in safety science.

STUDY QUESTIONS

1. Analyze a case study of your choice in terms of HRO and NAT. What would the two approaches focus on? Which factors would be analyzed? How would the outcome be explained?
2. Explain how the NAT system dimensions of complexity and coupling are, or are not, independent of each other.
3. What, from your point of view, makes NAT and HRO more competitive or complementary in nature?
4. In what sense is HRO complimentary to Hollnagel's ideas about safety not being the absence of catastrophic potential, but the presence of capacities to make things go right?

REFERENCES AND FURTHER READING

Baker, D. P., Day, R., & Salas, E. (2006). Teamwork as an essential component of high-reliability organizations. *Health Services Research, 41*(4 pt 2), 1576–1598. doi:10.1111/j.1475-6773.2006.00566.x

Baker, J. A. (2007). *The report of the BP U.S. refineries independent safety review panel.* Washington, DC: BP U.S. Refineries Independent Safety Review Panel.

Bierly, P. E., & Spender, J. C. (1995). Culture and high reliability organizations: The case of the nuclear submarine. *Journal of Management, 21*(4), 639–656.

Bourrier, M. (2011). The legacy of the high reliability organization project. *Journal of Contingencies and Crisis Management, 19*(1), 9–13. doi:10.1111/j.1468-5973.2010.00628.x

CAIB. (2003). *Report volume 1, August 2003.* Washington, DC: Columbia Accident Investigation Board.

Chi, M. T. H., Glaser, R., & Farr, M. J. (1988). *The nature of expertise.* Hillsdale, NJ: Lawrence Erlbaum Associates.

Costella, M. F., Saurin, T. A., & de Macedo Guimarães, L. B. (2009). A method for assessing health and safety management systems from the resilience engineering perspective. *Safety Science, 47*(8), 1056–1067.

Dekker, S. W. A. (2003). Failure to adapt or adaptations that fail: Contrasting models on procedures and safety. *Applied Ergonomics, 34*(3), 233–238.

Dekker, S. W. A. (2012). *Just culture: Balancing safety and accountability* (2nd ed.). Farnham, UK: Ashgate Publishing Co.

Dekker, S. W. A. (2014). Deferring to expertise versus the prima donna syndrome: A manager's dilemma. *Cognition, Technology and Work, 16*(4), 541–550.

Dekker, S. W. A., & Woods, D. D. (2009). The high reliability organization perspective. In E. Salas (Ed.), *Human factors in aviation* (pp. 123–146). New York, NY: Wiley.

Dismukes, K., Berman, B. A., & Loukopoulos, L. D. (2007). *The limits of expertise: Rethinking pilot error and the causes of airline accidents.* Aldershot, Hampshire, England, Burlington, VT: Ashgate Publishing Co.

Downer, J. (2013). Disowning Fukushima: Managing the credibility of nuclear reliability assessment in the wake of disaster. *Regulation & Governance, 7*(4), 1–25.

Farrington-Darby, T., & Wilson, J. R. (2006). The nature of expertise: A review. *Applied Ergonomics, 37*, 17–32.

Feynman, R. P. (1988). *"What do you care what other people think?": Further adventures of a curious character.* New York, NY: Norton.

Frederickson, H. G., & La Porte, T. R. (2002). Airport security, high reliability, and the problem of rationality. *Public Administration Review, 62*, 33–43.

Gaba, D. M. (2000). Structural and organizational issues in patient safety: A comparison of health care to other high-hazard industries. *California Management Review, 43*(1), 83.

Gephart, R. P. (2004). Normal risk technology, sense making, and environmental disasters. *Organization & Environment, 17*(1), 20–26.

Hale, A. R., & Hovden, J. (1998). Management and culture: The third age of safety. In A. M. Feyer & A. Williamson (Eds.), *Occupational injury: Risk, prevention and intervention* (pp. 129–166). London, UK: Taylor & Francis Group.

Hollnagel, E., Woods, D. D., & Leveson, N. G. (2006). *Resilience engineering: Concepts and precepts.* Aldershot, UK: Ashgate Publishing Co.

Hopkins, A. (1999). The limits of normal accident theory. *Safety Science, 32*(2), 93–102.

Hopkins, A. (2001). *Lessons from Esso's gas plant explosion at Longford.* Canberra, Australia: Australian National University.

Jensen, C. (1996). *No downlink: A dramatic narrative about the challenger accident and our time* (1st ed.). New York, NY: Farrar, Straus, Giroux.

Kates, R. W. (1986). *Normal accidents-living with high-risk technologies-Perrow, C.* Cambridge, MA: Blackwell Publishers.

La Porte, T. R. (1994). A strawman speaks up: Comments on the limits of safety. *Journal of Contingencies and Crisis Management, 2*(4), 207–211. doi:10.1111/j.1468-5973.1994.tb00045.x

La Porte, T. R. (1996). High reliability organizations: Unlikely, demanding and at risk. *Journal of Contingencies and Crisis Management, 4*(2), 60–71.

La Porte, T., Perrow, C., Rochlin, G., & Sagan, S. (1994). Systems, organizations and the limits of safety: A symposium. *Journal of Contingencies and Crisis Management, 2*(4), 205–206. doi:10.1111/j.1468-5973.1994.tb00044.x

La Porte, T. R., & Rochlin, G. (1994). A rejoinder to Perrow. *Journal of Contingencies and Crisis Management, 2*(4), 221–227. doi:10.1111/j.1468-5973.1994.tb00047.x

Lagadec, P. (1997). Learning processes for crisis management in complex organizations. *Journal of Contingencies and Crisis Management, 5*(1), 24–31.

LaPorte, T. R., & Consolini, P. M. (1991). Working in practice but not in theory: Theoretical challenges of "High-Reliability Organizations". *Journal of Public Administration Research and Theory, 1*(1), 19–48.

Lekka, C. (2011). *High reliability organisations: A review of the literature.* Derbyshire: UK: Health and Safety Executive.

Lekka, C., & Sugden, C. (2011). The successes and challenges of implementing high reliability principles: A case study of a UK oil refinery. *Process Safety and Environmental Protection, 89*(6), 443–451.

Leveson, N. G., Dulac, N., Marais, K., & Carroll, J. (2009). Moving beyond normal accidents and high reliability organizations: A systems approach to safety in complex systems. *Organization Studies, 30*(2–3), 227–249.

Libuser, C. B. (1994). *Organizational structure and risk mitigation.* Los Angeles, CA: University of California Press.

Mannarelli, T., Roberts, K. H., & Bea, R. G. (1996). Learning how organizations mitigate risk. *Journal of Contingencies and Crisis Management, 4*(2), 83–92. doi:10.1111/j.1468-5973.1996.tb00080.x

Marone, J. G., & Woodhouse, E. J. (1986). *Averting catastrophe: Strategies for regulating risky technologies.* Berkeley, CA: University of California Press.

Mezias, S. J. (1994). Financial meltdown as normal accident: The case of the American savings and loan industry. *Accounting, Organizations and Society, 19*(2), 181–192.

Mindell, D. A. (2008). *Digital Apollo: Human and machine in spaceflight.* Cambridge, MA: MIT Press.

Murray, C. A., & Cox, C. B. (1989). *Apollo, the race to the moon.* New York, NY: Simon and Schuster.

Perrow, C. (1984). *Normal accidents: Living with high-risk technologies.* New York, NY: Basic Books.

Perrow, C. (1994). The limits of safety: The enhancement of a theory of accidents. *Journal of Contingencies and Crisis Management, 2*(4), 212–220.

Perrow, C. (1999). *Normal accidents,* updated edition. New Jersey, NJ: Princeton University Press.

Perrow, C. (2011). Fukushima and the inevitability of accidents. *Bulletin of the Atomic Scientists, 67*(6), 44–52.

Perrow, C. (2013). Nuclear denial: From Hiroshima to Fukushima. *Bulletin of the atomic scientists, 69*(5), 56.

Pidgeon, N. F., & O'Leary, M. (2000). Man-made disasters: Why technology and organizations (sometimes) fail. *Safety Science, 34*(1–3), 15–30.

Rijpma, J. A. (1997). Complexity, tight–coupling and reliability: Connecting normal accidents theory and high reliability theory. *Journal of Contingencies and Crisis Management, 5*(1), 15–23.

Rijpma, J. A. (2003). From deadlock to dead end: The normal accidents-high reliability debate revisited. *Journal of Contingencies and Crisis Management, 11*(1), 37–45.

Roberts, K. H. (1989). New challenges in organizational research: High reliability organizations. *Organization & Environment, 3*(2), 111–125.

Roberts, K. H. (1990). Some characteristics of one type of high reliability organization. *Organization Science, 1*(2), 160–176.

Roberts, K. H., & Gargano, G. (1989). *Managing a high reliability organization: A case for interdependence: Managing complexity in high technology industries: Systems and people.* New York, NY: Oxford University Press.

Roberts, K. H., Madsen, P., Desai, V., & Van Stralen, D. (2005). A case of the birth and death of a high reliability healthcare organisation. *Quality and Safety in Health Care, 14*(3), 216–220.

Rochlin, G. I. (1996). Reliable organizations: Present research and future directions. *Journal of Contingencies and Crisis Management, 4*(2), 55–59.

Rochlin, G. I. (1999). Safe operation as a social construct. *Ergonomics, 42*(11), 1549–1560.

Rochlin, G. I., LaPorte, T. R., & Roberts, K. H. (1987). The self-designing high reliability organization: Aircraft carrier flight operations at sea. *Naval War College Review, 40,* 76–90.

Rosness, R., Grotan, T. O., Guttormsen, G., Herrera, I. A., Steiro, T., Storseth, F., ... Waero, I. (2010). *Organisational accidents and resilient organisations: Six perspectives, Revision 2* (SINTEF report A17034). Trondheim, Norway: SINTEF Technology and Society, Safety Research.

Rosness, R., Guttormsen, G., Steiro, T., Tinmannsvik, R. K., & Herrera, I. A. (2004). *Organisational accidents and resilient organizations: Five perspectives (Revision 1)* (STF38 A 04403). Trondheim, Norway: SINTEF Industrial Management.

Sagan, S. D. (1993). *The limits of safety: Organizations, accidents, and nuclear weapons.* Princeton, NJ: Princeton University Press.

Schwenk, C. R., & Cosier, R. A. (1980). Effects of the expert, devil's advocate, and dialectical inquiry methods on prediction performance. *Organizational Behavior and Human Performance, 26*(3), 409–409.

Vaughan, D. (1996). *The challenger launch decision: Risky technology, culture, and deviance at NASA.* Chicago, IL: University of Chicago Press.

Vaughan, D. (2002). Signals and interpretive work: The role of culture in a theory of practical action. In K. A. Cerulo (Ed.), *Culture in mind: Toward a sociology of culture and cognition* (pp. 28–54). New York, NY: Routledge.

Weick, K. E., & Sutcliffe, K. M. (2007). *Managing the unexpected: Resilient performance in an age of uncertainty* (2nd ed.). San Francisco, CA: Jossey-Bass.

Weick, K. E., Sutcliffe, K. M., & Obstfeld, D. (1999). Organizing for high reliability: Processes of collective mindfulness. *Research in Organizational Behavior, 21*, 81–124.

Weick, K. E., Sutcliffe, K. M., & Obstfeld, D. (2008). Organizing for high reliability: Processes of collective mindfulness. *Crisis management, 3*, 81–123.

Wildavsky, A. B. (1988). *Searching for safety.* New Brunswick, NJ: Transaction Books.

Wolf, F. G. (2001). Operationalizing and testing normal accident theory in petrochemical plants and refineries. *Production and Operations Management, 10*(3), 292–305.

Wolf, F. G. (2002). *Normal accident theory validated: Interactive complexity and resource availability as predictors of reliability.* Paper presented at the Reliability and Maintainability Symposium, Seattle, WA, 2002. Proceedings. Annual.

Woods, D. D. (1988). Coping with complexity: The psychology of human behavior in complex systems. In L. P. Goodstein, H. B. Andersen, & S. E. Olsen (Eds.), *Tasks, errors, and mental models.* New York, NY: Taylor & Francis Group.

Woods, D. D., & Patterson, E. S. (2001). How unexpected events produce an escalation of cognitive and coordinate demands. In P. A. Hancock & P. Desmond (Eds.), *Stress, workload and fatigue* (pp. 290–304). Mahwah, NJ: Lawrence Erlbaum Associates.

Woods, D. D., & Shattuck, L. G. (2000). Distant supervision-local action given the potential for surprise. *Cognition, Technology & Work, 2*(4), 242–245.

Wreathall, J. (2006). Properties of resilient organizations: An initial view. In E. Hollnagel, D. D. Woods, & N. G. Leveson (Eds.), *Resilience engineering: Concepts and precepts* (pp. 275–286). Aldershot, UK: Ashgate Publishing Co.

Xiao, Y., & Moss, J. (2001). *Practices of high reliability teams: Observations in trauma resuscitation.* Proceedings of the Human Factors and Ergonomics Society Annual Meeting, 45, Minneapolis/St. Paul, MN.

9 The 1990s and Onward
Swiss Cheese and Safety Management Systems

KEY POINTS

- By the late 1980s, a strong consensus had formed: human performance at the sharp end is shaped by local workplace conditions and upstream organizational factors.
- Lots of work from the previous decades point to this, including Taylor, Heinrich, human factors, and man-made disaster theory. All, in some way or other, considered a lack of safety at the front end to be attributable to what had preceded in the blunt, administrative, and engineered end of an organization.
- The Swiss cheese model became an iconic representation of this idea: problems experienced at the sharp end (or frontline) of an organization are not created there, but rather inherited from other imperfect parts.
- The Swiss cheese model is a defenses-in-depth model of risk, which suggests (as did Heinrich) that risk should be seen as energy that needs to be contained or channeled or stopped.
- This makes it difficult for Swiss cheese to be a true 'systemic' model, since it is not capable of explaining or portraying the complex emergence of organizational decisions, the erosion of defenses, drift, and a normalization of deviance.
- Swiss cheese conceptually aligns with safety management systems. These direct safety efforts and regulation at the administrative end of an organization, where assurance that safety is under control is sought in management systems, accountabilities, processes, and data.
- The gradual shift to 'back-of-house,' to organizational and administrative assurance of safety, had been long in the making. It is now intuitive that all work is shaped by the engineered context and workplace conditions and upstream organizational factors. If we want to understand or change anything, then that is where we need to look.
- This trend has also give rise to large safety bureaucracies and cultures of compliance. It has left us with a safety profession that broadly lacks purpose and vision in many industries, and with regulators whose roles have in certain cases been hollowed out and minimized.

9.1   INTRODUCTION

By the late 1980s, a strong consensus had formed: human performance at the sharp
end is shaped by local workplace conditions and upstream organizational factors.
Adequate accounts of accidents and disasters could not stop at the acts or omissions
of those who were closest in place and time. It was a long time coming for a wide
range scholarship to congeal and get turned into one of the most iconic 'brand'
images of these ideas: the Swiss cheese.

9.1.1   THINKING ABOUT THE SYSTEM HAD BEEN LONG IN THE MAKING

Back in the 1910s, Taylor did not think that frontline workers were simply lazy
or deficient. (Of course, he left that possibility open, and thought it should be
addressed by personnel selection and person-to-task matching.) Rather, Taylor
considered inefficiencies at the operational end to be the result of a lack of control
from above. This was a lack of effective supervision, a lack of coordination and
careful task design, deficiencies in engineering design or planning, or managerial
prework. You may recall Taylor's study of meat plant workers. The production
issues they had, the problems of balancing the line—those were inherited from
administrative, bureaucratic, planning levels, and amplified by the design of the
machinery. The problems that were experienced on the line were not created at the
level of the line.

Heinrich's initial work strongly echoed this position: by the time harm comes to a
worker, a number of dominoes—organizational, personal, and engineering antecedents that set the situation up for failure—have already toppled. Bird, too, realized
that opportunities for better control occur in all stages of an accident chain: precontact, during contact, and post-contact. He would affirm, as does the Swiss cheese
model, that precontact control—e.g., the development of programs and standards
and audit procedures, engineering measures, substitution of hazardous materials—
produces the most wide-ranging benefits. It may be difficult to motivate organizations to make the sacrifices needed to invest in such controls, but Bird argued that
it was better than trying to control the discharge of energy once released, and much
more predictable than what would be needed to contain the consequences once a
loss had occurred.

Early human factor pioneers, such as Fitts, Jones, and Chapanis, showed empirically
that 'human error' is not a cause, but a consequence of design and organizational
issues in the system in which people work. Human error is a consequence—a
symptom of trouble that gets designed into a system long before operators get to
interact with that system.

In the 1970s, Barry Turner looked further than just the engineered setting or
interface that created error traps. He studied the administrative, the bureaucratic, the
organizational layers that surround operational people and safety-critical technologies. That is where failures brewed; where accidents were incubated. The processes
that set failure in motion were not unmotivated workers or their errors at the sharp
end. Instead, the seeds of disaster needed to be sought among the normal processes
of organizational life, in everyday managerial processes of information gathering,

decision-making, and communicating. The potential for having an accident turned out tightly connected to the way we organize, operate, maintain, regulate, manage, and administer our safety-critical technologies.

If disasters and accidents are incubated inside an organization long before they actually happen, then normal accident theory showed how certain structural features make some systems more vulnerable than others to suffering such accidents. We need, said Perrow, to look at the interactive complexity and coupling of the systems we ask people to maintain and operate. Errors and failures in managing or operating risky technologies are less related to how people do their work, and much more the result of how these systems are planned, conceived, built, or left to grow in size, complexity, and safety-criticality (Perrow, 1984). High reliability theory then showed the extraordinary lengths that organizations, managers, supervisors, designers, and operational teams have to go to *not* have such accidents happen—despite their evident risk.

Looking upstream, rather than downstream had become *de rigeur*. To understand where a failure came from, and how its risk should be controlled, we should look back into the system, into the organization that helped produce it. An accident investigation today is barely considered complete if it does not somehow invoke organizational levels and their contribution to how things went wrong at the sharp end. In fact, investigations often go beyond the organization that suffered the accident, taking into account the regulations, inspections, and policy-making environments that surrounded it—layer after layer.

9.1.2 Impossible Accidents

In the late 1980s, two researchers at the University of Leiden in the Netherlands—Willem Wagenaar and Job Groeneweg—published a paper in the *International Journal of Man-Machine Studies*, entitled "Accidents at sea: Multiple causes and impossible consequences" (Wagenaar & Groeneweg, 1987). It presented an analysis of 100 maritime accidents, many of which had been attributed to 'human error.' The analysis was typical of the age it heralded, showing that any attempt to control or reduce 'the human error problem' was hopeless if it targeted only the human by selection, behavioral intervention, training programs, removal of the 'accident prone' people or attempts to increase people's motivation. Such measures, after all, assumed that the chains of events that lead to accidents can be fully overseen and controlled by the people who are deeply embedded in, and part of, such chains; and that not preventing the chain from becoming an accident was an act of stupidity or negligence. But that, the researchers concluded, was impossible:

> Accidents appear to be the result of highly complex coincidences which could rarely be foreseen by the people involved. The unpredictability is caused by the large number of causes and by the spread of the information of the participants... Accidents do not occur because people gamble and lose, they occur because people do not believe that the accident that is about to occur is at all possible.

(p. 42)

If we wanted to do something about human error, the researchers concluded, then human error itself is not to be the target of intervention. Instead, echoing Fitts and Jones 40 years earlier (1947), to control 'human error,' we needed to change the work environment, including hardware systems, machinery, tools, procedures, and schedules.

The 1990s saw a widespread adoption of such 'systems' thinking—not only in how we should investigate and understand accidents and disasters after the fact, but also how we should manage their risk before anything happened. The systems approach says that failure and success are the joint product of many factors—all necessary and only jointly sufficient. When something in an organization succeeds, it is not likely because of one heroic individual. And when something fails, it is not the result of one broken component, or one deficient individual. It takes teamwork, an organization, a system, to succeed. And it takes teamwork, an organization, a system, to fail.

Theoretical development during the 1990s was actually relatively limited. Fundamental ideas came mostly from the work of Heinrich, from human factors and Cognitive Systems Engineering, normal accidents theory, and high reliability theory. These theories offered most of the substance for thinking about the role of organizations in creating the conditions for failure and success. As you will see, the notion of frontline 'unsafe acts' as the deciding factor, of accidents as the end point of a linear chain—of a proportional cause–effect sequence of events—was sustained throughout, as was the idea that well-maintained defenses-in-depth were the way to stop them. But during this decade, Swiss cheese made a crucial difference in how the ideas were presented and popularized.

9.2   SWISS CHEESE

9.2.1   Defenses-In-Depth and Barriers

The origins of seeing accidents as a linear sequence of negative events that breach or topple layers of defense can be found in Heinrich's work (see Chapter 4). The safety barrier concept itself, though, came to the fore only in the 1960s and 1970s, in the work of, for example, Haddon (1973). The core assumption that drove the barrier concept is that accidents are the result of an uncontrolled release of harmful energy.

The introduction of fault tree and event tree analyses as system safety methods was a crucial contributor to barrier concepts and analysis. These methods were first proposed in 1961 as a way to assess the launch control system for the Minuteman ICBMs (see Chapter 6):

- Fault tree analysis starts from a chosen event (usually called the initiating event), and then considers the various ways in which this can possibly develop into an accident (Hollnagel, 2004);
- The opposite, called an event tree, starts with the unwanted outcome and works backwards, step-by-step, to find all the possible conditions that would lead to it. This produces a *tree* of events rather than a simple sequence (such as in Heinrich or Swiss cheese). After all, there is always more than one prior condition or event that triggers the next event (Hollnagel, 2004).

Both methods obviously open the way for a systematic (even quantifiable) consideration of the placement of barriers (of whatever kind) to halt the propagation of events and their escalation into an unwanted outcome. Barriers can be classified in many different ways (as will also become evident later in this chapter when you read about the various installments of the Swiss cheese model). A straightforward classification of barriers, which follows the linear accident trajectory and the various stages of things already having gone out of hand, is:

- Prevention
- Control
- Mitigation

Another way to look at it, Hollnagel (2004) proposed, was to consider barriers as either systems or functions. He classified barriers according to their nature, rather than their location or purpose in a particular imagined event sequence. His classification was:

1. *Material barriers*, which physically prevent an action or a harmful event to occur (this might be a wall, a railing, a concrete bunker);
2. *Funcational barriers* (also named 'active' or 'dynamic'). These are barriers that set up the preconditions for action. The precondition has to be met before the action can be taken (such as a lock, code, password, and gate).
3. *Symbolic barriers* are those which require somebody to interpret or understand their meaning and application. This would include instructions, signs, or permits.
4. *Incorporeal barriers* are based on the knowledge of the person who needs to embody and apply the barrier. An example is knowledge of the stall speed of an airplane, below which the airplane no longer flies.

The problem, of course (and Hollnagel recognized this), is that the prevention of one kind of accident might well fall into multiple barrier categories. Preventing an aircraft from stalling, for example, could be distributed across incorporeal barriers (knowledge of the flight manual and aircraft limitations at certain weights), symbolic barriers (a stall warning), functional barriers (a stick shaker that the pilot has to pull through to slow the aircraft down even further) and material barriers (a stick pusher that automatically pitches the nose of the aircraft down). When combined with an autopilot, the existence of these barriers could lead (and *has* led) to the new types of pathways to system disaster discussed in Chapter 5. In one accident, the autopilot was inadvertently programmed to conduct a go-around (which means the aircraft has to pitch up, add power, and fly away from the approach it was making), but the pilots were still intending to land (meaning the nose had to be pitched down, the power reduced and the aircraft brought down to the runway). Pilots and autopilot started 'fighting' each other—with every move by one countered by the other, until the aircraft was pitching up some 47°. At that point, it backslid and crashed (NTSB, 1994). As noted in Chapter 6, it has proven very difficult to meaningfully anticipate these kinds of 'going sour' scenarios in interaction with automated systems through traditional fault tree or event tree analyses.

This is just one example of the problems and weaknesses that afflict safety barriers as a way of thinking and of controlling, preventing, and mitigating risk. Recently, the following principles for better barrier management were proposed (Harms-Ringdahl & Rollenhagen, 2018, p. 81):

- Barriers should be made a simple as possible, and it should be easy to understand the assumptions behind their design:
- Barriers should cover as wide a range of initiating events as possible;
- Barrier design should consider the possible conflicts that can arise from interacting system goals and contributors;
- Barrier design has to take into account that the barrier can be seen as a hindrance, as something that gets in the way of getting the work done. People will otherwise avoid or circumvent such barriers;
- Barriers should be sensitive to human errors both in operation and in maintenance, and need to incorporate a way to be assessed for their (continued) effectiveness;
- Administrative barriers generally have low efficiency, especially over a longer time. They need to be carefully planned and often evaluated.

Although the Swiss cheese model is fundamentally a linear accident trajectory model that leans heavily on the concept of safety barriers, these analytic subtleties postdate its inception and evolution. And, as commented almost two decades after the first installment came out:

> the Swiss cheese model has an indisputable value as a means of communication, as a heuristic explanatory device. It has had a significant influence on the understanding of accidents, and therefore also on the practical approach to accident analysis and prevention. It has successfully been applied as a means of accident analysis and proactive measurements, although the level of resolution does not go as far as for other models.
>
> **Reason, Hollnagel, & Pariès (2006, p. 21)**

9.2.2 THE IMPETUS FOR SWISS CHEESE

James Reason was trained as a psychologist at the universities of Manchester and Leicester in the United Kingdom in the 1960s. He was appointed as professor of psychology at Manchester in 1977, investigating issues of ergonomics, stress, and motion sickness. In the late 1980s, while working on a book about human error, he got diverted from his original mission to provide a cognitive-psychological account, an account of mostly mental processes. He had been aiming the book at his peers: psychologists and scientists. But the 1970s and 1980s had driven large disasters and accidents back onto center stage, already inspiring researchers to formulate man-made disasters theory, normal accidents theory, and high reliability theory.

Tenerife, Flixborough, Challenger, Three Mile Island, Bhopal, Chernobyl, the Herald of Free Enterprise, and the King's Cross Underground fire—the disasters from that era were inescapable, as was the complicated role that 'human error' seemed to play in them. Each of these accidents was investigated in its own way by various authorities, but together they pointed to the many organizational and other contextual factors that shaped human understandings and decisions at the sharp end. It became obvious to Reason that an adequate account of human error was impossible without considering the relationship between people's assessments and actions on the one hand, and features of their organizational and operational world on the other (Reason et al., 2006).

9.2.3 RESIDENT PATHOGENS

As with many of the foundational ideas of Cognitive Systems Engineering (see Chapter 5), a crucial distinction in our thinking about systems disasters had been offered by Jens Rasmussen. In the late 1970s, Rasmussen (1980) conducted an analysis of 200 cases of nuclear power plant incidents, which were classified under the header of 'operational problems.' He wanted to know what we can actually learn from these sorts of reports. He found that 34% of all the incidents were attributed to human omissions (people failing to do something that needed to be done). Rasmussen realized that this was only the beginning of an organization's understanding of human error. Because those omissions came from somewhere, they were allowed by something, enabled, encouraged even. The suggestion was that we should distinguish between two kinds of 'human error':

- *Active* errors are those on or near the sharp end, whose effects are felt or visible almost immediately. These are errors that are associated with those active on the frontline of safety-critical systems.
- *Passive* errors lie dormant inside a system for a long time, to become evident only when they combine with other factors to breach the system's defenses. Such passive errors are incubated by people whose work is removed, in time and space, from the direct interface with the safety-critical system. They were also called *latent* errors, or, in a reference to a model of disease, *resident pathogens.*

Reason's review of disasters from the previous two decades, which made up part of the empirical material for his 1990 book, followed this distinction (Reason, 1990). His corpus of cases made it apparent that latent or passive errors, or resident pathogens, posed the greatest threat to the safety of complex systems. They nestled as root causes deep inside those systems long before any active errors or other issues at the sharp end brought them to the fore. Major disasters in well-defended systems were never caused by a single factor, but possible only if there was an unforeseen, and usually unforeseeable, concatenation of several diverse events. Perrow, of course, had already tried to capture this with the concepts of interactive complexity and tight coupling.

Each one of these events was necessary, and they were only jointly sufficient to push a system over the edge into failure. *Concatenation*, the word chosen by Reason, refers to a series of interlinked or connected events, a sequence, succession, or chain. This is important to remember when we discuss some of the implications of his theory for how we consider accidents in complex systems (see Section 4 below).

Reason always picked his words and phrases with care, and *resident pathogens* were one of them. It suggested an analogy between the breakdown of complex technological systems and the way multiple cause illnesses, such as cancer or cardiovascular disease, were triggered and developed. Reason found explicit similarities between latent failures in complex technological systems and resident pathogens in the body. These are causal factors which are already present before any accident sequence actually begins. They can have effects that are not immediately apparent, but can weaken a system's defenses and promote *unsafe acts* (Reason adopted the term for active errors originally used by Heinrich in the 1930s).

Breaching a system's defenses requires that resident pathogens and active triggering events come together, even though the precise nature of their interaction is hard to predict. For the most part, resident pathogens (if people in responsible positions recognize them as such) are tolerated or kept in check by other protective measures. Sometimes they are detected and corrected before they can create havoc: the very metaphor served as direct attention to indicators of 'system morbidity' prior to a catastrophic breakdown.

Reason was aware that suggesting the existence of resident pathogens was not the same as proposing a coherent or workable theory. Launching a mere search inside an organization for resident pathogens was indeed not very workable, because how were people going to decide what counted as one? And were the layers of defense as depicted in the original model sufficiently discriminate and comprehensive? After its original 1990 presentation, the Swiss cheese model underwent various stylistic modifications to accommodate developing insights, (Reason et al., 2006):

- The initial model from 1990 depicted a sequence of five 'planes' lying one behind the other: top-level decision makers, line management, preconditions, productive activities, and defenses. They were not categorically similar or consistent, but more an identification of the elements of production in order to describe how and why things might fail. There is no Swiss cheese resemblance yet, except for the last plane which shows a single hole. The term Swiss cheese had not yet been coined for this model either.
- The second incarnation of the model was developed in the early 1990s. It reduced the four productive planes to three (organization, workplace, person), and extended the single defensive layer to three layers. The model thus split into two: those things that contribute to the conditions for errors and violations on the one side, and those that help prevent or mitigate them on the other. The organization box included corporate culture and organizational processes as well as management decisions. This was consistent with the budding attention paid to safety culture at the time (see the next chapter).
- The third version of the model was developed in the mid-1990s and appeared in Reason (1997). Inspired by his work in the oil industry during that time, the model held that accident causation must have three basic elements: hazards, defenses, and losses. In it, the 'planes' were a bit more suggestive of Swiss cheese slices, but they still looked very mechanical and were not labeled. What Reason did add was an explanation of how the holes, gaps, or weaknesses arise. Short-term breaches may be created by the errors and violations of frontline operators, he believed. Longer-lasting and more

dangerous gaps are created by the decisions of designers, builders, procedure writers, top-level managers and maintainers. He now called these latent
conditions rather than latent errors or latent failures, where a condition is
not a cause, but it is necessary for a causal factor to have an impact.

- The fourth version of the model appeared (among other places) in his 2008
book. The model showed thick slices of cheese lined up and labeled according to elements from the case they served to exemplify. Reason used the
word Emmentaler in discussing his cheese slices, to more precisely locate
it in the birthplace of the hard cheese with many holes in it (Emmental is
a valley in west central Switzerland, not far from Bern). Taking the cheese
analogy to its logical conclusion, he introduced a mouse in the final cartoon.
The mouse represented an accumulation of small, niggling events, and failures that slowly nibbled away at the last slice: the heroic interventions of a
frontline operator. Audiences loved it.

Reason also realized that the resident pathogen metaphor shared a feature with the
largely discredited accident proneness theory (recall Chapter 3). Most importantly,
they share the assumption that an unequal liability to suffering an accident is due to
inherent and enduring features of the individual (in the case of accident proneness
theory) or the organization (in case of resident pathogens). Some organizations have
a greater accident liability because of their accumulation of resident pathogens. The
major difference lies of course in the target of remedies against the proneness to
having an accident: is it a particularly liable individual or the organization?

The human reliability community at the time had become increasingly aware that
attempts to discover and neutralize resident pathogens in an organization was going to
have a greater beneficial effect on system safety than chasing localized, active errors
by those at the sharp end (Rasmussen, 1985). Reason (1990), too, concluded that:

> Rather than being the main instigators of an accident, operators tend to be the inheritors
> of system defects created by poor design, incorrect installation, faulty maintenance
> and bad management decisions. Their part is usually that of adding the final garnish to
> a lethal brew whose ingredients have already been long in the cooking.
>
> *(p. 173)*

Human factors traditionally focused on the human at the sharp end: at the interaction
between the human and the system in that narrow, localized sense. We had to throw
the net of 'human factors' considerably wider, Reason argued, embracing a larger
range of individual and activities than those associated with the frontline operation
of a system.

In fact, Reason concluded from his review, it seemed that the further people were
removed from the frontline activities of their system, the greater the potential danger
they posed to it. Only responding to active errors would not lead to much system
improvement or accident prevention. The particular combination of causes and the
proximal acts that produced the latest accident, after all, were unlikely to repeat
themselves in precisely that way, so there was more benefit to trying to remove or
remedy underlying conditions responsible for a family of accidents.

Reason, together with others who had grown alarmed about the enthusiastic embrace of systems thinking, would rescind this view only a few years later. He called for revisiting the individual, because the net had been thrown too widely; the pendulum had swung too far (see Section 9.3.4). Leaving little doubt about where he ended up in his thinking about the individual versus the system, he called his 2008 book *The Human Contribution*, with a subtitle of 'Unsafe acts, accidents and heroic recoveries' (Reason, 2008). Unsafe acts—errors and violations—of people are "commonplace, banal even," he wrote, and commented that "the predominant mode of treating this topic is to consider the human as a hazard, a system component whose unsafe acts are implicated in the majority of catastrophic breakdowns" (p. 3).

In this book, the redemption of human fallibility did not come from turning back to the system that produced the conditions for it (as his 1990 and 1997 books had done to a much greater extent). It came instead from studying the other side of the coin. Humans were not only hazards, they were also (though on rare occasion) heroes. How was it that some people were capable of thwarting a disaster scenario? "Some people are born heroic," Reason said, "but most of us can acquire the skills necessary to give a better than evens chance" of heroically recovering from a doom scenario. In Chapter 11, you will see that the resilience literature is less interested in these two extremes (poor performance versus heroic performance) and focused much more on the huge portion of normal, quotidian work in the middle; of things that go right because of human effort—despite the obstacles, goal conflicts, design issues, and organizational resource constraints under which they work.

9.2.4 POROUS LAYERS OF SYSTEM DEFENSES

As said in the beginning of this chapter, the recognition has grown that mishaps (a refinery explosion, a commercial aircraft accident) are inextricably linked to the (mal-)functioning of that system—the surrounding organizations and institutions. The construction and operation of commercial airliners or upstream gas networks or healthcare or space shuttles or passenger ferries spawn vast networks of organizations to support it, to advance and improve it, to control and regulate it. People are not the instigators of failure; they are the recipients of it, the inheritors.

Complex technologies cannot exist without these organizations and institutions—carriers, regulators, government agencies, manufacturers, subcontractors, maintenance facilities, training outfits—that, in principle, are designed to protect and secure their operation. Their very mandate boils down to not having accidents happen. But since Barry Turner, safety science has increasingly realized that the very organizations meant to keep a technology safe and stable (human operators, regulators, management, maintenance) are actually among the major contributors to breakdown.

Yet most organizations, even if they are, in Reason's language, hosts to many resident pathogens, do not frequently produce organizational accidents. This is because safety-critical systems are generally too well protected against the vulnerability of single factors or processes. The protection against failure is to put in barriers. These barriers, or defenses, need to be put in place to separate the object to be protected from the hazard that might harm it.

These barriers, as explained in Section 9.2.1, are measures or mechanisms that protect against hazards or lessen the consequences of malfunctions or erroneous actions. These defenses come in a variety of forms. They can be engineered (hard) or human (soft), they can consist of interlocks, procedures, double-checks, actual physical barriers, or even a line of tape on the floor of the ward (that separates an area with a particular antiseptic routine from other areas, for example):

> The popular representation of the energy model leads us to think of barriers as very concrete physical structures or devices. However, a functional view may be more productive when it comes to systematic loss control. A functional view implies that we think in terms of goals and means. We may think of a function as a task which is defined by one or more objectives to be achieved under specified conditions, for instance 'prevent ignition of hydrocarbons after an uncontrolled hydrocarbon release in the process module.' By taking a functional view, we thus focus on the tasks that are necessary to adequately control a specific hazard. These tasks may be performed by passive physical structures (e.g. fire proof walls), by active technical systems (e.g. the gas detection and emergency shutdown system on a production platform), or by humans, usually in interaction with technology and supported by procedures (e.g. the control of hot work so as to keep it separate from inflammable objects and substances). Thinking in terms of functions invites us to consider alternative means to implement a loss reduction strategy. For instance, if a gas detection system has to be inoperative during maintenance, an operator with a portable gas meter and radio communication with the control room operator may perform its task. Moreover, we need to consider that barriers can deteriorate and need to be monitored and maintained ... The term 'safety function' is sometimes used in a sense similar to 'barrier function.'
>
> **Rosness et al. (2010, pp. 36–37)**

According to Reason, the best chance of minimizing accidents is by identifying and correcting delayed action failures (latent failures) before they combine with local triggers to breach or circumvent the system's defenses. This is consistent with ideas about barriers and the containment of energy or the prevention of uncontrolled release of energy. All these defenses are designed to serve one or more of the following functions (Reason, 1997, p. 7):

- to create understanding and awareness of local hazards;
- to give clear guidance on how to operate safely;
- to provide alarms and warnings when danger is imminent;
- to restore the system to a safe state in an off-normal situation;
- to interpose safety barriers between the hazards and the potential losses;
- to contain and eliminate the hazards should they escape this barrier;
- to provide the means of escape and rescue should hazard containment fail.

Together, these defenses generally protect a system from the hazards it embodies. At the same time, these defenses are not perfect. They are porous. Recall that the first drawing of the model (from 1990) only had a single hole in the last layer of defense: all others simply were solid panes stacked in a row. But from the next incarnation onward, the holes in these layers of defense took on greater significance and turned the model into the Swiss cheese we know it as today. The term 'Swiss cheese model,' incidentally, was not coined by Reason himself, but probably by Rob Lee, who was then director of the Bureau of Air Safety Investigation in Australia (Reason et al., 2006).

What are these holes, then? Layers of defense get undermined by the human and organizational flaws and processes which Turner also studied. An interlock can be bypassed, a procedure can be ignored, a safety valve can begin to leak in a way that is not detected. An organizational layer of defense, for example, involves such processes as goal setting, organizing, communicating, managing, designing, building, operating, and maintaining. All of these processes are fallible, and produce the latent failures that reside in the system. This is not normally a problem, but when combined with other factors, they can contribute to an accident sequence (Figure 9.1).

According to the Swiss cheese model, accidents happen when all of the layers are penetrated. Incidents, in contrast, happen when the accident progression is stopped by a layer of defense somewhere along the way. The Swiss cheese model indeed got its name from the image of multiple layers of defense with holes in them. Only a particular relationship between those holes will allow hazard to reach the object that was supposed to be protected. If these holes all 'line up,' then they together create a linear window of opportunity for an 'accident trajectory.' For Reason, defenses are weakened by:

- line management deficiencies
- supervisory shortcomings
- fallible organizational decisions
- unsafe acts at the sharp end.

Unsafe acts, as they were for Heinrich, are still central to the Swiss cheese model. Without them, a full accident cannot occur, or at least it cannot breach the final layer of defense. Unsafe acts by people at the sharp end are still necessary to make the model

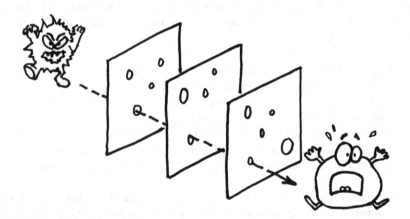

FIGURE 9.1 The defenses-in-depth or Swiss cheese model: layers of defense separate the object to be protected from the source of harm. But the layers are porous and can be penetrated. (Drawing by author.)

work at all. Of course, an unsafe act can only be defined, or only be meaningful, in the presence of a particular hazard and the potential that it can turn into a risk—otherwise there is nothing unsafe about the act. So for people to commit unsafe acts, a complex interaction was necessary between intrinsic system influences and outside factors.

Following in the footsteps of Heinrich, Reason (1990) argued that an unsafe act is "an error or a violation committed in the presence of a potential hazard: some mass, energy or toxicity that, if not properly controlled, could cause injury or damage" (p. 206). He suggested the following classification of unsafe acts:

- *Unintended* action: This could be a slip, a lapse, or a mistake, respectively, related to attentional failures, to memory failures or to rule- or knowledge-based issues (e.g., misapplication of an otherwise good rule because of misunderstanding or buggy knowledge).
- *Intended* action: This could be either a mistake or a violation. Reason (1990) distinguished between exceptional and routine violations. Violations are exceptional when they occur only occasionally to deal with a specific situation that could not otherwise be safely or adequately managed. Violations are routine if they occur every time a particular situation shows up.

Defenses themselves can become dangerous, too, as Reason pointed out in his 1997 book. The idea, for example, that an extra barrier can help hospital wards prevent medication errors, has been proven quite false. One barrier in the form of a red vest with 'don't interrupt!' warnings on it, warn by a nurse preparing medication, actually led to more interruptions, because colleagues could now more easily find the nurse and often wanted to ask just a small thing to not have to interrupt with something bigger later.

Another barrier, in the form of double-check of prescriptions and medication preparations, led to an increase in medication errors, not a decrease. The problem was the illusion of social redundancy, that is, the false idea that two people are entirely independent monitors or checkers of the same process. Given that colleagues know each other, and get to rely on each other (and, as you learned under groupthink, probably wish to preserve some social harmony) the double-check is not only ineffective, but riskier than not having one. People do rely on each other to get things right, and thus do not offer a robust double-check.

In an example from a remote mine site, a caterer had been brought in to provide meals to the workers. The contract was up for renewal, and the caterer had had a share of hand injuries as its workers prepared the food for the miners. In order to qualify for renewal of the contract, the number of injuries needed to go down. The caterer invested in steel-mesh gloves for its workers, and hand injuries were all but eliminated. But some time later, gastro-enteritis broke out among the miners. Contamination from uncooked food had spread through the gloves. This is what Reason would have called a 'dangerous defense.'

9.2.5 SHARED ASSUMPTIONS BETWEEN REASON, HEINRICH, AND BIRD

Swiss cheese has been credited with affirming an interest in organizational and managerial control of safety, and in the precursors and latent conditions that undermine it. It has also reminded people that saying that 'human error' caused the accident is inadequate. It emphasized and visualized—like Heinrich's and Bird's models—that frontline workers are not just the instigators of trouble, but the recipients of trouble that is present elsewhere in the organization.

Then there are other assumptions that Swiss cheese has in common with Heinrich's and Bird's models:

- *The necessity of 'unsafe acts' by people on the frontline* as a breach of the final defense. These unsafe acts are required to make the whole model work. Without them, accumulated risks that penetrated the other barriers cannot become a full accident;
- *Accidents and losses are the result of a linear, single sequence of events or factors.* The order in which one thing affects the other is invariable. And there is only one track along which failure travels: it does not bifurcate, branch, or run along multiple parallel tracks. Any factor or event removed from the fixed order will prevent the outcome, though some factors or events are more difficult to remove than others;
- *A failure is necessarily preceded by a failure.* For something to go wrong in one stage of the domino model or defenses-in-depth model, something must have gone wrong before it. And something before that, until the original failing domino or breached layer of defense is reached;
- *Risk and hazard is modeled as the uncontrolled transfer of energy.* An injury or incident is the result of such a transfer that exceeds human, machine, or containment limits. To contain hazards and risks, energy transfers need to be controlled or stopped by dominos that do not fall or layers of defense that do not have holes in vulnerable places;
- *The common cause hypothesis is true.* That is, incidents and accidents are caused by similar or identical sequences of events. No matter how severe the outcome, the causal trajectory that got you there is the same. Remember from Chapter 4 that this is of course a very important assumption that needs to be true for the idea of the pyramid or triangle to work, or for the Swiss cheese to be able to show how incidents relate to accidents. In both cases it has to be true, otherwise you cannot prevent severe outcomes by focusing on less severe ones (in the case of Swiss cheese, incidents are stopped before they have breached all the defenses, whereas accidents represent the breach of all of them).

These basic assumptions are shared between Swiss cheese, Heinrich, and Bird. Although, two aesthetic features changed:

- the eyes of the dominos are transformed into holes through which an accident trajectory can pass;
- the dominos no longer topple each other in succession but rather represent imperfect layers of defense.

The differences are subtle and not substantive. In the Swiss cheese model, an accident trajectory travels through the layers of defense while these remain upright (albeit porous), allowing a hazard which is otherwise barred, to reach the place where it can do harm. In the Swiss cheese model, Bird's 'inadequate control' is spread across and made visible in each layer (or slice): the holes are the evidence for that.

The resemblance between prewar safety models and Swiss cheese is testimony to the relevance and resilience of the ideas in it. The assumptions they have in common can also be linked to some of the critique of Swiss cheese and lack of theoretical progress during the 1990s.

9.3 LINEARITY, JUDGMENTS, AND BUREAUCRATIC ORDER

In this section, we will reflect on some of the critiques and calls for further development that have been directed at the Swiss cheese model. We will consider, in this order:

- The assumptions of linearity and proportionality (or Newtonian assumptions) that may have held back further theorizing about the complexity and emergence of accidents;
- The use of judgmental labels for human actions (or inaction) that may sometimes take the place of explanation and understanding;
- The extent to which safety has become a matter of demonstrating that risks are under control through administrative ordering has contributed to an increasing bureaucratization of safety, and a focus on the 'work of safety,' rather than the 'safety of work.'

9.3.1 LINEARITY AND PROPORTIONALITY

Dominos and breached layers of defense are cause–effect, or sequence-of-events models. One thing leads to the other which then leads to the next, and so on. Seeing an accident in that way of course allows us to trace back what contributed to a failure at the frontline. But it comes at a cost. Most fundamentally, you see some of Newton's thinking reflected in the assumptions shared by these models:

1. Something can go wrong *only* if something else goes wrong. For every effect, there has to be a cause;
2. One thing typically leads to one other thing (one more domino, or the next breach in the subsequent layer of defense). One thing cannot really lead to lots of other things. The model is linear, in some sense: there is one line from inception to accident.
3. Risk is seen as energy that is not contained or channeled. Also, there is a fixed amount of energy and proportionality in the sequence: something small cannot cause something big (a small domino cannot topple a big one).

MACONDO AS A SERIES OF DEFENSE FAILURES

The BP report on the Macondo blowout, for example, modeled the disaster as a linear series of breaches of layers of defense, finding the following ones (BP, 2010):

1. The cement barrier did not isolate the hydrocarbons;
2. The shoe track barriers did not isolate the hydrocarbons;
3. The negative pressure test was accepted although well integrity had not been established;
4. Influx was not recognized until hydrocarbons were in the riser;
5. Well-control response actions failed to regain control of the well;
6. Diversion to the mud gas separator resulted in gas venting onto the rig;
7. The fire and gas system did not prevent hydrocarbon ignition;
8. The blowout preventer emergency mode did not seal the well.

Of course, in hindsight, a plausible sequence of events (of linear, successive failures) may well look like that. And finding faults and failures is fine because it gives us something, or lots of things, to fix. Again, that is the power of these kinds of models. The problem is that the sequence of events only represents the last moments before the accident. It traces the cause–effect series from the sea floor upward to the rig and into the air around it. The holes found in the layers of defense (the cement, the engineers, the regulator, the manufacturer, the operators, the emergency systems) are easy to discover once the accident has happened.

And it leaves the organizational, human story entirely unexamined. Why did nobody at the time see these now so apparent faults and failures for what they (in hindsight) were? At the time, people did not see holes, or act on them. Their sensemaking; their creation of local rationality in the thousands of little and larger trade-offs and decisions they make every day—that is impossible to visualize in this model. People were doing normal work leading to outcomes, mostly without anything having been breached relative to how they understood the system at the time. The holes, deficiencies, and failures were not seen as such, nor easy to see as such, or act on, by those on the inside (or even those relatively on the outside, like the regulator) before the accident happened. Because failures are a precondition for failure in the Swiss cheese model, it has no way of incorporating this insight:

> The weak point is that the SCM does not propose a more detailed account of how latent failures, intervening factors, and active failures may combine, i.e., that it does not provide a more detailed accident model. It therefore does not fully support the contemporary view of accidents as the result of a conjunction or aggregation of conditions, none of which need to represent a failure as such.
>
> **Reason et al. (2006, p. 20)**

Man-made disaster theory did a better job at this, as you will recall from Chapter 7. Accidents are preceded by a gradual but growing mismatch between how risk is actually under control and how whole systems of people and bureaucracies believe it to be under control. Modeling that into Swiss cheese is difficult. You can of course represent the various organizations and their parts (contractors, regulators, engineering) as successive layers of defense. But still, the buildup of latent failures, if that is what you want to call them, is not modeled. The process of erosion, of attrition of safety norms, of drift toward margins—none of that can be captured well by defense-in-depth models. They are inherently metaphors for resulting forms, not models oriented at processes of formation (or deformation). The models are static. They themselves cannot really *account* for the many influences that weaken defenses.

The Deepwater Horizon commission, which reported to the U.S. President, seemed to realize the limits of modeling the accident as a series of linear failures or a concatenation of latent and active conditions that were visible only in hindsight. The commission quoted the Columbia Space Shuttle Accident investigation when it warned that "complex systems almost always fail in complex ways" (Graham, 2011, p. viii). Drawing up a linear story of such a complex event does it no justice. Even BP itself, in its 2010 report, acknowledged that what accounted for the accident was:

> ...a complex and interlinked series of mechanical failures, human judgments, engineering design, operational implementation and team interfaces [that] came together to allow the initiation and escalation of the accident. Multiple companies, work teams and circumstances were involved over time.
>
> *(p. 11)*

Any linear line-up of breached layers of defense has difficulty with portraying of a multitude of complexly interactive pathways and tight couplings that come together.

Another aspect typical of complex system failure is difficult to model as well. Remember from Chapter 8 how Perrow (1984) related the pressurization incident of an airliner that started with a small crack in a water tank. This is a characteristic of complex systems: small things can lead to big events. There is seldom a linear or proportional relationship between cause and effect. Things get amplified; they escalate or spin out of control.

This has been described in other books (Dekker, 2011), and stands out from a number of big accidents and disasters. A stretched lubrication interval led to a worn screw in the tail of MD-80 passenger jets, which led to the loss of a jet and all its passengers and crew. Changing the bureaucratic labeling of a foam strike from a safety issue to a maintenance problem led to different organizational treatment, categorization and budgeting of that problem, which eventually led to a disaster with a returning space shuttle (CAIB, 2003).

Although Swiss cheese is sometimes called a 'system model' or a 'systemic model' (Reason, 1995), it is not what is considered systems thinking. The systems part of Swiss cheese is largely limited to identifying, and providing a vocabulary for, the upstream structures (the blunt end) behind the production of errors at the sharp end. The systems part of these models is a reminder that there is context;

that we cannot understand errors without going into the organizational background from which they hail. All of this is necessary, of course, as errors are still all too often taken as the legitimate conclusion of an investigation. The Swiss cheese model contributes to this search of the wider organizational context not by predicting how specific accidents might happen (in the way that an event tree or fault tree would try to). Instead, it reminds people where they should look for the places where their systems and organization may be weakened:

> Rather than trying to predict the likelihood of specific types of accidents, it uses extensive practical experience to point to system functions that, if damaged, often are associated with the occurrence of incidents and accidents. This clearly has a considerable pragmatic value.
>
> **Reason et al. (2006, p. 20)**

Yet reminding ourselves of the context from which problems can stem is of course no substitute for beginning to explain the dynamics, the subtle, incremental processes that lead to, and normalize, the behavior in the system that we can eventually observe. This requires a different perspective for looking at the messy interior of organizations, and a different language to cast the observations in (see Chapters 7 and 11).

The central issue in the examples above is not necessarily that defenses are a bad or dangerous thing. Rather, it is that we should not limit ourselves to thinking about risk only as energy (heat, momentum, speed, or something caustic, pharmacological, toxic or radiological) that needs to be contained or carefully channeled. The popularity of the Swiss cheese model, with its image of the layers of defense between the source of hazard and the object to be protected, may have unwittingly contributed to a fixation on risk-as-energy-to-be-contained, which may or may not work well for the system in question:

> The energy perspective may be most relevant for systems where the technical core and the hazard sources are well defined, physically confined and stable, for instance nuclear power plants or offshore oil production platforms. The scenarios following the release of a major hazard in such systems are usually confined to one or a few paths (e.g. fire/explosion or structural collapse on an oil platform). In this case, quantitative risk analyses may emphasize the reliability of barrier functions.
>
> In contrast, air transport is a distributed large-scale system. The functional technical core is divided among aircraft and infrastructures. Safe operations depend on the co-ordination of decentralized activities. Moreover, it is simply not feasible to design an aircraft strong enough to withstand a head-on mid-air collision. For these reasons, risk reduction efforts should emphasize preventing the release of hazardous energy, for instance by ensuring that critical systems are operative when they are needed.
>
> Road transport could, however, benefit from more use of the energy/ barrier model. Haddon developed the energy barrier model for the road safety. And we have seen that the Swedish road authorities use it extensively, for instance by introducing mid-barriers on rural roads.
>
> **Rosness et al. (2010, p. 42)**

Indeed, it is always good to remind ourselves that other possibilities exist too. Risk can be constructed in many different ways, and the foundational models discussed throughout this book attest to a number of them:

- *Risk as energy to be contained.* As depicted in the Swiss cheese model, the sources of hazard need to be kept apart from the objects to be protected by layers of defense. These contain or channel the energy from the sources of hazard.
- *Risk as the structural property of a system's coupling and interactive complexity.* This was articulated in Perrow's normal accident theory: the way to deal with risk is to reduce coupling and complexity (and extra layers of defense may well add coupling and complexity).
- *Risk as a gradual drift in what it is seen as normal operations,* or a normalization of deviance. This conceptualization of risk can be seen in drift models and in Vaughan's and Snook's ideas, as well as Barry Turner's of course. Risk, in this case, lies in people's reading or interpretation of their operations and the gradual shifts that take place in that reading.
- *Risk as an exportable property.* This was articulated by Beck (1992), who recognized that those who do not want the risk can typically sell or give risk to those who want to have it. This happens in financial markets and insurance, but also in the export of toxic waste, for instance. Contracting work out is also a form of exporting risk. The arrangement can work as long as people actually know what they are buying (and selling). When the mortgage crisis hit the United States in 2008, it turned out that such knowledge had got lost (or perhaps deliberately obfuscated by some) in the ways risk had been packaged, chopped up, and repackaged (Dekker, 2011).
- *Risk as the absence of adaptive capacity.* This is the conceptualization of risk favored by cognitive systems engineering and particularly resilience engineering. Variation in operations itself is not risky, but systems (including people, organizations, and human-machine systems) which have become stale or miscalibrated in their strategies for absorbing such variation do represent risk.

The conceptualization of risk as energy to be contained or managed has its roots in efforts to understand and control the physical (or purely technical) nature of accidents. This of course goes back to Heinrich. It also spells out the limits of such a conceptualization: it is actually not well suited to explain the organizational and socio-technical factors behind system breakdown. The Swiss cheese model is not equipped with a language that can meaningfully handle processes of gradual adaptation, risk management, and decision-making. Perhaps it was never meant to:

> The Swiss cheese model does not provide a detailed accident model or a detailed theory of how the multitude of functions and entities in a complex socio-technical system interact and depend on each other. That does not detract from its value as a

means of communication, but may limit is use in analysis and as a support for proactive measurements. On the other hand, the Swiss cheese model was never developed to be a detailed model in this sense.

Reason et al. (2006, p. 21)

Perhaps Swiss cheese had increasingly been asked to account for more than it actually could. In 2004, a few years after a mid-air collision over Überlingen in southern Germany showed again how much hugely intertwined and complex socio-technical processes matter for both containing and creating major system disasters, Reason himself asked at a workshop convened for that purpose, "Is Swiss cheese past its sell-by date?" (Reason et al., 2006, p. 1).

9.3.2 JUDGMENTS RATHER THAN EXPLANATIONS

When the Swiss cheese model does talk about human action, it relies on a judgmental vocabulary. Recall the terms chosen by Reason to explain how defenses can become porous or breached:

- line management deficiencies;
- supervisory shortcomings;
- fallible decisions;
- unsafe acts;
- violations.

Others who have relied on Swiss cheese for their own models have added to this judgmental vocabulary. For example, HFACS (a human factors analysis and classification system based on Swiss cheese) uses 'poor leadership' as one of those labels (Shappell & Wiegmann, 2001). These are all quite judgmental indeed, in the same way that Turner's labels for flawed processes in the organizational bureaucracy were.

A question we need to ask is to what extent this actually *explains* human action and whether it offers us levers to do anything about it. The terms above judge people for not doing something that (in hindsight) could have avoided the outcome, such as not being deficient in managing, not making fallible decisions, avoiding supervisory shortcomings, not committing unsafe acts or violations. Of course, we can perhaps get some short-term leverage by (re-)imposing rules or regulations. We can put up posters and make moral appeals for people to care about safety, but the effects are generally short-lived. Actual work is always under pressure to deal with goal conflicts, and adapt to conditions of scarcity and competition. It will likely again drift into niches that generate greater operational returns at no apparent cost to safety.

From the point of view of people inside the situation, as well as potential other observers, those 'errors' and 'deficiencies' and 'shortcomings' are little more than normal work. If we want to understand why it made sense for people to do what they did, we have to reconstruct their local rationality. What did they know? What was their understanding of the situation? What were their multiple goals, resource constraints, pressures? Why did it make sense for them to do what they did? People do not come to work to do a bad job.

If we take our guidance from a defense-in-depth model, however, to understand why it made sense for people in one layer to do what they did, it can quickly point to the layer before it. Problems at the sharp end can automatically get traced back to problems elsewhere in the organization—elsewhere along the layers of defense. Sharp-end errors can get attributed it to pathogens deeper in the organization). The issue is that rather than explaining error, it displaces the error onto another person: a supervisor, a manager, a maintenance technician, a board member.

9.3.3 ADMINISTRATIVE ORDERING AND SAFETY BUREAUCRACIES

Safety is increasingly seen as a matter of administrative ordering and managerial control. The Swiss cheese model may have contributed to this trend: it shows an outcome failure as the result of many smaller, prior failures in organizational and administrative layers upstream from the sharp end. Again, it is empowering and emancipatory because it provides a strong reminder of context, of the system surrounding people, not just the person at the sharp end.

Yet as much as the model encourages decision makers to think about how to enable and empower people at the sharp end to do the right thing, and to avoid leaving them to hold the bag of an imperfect organization, some might have seen it as an affirmation that people are a problem to control. The frontline, after all, not only inherits latent pathogens from the organization; many believe that it could not function without all the administrative, managerial, and supervisory control that hangs over it. And if layers of organizational protection are imperfect and porous, this is the result not only of bad procedures or unsuitable designs, but (in the language of the model) of violations, unsafe acts, line management deficiencies, fallible decisions and deficient supervision. That means that there are now lots of other people, higher up in the organization, who are also a problem to control.

The Swiss cheese image conveys the idea that safety at the sharp end is assured by good administrative and technological ordering of the upstream system. We need to find and fix the holes or failures there to prevent outcome failures. This suggests that we might put our faith in bureaucracy and administration, as well as science and technology. It can help entrench the idea that safety is generated chiefly through planning, process, paperwork, audit trails, and administrative work. It may have resulted in a new focus on safety management systems that count and tabulate negative events, on procedural compliance, and surveillance and monitoring, and it puts new procedural and regulatory limits on the people who do work at the sharp end.

If the image of Swiss cheese suggests that better hierarchical ordering of the organization will translate into better risk management, then you can see this belief in the proliferation of safety management systems, auditing systems, or loss prevention systems across industries. We have to make sure layers of defense are in good order, and patch the holes in them. Some even read the function of a safety management system to be just that. It suggests tracking down 'failures' of risk management in the wider organization (e.g., deficient supervision, poor leadership), and other deficiencies related to design, procedures, technologies, checklists, or the organization of work.

Cultures of safety bureaucratization and far-reaching compliance have become pervasive and difficult to contain (Dekker, 2014, 2018). Here are some of the implications:

- Efficiency and safety at the operational sharp end (as they were in Taylor's time) are largely dependent on the managers, planners, engineers higher up in the organization (at the 'blunt' rather than 'sharp' end). Problems at the sharp end are inherited from a lack of administrative and organizational control at the blunt end. While it is emancipatory to see frontline or sharp-end workers as the recipients rather than instigators of trouble, it also makes clear that the solutions for this lie with Taylor's 'smart' people: the planners and managers at the top.
- This removes a deference to craftsmanship and expertise at the sharp end. Expanding machineries of safety can make an organization listen less to technical expertise and operational experience, and disempower middle management and supervisory layers. People may no longer feel as able or empowered to think for themselves. This could stifle innovation, choke off initiative, and erode problem ownership;
- A belief in the administrative ordering of work to ensure it is both safe and efficient not only involves several layers of line management and supervision, but extensive safety bureaucracy. These sometimes formalize practices in which accountability for safety is generated chiefly through process, paperwork, audit trails, and administrative work—all at an increasing distance from the operation.
- Rather than safety as a practical and ethical responsibility downward, it can become a bureaucratic accountability upward. In many industries, safety has morphed from an operational value into an administrative machinery. Burgeoning safety bureaucracies, which work at an increasing distance from the operation, are usually organized around lagging, negative indicators. They tend to establish vocabularies of deficit, control, and constraint. People are once again a problem to control.
- A machinery for the surveillance and monitoring of human behavior is necessary, then, to manage safety. It happens through safety management systems, error counting and classification systems, data recording and monitoring systems, operational safety audits, and other auditing systems.
- Better organization of the blunt-end will enhance the possibility of an accident-free future. According to Swiss cheese thinking, the best chance of minimizing accidents is by identifying and correcting delayed action failures (latent failures) before they combine with local triggers to breach or circumvent the system's defenses.
- Accidents (i.e., accident trajectories that penetrate all the layers of defense) point to risk that was not managed well. And behind that, a person or persons can be found culpable or at least responsible (and held liable).

The last bullet solves an important problem. Because some, including Reason himself, have commented that the increased focus on upstream factors, aided by his own model, has possibly diluted responsibility and accountability (Young, Shorrock, Faulkner, & Braithwaite, 2004). Some people started fearing that Swiss cheese thinking got people off the hook too easily. That it might allow them to 'blame the system' whenever things go wrong, even when all system provisions to do the right thing were thought to be in place (Sharpe, 2003).

Reason asked in 1999: "Are we casting the net too widely in our search for the factors contributing to errors and accidents?"(p. 210). He argued that the pendulum may have swung too far, and that it was time to revisit the individual. The Swiss cheese approach, after all, should not make it impossible to hold people accountable for poor behavior. Were we no longer allowed to blame anyone? He wasn't alone. Patient safety researchers Wachter and Pronovost observed (2009):

> ...beginning a few years ago, some prominent ... leaders began to question the singular embrace of the 'no blame' paradigm. Leape, a patient-safety pioneer and early proponent of systems thinking, described the need for a more aggressive approach to poorly performing [practitioners].

> *(p. 1401)*

Their proposal made it into the *New England Journal of Medicine*. A swift response by a colleague and me was published there soon as well (Dekker & Hugh, 2010). The appeal to punish noncompliant practitioners (or 'hold them accountable') is part of a string of such calls during the 1990s and 2000s. Aviation safety researchers, concerned at the enthusiasm of investigators to find all sorts of system precursors and 'latent conditions,' called at one of their conferences for a refocus on active 'errors' by frontline workers rather than mitigating factors in the system (Young et al., 2004). Sometimes, they argued, accidents *are* caused by people at the sharp end—it is as simple as that! Perhaps the desire to simplify and blame remains alive and well (Alicke, 2000; Catino, 2009; Douglas, 1992).

9.4 SWISS CHEESE AND SAFETY MANAGEMENT SYSTEMS

9.4.1 DIRECTING ATTENTION AWAY FROM THE SHARP END ALONE

Rather than controlling individual behavior or human factors on the frontline of an organization, Swiss cheese became a popular image that helped direct attention back into the organization itself. By layering the organizational contribution to safety (through its defenses), it also systematized the context from which frontline problems stem. It got decision makers to consider more critically how to direct safety efforts and resources at entire systems, at their own organization. It has once again put goal conflicts, procedural imperfections, design issues, and operational and organizational constraints on the agenda, and has made people aware that these are the sorts of things that create problems and safety risk for everybody.

The issue, Swiss cheese clearly communicates, is not just frontline behavior. Problems cannot simply be adjusted with clever incentives or sanctions targeted at persons there. Swiss cheese literally illustrates that there is a whole organizational and engineered context that surrounds or at least precedes human action at the sharp end. This is the system that sets the conditions for their success or failure. And a safety management system is intended to assure and formalize the way in which it does so.

9.4.2 Demonstrating That Safety Risks Are Well Managed

A safety management system is a systematic approach to managing safety. Among other things, it:

- Defines how the organization is set up to manage risk;
- Identifies operational risks and implements suitable controls;
- Develops effective communications across all levels of the organization about these risks and their controls;
- Implements a process to identify and correct nonconformities;
- Supports a continuous improvement process.

To achieve this, a safety management system includes, for example:

- organizational structures related to safety functions;
- people's responsibilities and accountabilities related to safety;
- processes for gathering, analyzing, and storing safety-related information;
- safety policies and procedures.

Safety management systems are consistent with the position popularized by the Swiss cheese model: safety can be assured by assessing and securing the quality of the organizational layers of defense that control hazards and keep them away from the places where they can do harm.

It is difficult to pinpoint an exact origin of the safety management system. Many historical developments, among them the conceptual ones laid out in this book, point in the direction of safety management systems as a plausible and almost to-be-expected late 20th-century outcome. Developments in our understanding of the organizational and engineered context of safety and disaster prevention laid the theoretical groundwork, as did emerging insights about resilience (see Chapter 11). The latter fits the idea that an organization should demonstrate that safety risks are well managed not by showing the absence of those risks, but by the presence of organizational, managerial, and other capacities and processes.

It is also possible to draw a link with total quality management, which preceded safety in its systematic, organization-wide approach to assuring quality. Safety management systems are also a natural companion to any business management system, particularly in safety-critical industries. The promotion of safety cases (see Chapter 6) was yet another contribution to the rigorous analysis, documentation, and demonstration of the safety of a system. Various industries began promoting the adoption and use of safety management systems in the 1990s. This sometimes happened as a result of an inquiry into a large disaster (Gephart, 1984). The case of Piper Alpha was one of the first ones. Using a phrase for SMS that has since been lost to history, the Cullen inquiry ended up arguing that the responsibility for safety should be more on companies than on the regulator, and that a formal management system would be the way to make the link between them:

> The recommendations relating to safety assessment and the regulatory system stem from Cullen's view that existing detailed and prescriptive regulations were inflexible. He considered that the responsibility for offshore safety should be put more clearly on the companies rather than the regulator. By requiring a formal safety assessment (FSA) from the operator of any offshore installation, mobile or fixed, a more objective framework for regulation could be established. The FSA encompasses the whole life cycle of a project, from feasibility study through design, construction, operation, and abandonment. Its need arises because the combinations of mechanical and human failures are so numerous that a major accident hardly ever repeats itself. The techniques used include hazard and operability (HAZOP) studies, quantitative risk assessment (QRA), fault tree analysis, human factors analyses, and safety audits. … A Formal Safety Assessment (FSA) is an essential element in a modern safety regime for major hazard installations. FSA is the identification and evaluation of hazards over the life of a project… It is a demonstration that so far as is reasonably practicable the risks to personnel have been minimized.
>
> *COSOP (1991, pp. 3–19)*

An FSA, and later SMS, was deemed essential for demonstrating that risks were under control. It did not ask an organization to demonstrate that it was safe, but rather to present a formal system that it would use to demonstrate that it was trying to be as safe as it could be:

> Safety management refers to the activities of a safety-related character concerning organization, responsibility, processes, and resources that are required to direct and manage railway operations. Safety management is an organizational process that encompasses many steps, from strategic goals to evaluation of results. Safety management includes both the daily work, with checking that everything functions as it should, as well as a comprehensive assessment of risk and changes. These two forms are of different character. The daily work is of a practical nature and characterized by the need for somebody to be present all the time for safety to be adequate. The comprehensive assessment or the risk analysis is abstract and characterized by a comprehensive view and assessment of changes.
>
> **Almklov, Rosness, & Storkersen (2014, pp. 28–29)**

And safety management is more than that. A safety management system is also the way for a major hazard facility (and other organizations) to show that it was transparent to the community about its efforts to manage the risks associated with its operations. The European Council directive of 1996 (also known as the Seveso II Directive), which grew out of the 1976 Seveso chemical disaster in northern Italy, is aimed at doing just this, and at improving the safety of sites across Europe that contain large quantities of dangerous substances.

Two other trends converged at the end of the 20th century that boosted the wide embrace of safety management systems. One was the increasing adoption of neoliberal governance across much of the West (and, in fact, elsewhere). In 1981, it was noted how the Reagan administration in the United States (which, together with Thatcher's in the United Kingdom was seen as the forerunner of neoliberalism) believed that safety regulation had gone too far. Terms and standards had been set so strictly that costs easily outweighed benefits (Mendelhoff, 1981). A pushback on this was deemed necessary and politically expedient. Neoliberalism promoted deregulation, pro-market polices and in certain cases a reduction of public expenses on regulatory agencies (Blanton & Peksen, 2018; Saull, 2015). The burden of safety proof started shifting to the regulated organization, which now needed to demonstrate to a regulator that it knew what it was doing.

The second, related, trend has been that of increasing technological sophistication and specialization, and the concomitant expertise required to play a meaningful role as a government inspectorate (Wynne, 1988). It was simply getting too expensive and difficult to train and retain inspectorate staff to keep covering the technical nuances that needed to be deeply understood to be able to regulate a particular activity or innovation (OECD, 2014).

From the 1990s onward, it was becoming more efficient and cost-effective to target regulatory inspections at the administrative and management systems that were supposed to keep risk under control, rather than at the actual installation or technology itself. This has helped accelerate the merging of regulatory agencies too. Instead of separate regulators for different transport sectors, for instance (aviation, road, rail, shipping), a number of countries now have single transport regulators. Due to the generic basis of safety management systems, it is possible (at least in principle) to go across sectors and conduct audits and inspections partially independent of technical background.

9.4.3   THE SAFETY OF WORK, OR THE WORK OF SAFETY?

There is an important and far-reaching consequence of the shift of safety focus to the organizational 'back-of-house,' and the administrative activities that are intended to demonstrate risk control and compliance. Organizations now typically need to have an internal apparatus to develop, implement, change and update rules, and to notify, keep records, and report. It creates a lot of extra and even unnecessary work that does little to improve or assure safety. As Almklov and colleagues explained the dynamic, this can move an organization to:

> ...spelling out institutional procedures and decision rules that would otherwise be implicit, and establishing paper audit trails or their electronic equivalents. Those developments allow auditors and inspectors of various kinds—the exploding world of 'waste-watchers, quality police and sleaze-busters'—to verify that the written rules, procedures and protocols have been followed.

(2014, p. 26)

Safety management systems are meant as the catchment for all of this, including the necessary organizational structures, roles, responsibilities and accountabilities, policies and procedures, and documentation (lots of documentation). Almklov et al. (2014) explain:

> Safety management has become subsumed by the more generalized accountability-based mechanisms of governance that dominate today. An example is the trend towards increased reliance on internal control and self-regulation, where companies are expected to have transparent standardized systems for control. For external auditors and authorities, it is primarily the systems that are subject to control and regulation... safety standards should be seen not only as attempts to ensure safety and interoperability but also as a means of making safety work transparent across contexts. If workers perform tasks as the standards prescribe, they are compliant, at least from an accountability perspective, and this compliance is transparent to regulators and others without having to further investigate details of the local setting... [Yet] the rules, which are made to be applicable in several different settings, are more complex, more abstract, and less locally relevant than what is optimal for each setting... Standards are a means of making information mobile across contexts. Decisions and activities enter the systems of accountability by being performed and described according to standards. The bureaucratic methods of accountability depend upon activities and situations of each local context being translated into slots on the accountants' sheets.
>
> *(pp. 26–27)*

Evidence suggests that safety management systems and the shift in bureaucratic administration of safety duties are still captured by a kind of 'rule homeostasis.' The total number of rules remains high, even if the administrative center of gravity for those rules has changed from the government to the organization itself. Fear of not meeting due diligence requirements tends to make organizations overreact and overspecify their safety rules. This makes deregulation or self-regulation a very limited answer to bureaucratic safety clutter. Deregulation and safety management systems, in a typical neoliberal sense, 'make the customer do the work.' And the customer tends to find help in the market to get that work done. As commented by a cargo ship captain in Almklov's study:

> Consultant companies have never earned as well... I know many competent people in [this consultant company], but everything is going on paper to be documentable. I have written deviations and commented the formulations on the deviations, and they are sent back and forward. It's silly.
>
> *(p. 27)*

Consistent with neoliberal governance, the amount of safety bureaucracy demanded by regulators typically accounts for only 40% of this. About 60% is generated by the organization itself and imposed on employees and contractors alike (Saines et al., 2014). This is, paradoxically, a typical response to deregulation and the need to now demonstrate compliant internal safety management (Storkersen, Antonsen, & Kongsvik, 2016). The false belief is that the more that can be demonstrated (the more internal rules, processes and procedures), the more compliance is assured. The opposite turns out to be true, and would seem pretty intuitive: the more rules there are, the easier it becomes to show that at least some of them are not complied with (Tooma, 2017). At the heart of this, though, is a dilemma for all stakeholders:

> Safety specialists are often agents in relationships characterized by principal-agent dilemmas: The agents hired to help a company with the safety systems do not necessarily have the exact same interests as their principal. We have suggested that at least in some cases, it can be in the interest of the hired safety specialists (the agent) to work with more standardized systems and systems that require less local adaptation.
>
> *(p. 33)*

This is the market at work in driving the bureaucratization of safety (Dekker, 2014). Bureaucratization means the administrative governing of the relationship between the means an organization dedicates to safety and the ends it hopes to achieve with them. This is typically done by organization members who are not representative of frontline workers. According to sociologist Max Weber (1864–1920), bureaucratization involves hierarchy, specialization, and division of labor, and formalized rules:

- Safety becomes a bureaucratic accountability to be demonstrated 'up' the organizational hierarchy, more than an ethical responsibility for those who do safety-critical work at the frontline. Safety and operational managers need to show low numbers of injuries to their superiors, for instance, and the most important question at pre-start work meetings in one large organization was whether everybody had signed the attendance sheet, so that compliance could be documented and demonstrated;
- The specialization and division of labor: not only has safety work become more of a specialization separate from operational labor, it also has further differentiations and divisions within (e.g., from occupational hygienists, biohazard managers, emergency response planners to process safety specialists);
- Bureaucratic work is characteristically conducted by nonrepresentative members of an organization. In this case, those who do safety work are not necessarily chosen or elected to speak or act on behalf of a constituency (e.g., the operators doing safety-critical work), nor do they necessarily have experience or expertise at doing frontline work.
- Formalized rules refer to standardized responses to known problems and fixed procedures that govern the collection, analysis, and dissemination of information as well as the processes by which decisions are arrived at, and how both authority and responsibility for decisions are distributed, upheld, and accounted for.

As Weber warned, the creation of additional internal bureaucracy—with a slew of external stakeholders orbiting around it to supply services that help feed and grow the bureaucracy—has secondary effects that run counter to the organization's objectives, and probably counter to the whole idea of deregulation. As Hasle and Zwetsloot observed:

> ...critics have used harsh words to describe management systems, such as 'scam', 'fraud', 'bureaucracy' and 'paper tigers' and pointed out that workers lose influence. Other issues are their usefulness and cost for small and medium enterprises, and their relevance in the 'changing world of work', where production is increasingly outsourced, and risk can be easily shifted to partners in the supply chain, or to contingency workers. The certification regimes associated with [safety management] systems have also been criticized, e.g. for increasing the cost to businesses and for becoming an aim in themselves.
>
> *(2011, p. 961)*

A considerable amount of 'safety clutter' now pervades workplaces, with a requirement to document and demonstrate a process for even the smallest of tasks. This is the type of bureaucracy that is surprisingly easy to accumulate. It is persistent, and it does not contribute to frontline safety (Rae, Weber, & Dekker, 2018). In fact, it typically gets in the way of getting work done (Dekker, 2014). Hasle and Zwetsloot point to a number of other 'ills' as well, as do other studies:

- The rise of a costly safety industry that helps convert safety management into a purpose in itself, with an expansion of internal bureaucratic clutter that is of little use to small- and medium-sized companies (if to large ones).
- Safety management systems might not have the sensitivity to respond to changing work arrangements (and in fact might accelerate such changes), e.g., the increasing reliance on contingent workers with precarious employment conditions (Valluru, Dekker, & Rae, 2017).
- Safety management systems cannot offer assurance, as Beck (1992) flagged decades ago, that risk does not get transferred to other parts of the world or move along the production chain to where there is less resistance and greater economic benefit to the organization.
- In one study, the adaptive capacity of operational people was shown to be eroded under this type of governance, with frontline operators retreating into formal rules and roles just to show compliance—even in escalating, non-normal situations that demanded adaptation and innovation (Bergström et al., 2009).
- As Hasle and Zwetsloot observe, 'workers lose influence' as a result of this market-driven arrangement. This has been backed up by research elsewhere, which shows that the change to government deregulation, which yields more internal bureaucracy, has taken influence away from workers, yet blames them more frequently when things go wrong (Gray, 2009).

Sociologists, following Weber, would probably not be surprised. They saw it happen before: a society dominated by bureaucratic organizations—with governments abetting them and small and medium enterprises supporting them—can erode the autonomy and power of many on the inside:

> Tracing that historic transformation, Coleman affirmed Weberian pessimism. He observed that this change altered social relations: Individuals not only interacted with individuals as before, they also interacted with organizations, and organizations interacted with other organizations. Coleman's primary insight was that this structural transformation produced both perceived and real loss of power for individuals.
>
> **Vaughan (1999, pp. 271–272)**

It has also produced a dilution of the duties of safety people, and a diffusion of the goals they purportedly need to help the organization achieve. A recent study showed that safety professionals were often unable to articulate a clear goal for their work activities (Provan, Dekker, & Rae, 2017, 2018). The underlying purpose had to be teased out and observed through their practice: it is not something that is typically made explicit by those who do the work. When safety people did say why they were performing the work activity, they expressed their goal in overly generic ways, for example:

- 'to implement the safety management system';
- 'to support the achievement of the organization's goals';
- 'to simplify activities';
- 'to improve safety'.

This work of safety rarely had a clear and specific goal that was related to a current risk exposure facing frontline workers. In other words, the work of safety which they do actually has little to do with the safety of work. Safety professionals are arguably best placed in the organization to monitor the ever-changing nature of safety risk, but this is not a role that they play in current systems of safety governance. In fact, as their roles have moved closer toward supporting management needs and increased bureaucratization, it has moved further away from the safety risk facing the workforce.

Safety people often lack the legitimacy to intervene in the tools and tasks of people performing work in which the safety professional has no direct experience (Weber, MacGregor, Provan, & Rae, 2018). Assuring and assisting in the safety of work turns out to be the hardest part of a safety person's task, for which safety people hardly find time among their 'blunt-end' administrative duties. This means that safety people could be argued to not be acting in the best interests of the organization, its workers, or other stakeholders, including owners, regulators, or the surrounding community and environment (Figure 9.2).

The gradual shift to 'back-of-house,' to organizational and administrative assurance of safety, had been long in the making. It was popularized by the Swiss cheese model in the late 1980s. The contributions and shifts that this trend has brought to safety science are huge. No longer can we just stop at the human factor at the sharp end (Woods, Dekker, Cook, Johannesen, & Sarter, 2010). No longer can we deny that all work is shaped by the engineered context and workplace conditions and upstream organizational factors, and that if we want to understand or change anything—*that* is where we need to look.

FIGURE 9.2 Safety people often lack the legitimacy to intervene in the tools and tasks of people performing work in which the safety professional has no direct experience (Weber et al., 2018). (Drawing by author.)

But this trend has also left us with a safety profession that now broadly lacks purpose and vision in many industries, with regulators whose roles have in certain cases been hollowed out and minimized, and with an overwhelming safety bureaucracy that has little capacity to understand or administratively accommodate the nuances, innovations, and adaptations of frontline work (Dekker, 2018). It is from this setting that a newly totalizing concept entered safety science and popular imagination: that of 'safety culture.' The next chapter is dedicated to it.

STUDY QUESTIONS

1. In what way is the Swiss cheese a 'systemic model,' and in which ways isn't it one?
2. What is the conceptual relationship between Frederick Taylor and the Swiss cheese model?
3. What are 'resident pathogens' in the Swiss cheese model? How are accidents caused, according to the model?
4. What is an 'impossible accident,' as opposed to a 'normal accident'?

5. What aspects of modeling (and controlling) risk do Heinrich and Swiss cheese have in common? Name at least three.
6. In what ways did Barry Turner lay the groundwork for Swiss cheese?
7. How can you see Swiss cheese strongly reflected in investigation methods used in your own organization or industry? What might applying a method like that typically bring out *and* miss as a result of that?
8. There are various ways to define and thus model 'risk.' Which of the ways you have learned about in this book (and this chapter in particular) is the Swiss cheese good at? Which ones not so much, and why not?
9. What is 'safety clutter'?
10. Explain how the concept of 'safety culture' would be a logical next installment of a safety-scientific trend that has increasingly highlighted the role of administration and organization in assuring safety.

REFERENCES AND FURTHER READING

Alicke, M. D. (2000). Culpable control and the psychology of blame. *Psychological Bulletin, 126*(4), 556–574.

Almklov, P. G., Rosness, R., & Storkersen, K. (2014). When safety science meets the practitioners: Does safety science contribute to marginalization of practical knowledge? *Safety Science, 67,* 25–36.

Beck, U. (1992). *Risk society: Towards a new modernity.* London, UK: Sage Publications.

Bergström, J., Dahlström, N., Van Winsen, R., Lützhöft, M., Dekker, S. W. A., & Nyce, J. M. (2009). Rule- and role retreat: An empirical study of procedures and resilience. *Journal of Maritime Studies, 6*(1), 75–90.

Blanton, R. G., & Peksen, D (2018). Pro-market policies and major industrial disasters—a dangerous combination? *Sociological Forum, 33*(1), 5–29.

BP. (2010). *Deepwater horizon accident investigation report.* London, UK: DIANE Publishing.

CAIB. (2003). *Report volume 1, August 2003.* Washington, DC: Columbia Accident Investigation Board.

Catino, M. (2009). Blame culture and defensive medicine. *Cognition, Technology and Work, 11,* 245–253.

COSOP. (1991). *Report of the consultative committee on safety in the offshore petroleum industry (OR 0935).* Canberra, Australia: Committee on Safety in the Offshore Petroleum (Australia), Mines Department Library.

Dekker, S. W. A. (2011). *Drift into failure: From hunting broken components to understanding complex systems.* Farnham, UK: Ashgate Publishing Co.

Dekker, S. W. A. (2014). The bureaucratization of safety. *Safety Science, 70*(12), 348–357.

Dekker, S. W. A. (2018). *The safety anarchist: Relying on human expertise and innovation, reducing bureaucracy and compliance.* London, UK: Routledge.

Dekker, S. W. A., & Hugh, T. B. (2010). Balancing "No Blame" with accountability in patient safety. *New England Journal of Medicine, 362*(3), 275.

Douglas, M. (1992). *Risk and blame: Essays in cultural theory.* London, UK: Routledge.

Fitts, P. M., & Jones, R. E. (1947). *Analysis of factors contributing to 460 "Pilot Error" experiences in operating aircraft controls (TSEAA94-12).* Dayton, OH: Aero Medical Laboratory, Air Material Command, Wright-Patterson Air Force Base.

Gephart, R. P. (1984). Making sense of organizationally based environmental disasters. *Journal of Management, 10,* 205–225.

Graham, B., Reilly, W. K., Beinecke, F., Boesch, D. F., Garcia, T. D., Murray, C. A., & Ulmer, F. (2011). *Deep water: The Gulf oil disaster and the future of offshore drilling (Report to the President)*. Washington, DC: National Commission on the BP Deepwater Horizon Oil Spill and Offshore Drilling.

Gray, G. C. (2009). The responsibilization strategy of health and safety. *British Journal of Criminology, 49,* 326–342.

Haddon, W. (1973). Energy damage and the ten counter-measure strategies. *Human Factors, 15*(4), 355–366.

Harms-Ringdahl, L., & Rollenhagen, C. (2018). Safety barriers. In N. Moller, S. O. Hansson, J. E. Holmberg, & C. Rollenhagen (Eds.), *Handbook of safety principles* (pp. 63–86). Hoboken, NJ: John Wiley & Sons.

Hasle, P., & Zwetsloot, G. I. J. M. (2011). Editorial: occupational health and safety management systems: issues and challenges. *Safety Science, 49*(7), 961–963.

Hollnagel, E. (2004). *Barriers and accident prevention.* Aldershot, UK: Ashgate Publishing Co.

Mendelhoff, J. (1981). Does overregulation cause underregulation? The case of toxic substances. *Regulation, 5*(5), 47.

NTSB. (1994). *Safety recommendations A-94-164 through 166 concerning China Airlines Airbus A-300-600R accident at Nagoya, Japan, April 26, 1994.* Washingtion, DC: National Transportation Safety Board.

OECD. (2014). *Regulatory reform and innovation* (2102514). Paris, France: Directorate for Science, Technology and Industry, Organisation for Economic Cooperation and Development.

Perrow, C. (1984). *Normal accidents: Living with high-risk technologies.* New York, NY: Basic Books.

Provan, D. J., Dekker, S. W. A., & Rae, A. J. (2017). Bureaucracy, influence and beliefs: A literature review of the factors shaping the role of a safety professional. *Safety Science, 98,* 98–112.

Provan, D. J., Dekker, S. W. A., & Rae, A. J. (2018). Benefactor or burden: Exploring the professional identity of safety professionals. *Journal of Safety Research, 66,* 21–32.

Rae, A. J., Weber, D. E., & Dekker, S. W. A. (2018). Safety Clutter: The accumulation and persistance of 'safety' work that does not contribute to operational safety. *Policy and Practice in Health and Safety, 16*(2), 194–211.

Rasmussen, J. (1980). What can be learned from human error reports? In K. Duncan, M. Gruneberg, & D. Wallis (Eds.), *Changes in working life* (pp. 97–113). London, UK: John Wiley & Sons.

Rasmussen, J. (1985). Trends in human reliability analysis. *Ergonomics, 28*(8), 1185–1195.

Reason, J. T. (1990). *Human error.* New York, NY: Cambridge University Press.

Reason, J. T. (1995). A systems approach to organizational error. *Ergonomics, 38*(8), 1708–1721.

Reason, J. T. (1997). *Managing the risks of organizational accidents.* Aldershot, UK: Ashgate Publishing Co.

Reason, J. T. (1999). Are we casting the net too widely in our search for the factors contributing to errors and accidents? In J. Misumi, B. Wilpert, & R. Miller (Eds.), *Nuclear safety: A human factors perspective* (pp. 210–223). London, UK: Taylor & Francis Group.

Reason, J. T. (2008). *The human contribution: Unsafe acts, accidents and heroic recoveries.* Farnham, UK: Ashgate Publishing Co.

Reason, J. T., Hollnagel, E., & Pariès, J. (2006). *Revisiting the "Swiss Cheese" model of accidents (EEC Note No. 13/06).* Brussels, Belgium: Eurocontrol.

Rosness, R., Grotan, T. O., Guttormsen, G., Herrera, I. A., Steiro, T., Storseth, F., ... Waero, I. (2010). *Organisational accidents and resilient organisations: Six perspectives, Revision 2* (SINTEF report A17034). Trondheim, Norway: SINTEF Technology and Society, Safety Research.

Saines, M., Strickland, M., Pieroni, M., Kolding, K., Meacock, J., Nur, N., ... Gough, S. (2014). *Get out of your own way: Unleashing productivity.* Sydney, Australia: Deloitte Touche Tohmatsu.

Saull, R. (2015). Capitalism, crisis and the far-right in the neoliberal era. *Journal of International Relations and Development, 18*(8), 25–51.

Shappell, S. A., & Wiegmann, D. A. (2001). Applying reason: The human factors analysis and classification system. *Human Factors and Aerospace Safety, 1,* 59–86.

Sharpe, V. A. (2003). Promoting patient safety: An ethical basis for policy deliberation. *Hastings Center Report, 33*(5), S2–S19.

Storkersen, K., Antonsen, S., & Kongsvik, T. (2016). One size fits all? Safety management regulation of ship accidents and personal injuries. *Journal of Risk Research, 20*(7), 1–19. doi:10.1080/13669877.2016.1147487

Tooma, M. (2017). *Safety, security, health and environment law* (2nd ed.). Annandale, NSW: Federation Press.

Valluru, C. T., Dekker, S. W. A., & Rae, A. J. (2017). How and why do subcontractors experience different safety on high-risk work sites? *Cognition, Technology and Work, 19,* 785–794.

Vaughan, D. (1999). The dark side of organizations: Mistake, misconduct, and disaster. *Annual Review of Sociology, 25*(1), 271–305.

Wachter, R. M., & Pronovost, P. J. (2009). Balancing "No Blame" with accountability in patient safety. *New England Journal of Medicine, 361,* 1401–1406.

Wagenaar, W. A., & Groeneweg, J. (1987). Accidents at sea: Multiple causes and impossible consequences. *International Journal of Man-Machine Studies, 27*(5–6), 587–598.

Weber, D. E., MacGregor, S. C., Provan, D. J., & Rae, A. R. (2018). "We can stop work, but then nothing gets done." Factors that support and hinder a workforce to discontinue work for safety. *Safety Science, 108,* 149–160.

Woods, D. D., Dekker, S. W. A., Cook, R. I., Johannesen, L. J., & Sarter, N. B. (2010). *Behind human error.* Aldershot, UK: Ashgate Publishing Co.

Wynne, B. (1988). Unruly technology: Practical rules, impractical discourses and public understanding. *Social Studies of Science, 18*(1), 147–167.

Young, M. S., Shorrock, S. T., Faulkner, J. P. E., & Braithwaite, G. R. (2004). *Who moved my (Swiss) cheese? The (r)evolution of human factors in transport safety investigation.* Paper presented at the International Society of Air Safety Investigators, Gold Coast, QLD, Australia.

10 The 2000s and Onward
Safety Culture

KEY POINTS

- Previous decades solidified our focus on the sorts of things that can be found and fixed in an organization before they can create or contribute to an accident. Encouraging organizations to build a 'good' safety culture is the logical continuation of this trend.
- Safety culture has given organizations an aspiration, getting leaders and others to think about what they want to have rather than what they want to avoid. Researchers and practitioners became concerned with specifying what is necessary inside an organization and its people to enhance safety.
- In the last two decades, much research has been published on safety culture, but little consensus has been reached on what it is. There is also a problem of predictive capacity. Organizations with observed deficiencies of their safety culture have not been proven to generate more accidents. Organizations with a great safety culture as measured by attitude surveys, on the other hand, have been involved in high-potential incidents and accidents.
- A functionalist approach to safety culture sees and measures it as something that an organization 'has.' A culture can be taken apart, redesigned and formed. Management can 'work' on parts of that culture (e.g., hazard reporting, procedural compliance). It assumes that values drive people's attitudes and beliefs, which in turn determine their behavior.
- The interpretivist, or qualitative, approach defines culture as something that an organization 'does.' It considers culture as a bottom-up, complex, emergent phenomenon; greater than the sum of its parts; resistant to reductionist analysis, measurement, and engineering. For this reason, it cannot be trained and injected into individual minds.
- Critiques of 'safety culture' have been targeted at (1) the normative idea that some cultures are 'better' than others, (2) the implication that cultures are consistent and coherent rather than full of conflict and contradiction, (3) the avoidance of any mention of power in most models of safety culture, (4) the methodological individualism that sees culture as the aggregate of measurable individual attitudes, beliefs, and values, (5) the lack of usefulness of the concept in safety regulation and investigations, (6) the lack of predictive value, and (7) the fact that the 'container term' of 'safety culture' tries to say so much that it ends up saying very little.

10.1   THE ORIGINS OF SAFETY CULTURE

10.1.1   CONTINUING THE TREND INTO THE BLUNT END

Up to 2000, a typical literature search yielded only a couple of hundred references to safety culture. From 2000 onward, you will find thousands—many thousands. This growing focus should not come as a surprise. High-visibility accidents from the 1970s onward (e.g., Tenerife, Three Mile Island, Chernobyl, Herald of Free Enterprise, Bhopal, Clapham Junction, Piper Alpha) now form the canon of cases in the safety world. Back when they happened, these accidents helped drive the search for causes at the blunt end of organizations. For most people, and consistent with the discoveries about 'human error' during World War II (WWII) (see Chapter 5), it was obvious that these accidents were not just the result of omissions or violations at the sharp end. They could be seen as evidence of a massive mismanagement of risk by organizations, institutions, even governments. Clearly, there were severe problems of control, created by administrative failures in complex socio-technical systems.

Man-made disaster theory continued and enriched the trend (see Chapter 7). Rather than casting accidents in mechanical or technical terms, it suggested we consider them in administrative, managerial, organizational, or indeed even cultural terms. Subsequent theorizing about high reliability organizations and normal accidents (Chapter 8) affirmed that there was much to find—and much to fix—in the way organizations build and run safety-critical systems. Swiss cheese (Chapter 9) became the icon of our preoccupation with everything that could be found and fixed in an organization and its surroundings *before* bearing down on the specific time and place of an accident at the sharp end. Encouraging organizations to build a 'good' safety culture was the logical continuation of this trend, and the explosion of research into it both answered and fed the growing popularity of the label.

Safety culture has given organizations an aspiration: something positive to aim for. Accident prevention (partly the title of Heinrich's 1931 book), after all, was organized around what organizations should avoid. It was about what they did not want to have. Safety culture flipped that around. If accidents, said Pidgeon,

> ...could indeed be put down to a 'poor' safety culture, then there might surely be something, for want of a better word, called a 'good' safety culture, which safety managers might then promote, design, or encourage in order to head off some of the worst consequences of organizational-system failure.

> **Pidgeon (1998, p. 203)**

Safety cultures could get organization leaders and others to think about what they wanted to have rather than what they wanted to avoid. Instead of modeling organizational preconditions for accidents as failures, risks or hazards, researchers and practitioners became concerned with specifying what is necessary inside an organization and its people to enhance safety (Pidgeon, 1998). It identified what organizations might want *more* of, what they might want to invest in and build up—not what they should have less of.

10.1.2   POLITICAL ORIGINS

You might recall from Chapter 1 how concern with safety was born out of politics. The previous chapter showed how safety management systems also fit with a particular form of governance. Pro-market policies and neoliberalism promoted a partial withdrawal by government regulators from the technical details of running an operational system. Concern for safety *culture* was born out of politics too. More specifically, you could say that the very legitimacy of the label 'safety culture' grew out of the ideological clash between different systems of governance. The label 'safety culture' was not used in the analysis of the 1979 Three Mile Island incident (see Chapter 9). But the investigation of the incident by the Kemeny Commission, established by then President Carter, pointed to much of what we would put under 'safety culture' today:

> During the investigation into the root causes of the TMI accident it was found that, while some design deficiencies and system malfunctions were contributory, the main causes were human-related. The [investigation] report presents the results of one of the main investigations into the TMI accident. It does not use the term safety culture; however, it fully identifies all relevant aspects of safety culture, including the important role of the mentality (culture) existing in organizations from the highest level of management down to the individual workers.
>
> **van Erp (2007, p. 157)**

The Kemeny Commission said that the nuclear industry had to dramatically change its attitude toward safety and regulations. It should establish a program that would specify appropriate safety standards including those for management, quality assurance, and operating procedures and practices. And there should be a systematic gathering, review, and analysis of operating experience at all nuclear power plants, coupled with an industry-wide international communications network to facilitate the speedy flow of information to affected parties. Nuclear plants were also encouraged to have separate groups that could report about safety directly to high-level management and clearly define roles, responsibilities, and accountabilities for operating procedures and practices.

In 1986, a catastrophic nuclear accident happened at Chernobyl in Ukraine, then a republic of the USSR. During a test of a safety emergency core cooling feature, a power surge produced an explosion of reactor number 4. The first analysis of the causes of the accident was published only a few months later. The International Atomic Energy Agency, linked to the United Nations, had created a group known as the International Nuclear Safety Advisory Group (INSAG). Their INSAG-1 report, based on data and analyses supplied by Soviet members, blamed operator errors and violations of rules and regulations. There was little else for INSAG to go on. During those years, it was difficult for Westeners to get independent access to sources behind the iron curtain, particularly in politically sensitive and safety-critical areas. INSAG did point out that the errors and violations could themselves be traced to a lack of knowledge of nuclear physics, engineering and the reactor's technical features, and a lack of experience and training. The primary cause of the accident, INSAG-1 concluded, was the extremely improbable combination of rule infringement plus the operational routine allowed by the power station staff. Deficiencies in the reactor design and the regulations that made the accident possible were mentioned only casually. The absence of safety culture was a major criticism in the report, but it was kept at the operator and plant level and focused on the operators' intentional disabling of safety systems on the evening of the test.

In 1992, INSAG admitted that it had been misled. By then, the Berlin wall and iron curtain had come down, Ukraine had once again become independent, and communism itself had all but crumbled across Eastern and Central Europe. INSAG produced another report in 1992 (called INSAG-7) in which the group said that their previous report had

> concentrated on the immediate issues of the Chernobyl accident and made little reference to the regulatory and general safety framework within which the plant was operated. A number of matters have since come to light ... that make it right at this time to present broader views.

(INSAG-7, p. 20)

What were the matters that had come to light? INSAG (1992, pp. 24–25) summed them up:

- A plant that fell well short of the safety standards in effect when it was designed, and even incorporated unsafe features;
- Inadequate safety analysis;
- Insufficient attention to independent safety reviews;
- Operating procedures not founded satisfactorily in safety analysis;
- Inadequate and ineffective exchange of important safety information, both between operators and between operators and designers;
- Inadequate understanding by operators of the safety aspects of their plant;
- Insufficient respect on part of the operators for the formal requirements of operational and test procedures;
- An insufficiently effective regulatory regime: it was unable to counter pressures for production;
- A general lack of safety culture in nuclear matters, at the national level as well as locally.

Many of the issues above were related to the management and governance of nuclear power in the USSR. The Soviets themselves had weighed in as well. By 1991, the political climate had opened somewhat, and evidence of the size and international scope of the catastrophe had become incontrovertible over the 5 years since. A commission of the USSR State Committee for the Supervision of Safety in Industry and Nuclear Power reassessed the causes and circumstances of Chernobyl produced new insights and conclusions. The chief reasons for the accident were now determined to lay in issues of the design and physics of the reactor type used at Chernobyl, especially those related to the cooling system and control rods.

Two earlier accidents with the same kind of reactor (one at Leningrad in 1975 and an earlier failure at Chernobyl in 1982) could in their own ways be considered precursors of the 1986 Chernobyl accident. But they had led to only modest design modifications and few improvements in operating practices. What is more, there was no sharing of incident data across sites, so Chernobyl was unaware of what had happened at Leningrad (today's St. Petersburg, some 1,000 km to the north). Acknowledging that their earlier conclusions had been based on erroneous, incomplete, and premature information in August 1986, INSAG overturned its focus on operators at the sharp end. To be sure, human factors remained crucial to understanding what happened at Chernobyl, INSAG thought. Operators' turning off the emergency core cooling system, interfering with the settings of protective equipment and various other actions were still considered as "most ill judged" (INSAG, 1992, p. 24). But INSAG no longer saw them as contributing to the original cause of the accident, even if they could be construed as breaches of regulations. Crucially, turning off the emergency system that was designed to prevent the two turbine generators from stopping was not a violation of regulations.

What INSAG would not say directly, was what the Soviets had said about Three Mile Island a decade earlier. That sort of nuclear incident, Soviet engineers claimed at the time, could only happen in a capitalist culture that emphasized production and short-term return on capital over safety (Pidgeon, 1998). Of course, each way of organizing a society and its resources creates different trade-offs and carries different strengths and weaknesses. Adversaries are usually happy to point out the weaknesses, particularly in hindsight. Now some insiders blamed the Soviet system of governance for Chernobyl. But they could not do so in public. Valerie Legasov, who had been a deputy director of the Kurchatov Institute of Atomic Energy, was a key member of the Chernobyl investigative commission and the one who delivered its findings to INSAG in August 1986. The limits on what he was able to divulge must have caused him great distress. Knowing that the report he handed over had been politically censored, he was not able to talk, for instance, about the risk secrecy pervading the Soviet nuclear industry, which suppressed knowledge of design flaws and incidents. In 1988, Legasov committed suicide by hanging himself in the stairwell of his Moscow apartment. Reportedly, on tapes he recorded, he called Chernobyl "apotheosis of all that was wrong in the management of the national economy and had been so for many decades" (transl. *Pravda* 20 May 1988, a month after his death). INSAG did come close to repeating Legasov's charge, when in 1992 it admitted that "the Soviets blamed the culture at the plant, and our first report paid too much attention to that. Now we blame the culture all the way to the top." It continued:

> The accident can be said to have flowed from a deficient safety culture, not only at the Chernobyl plant, but throughout the Soviet design, operating and regulatory organizations for nuclear power that existed at that time. Safety culture requires total dedication, which at nuclear power plants is primarily generated by the attitudes of managers of organizations involved in their development and operation. An assessment of the Chernobyl accident in this respect demonstrates that a deficit in safety culture was inherent not only to the stage of operation, but also and to no lesser extent to activities

at other stages in the lifetime of nuclear power plants (including design, engineering, construction, manufacture and regulation).

INSAG (1992, pp. 23–24)

Much of this could be recognized in what the U.S. nuclear industry had been told in the wake of Three Mile Island. Share information; learn from incidents. Make safety a pervasive concern of all involved, from top to bottom. Do not think that technical prowess alone can insulate you against nuclear disasters. But 'safety culture' was not used in the description. Perhaps it was not needed as container term. Or perhaps it was not deemed necessary as cover for a larger charge about the system of governance under which the incident had brewed and sprung into view.

To complete the path to their 1992 report, INSAG published a 1991 report dedicated entirely to safety culture. This report (commonly referred to as INSAG-4) helped prepare the way for INSAG-7. INSAG-4 explained that safety culture represented a way of conceptualizing processes of risk handling and management. It was a way to characterize some of the common behavioral, managerial, and organizational preconditions to disasters and also offered means to do something about it: which management or other strategies could enhance or complement existing risk control practices (Pidgeon, 1998). Acknowledging the intangible but critical nature of safety culture, INSAG defined it as follows (emphasis in the original):

> *Safety culture is that assembly of characteristics and attitudes in organizations and individuals which establishes that, as an overriding priority, nuclear plant safety issues receive the attention warranted by their significance.* This statement was carefully composed to emphasize that Safety Culture is attitudinal as well as structural, relates both to organizations and individuals, and concerns the requirement to match all safety issues with appropriate perceptions and action. The definition relates Safety Culture to personal attitudes and habits of thought and to the style of organizations. A second proposition then follows, namely that such matters are generally intangible; that nevertheless such qualities lead to tangible manifestations; and that a principal requirement is the development of means to use the tangible manifestations to test what is underlying.

INSAG (1991, p. 1)

Some have speculated that by using safety culture, INSAG could avoid a direct political accusation at the address of a—by then—crumbling Soviet system of governance. As worded by Pidgeon (1998), the safety culture "hypothesis stemmed much more from a rhetorical attempt to reassure Western publics that Chernobyl could not happen here" (p. 203). Depending on how it was read, though, pointing to a deficient safety culture either nullified the claim of infallibility made by Soviet engineers more than a decade before (in politically feasible terms), or showed that no system of governance was immune to the challenges and problems of building and maintaining a safety culture.

To pull off this political hat trick, INSAG appealed to a major strength of the concept of safety culture: its all-encompassing width and vague definition. Responsibility for the disaster could be put on a whole system supplying and surrounding and monitoring a safety-critical activity. The risks had sat in, and eaten through, all the ways in which a system was designed, operated, managed, governed, and regulated. These same strengths, however, could eventually become a weakness of the concept of safety culture. As you will see, blaming a deficient safety culture runs the same risk as blaming human error (see Chapter 5). That kind of conclusion—whether human error at the level of the individual or lack of safety culture in a whole system—can stop further inquiry, stick a large label on something that is poorly understood, and generate recommendations that do little more than exhorting stakeholders to try a little harder.

10.1.3 THEORETICAL ORIGINS

It is impossible to identify one point where 'safety' and 'culture' flowed together for the first time. But prior to Chernobyl, an important substratum of knowledge was already available for the concept of safety culture to emerge and solidify. It certainly did not appear in scientific theory as a newly discovered object. The development of safety thinking in the decades before Chernobyl had already prepared the ground, and made it possible for INSAG to use, promote, and help push safety culture onto the agenda of safety researchers and practitioners alike. This development had started almost a century earlier.

Recall from the first chapter of this book that safety performance in, and of, organizations had been a concern since the Industrial Revolution. More recently, though, the roles of organizational management, leadership, administration, and governance all started to figure prominently in safety-scientific theory during the 1970s and 1980s. Recall, for instance, from Chapter 7 how man-made disaster theory found these ingredients of the incubation of disasters:

- Rigidities in institutional beliefs;
- Distracting decoy phenomena;
- Neglect of outside complaints;
- Multiple information-handling difficulties;
- Failure to comply with regulations;
- Tendency to minimize emergent danger.

If you would not know better, then this could be a typical list of ailments of an organization which has a deficient safety culture. The links between man-made disaster theory and safety culture go deeper than that, though. What was groundbreaking about this theory was that it defined accidents not by their physical characteristics or impacts at all, but in sociological terms. Turner observed that organizations operate with a variety of cultural beliefs and norms related to risks and hazard. These impact the way risks and hazards are considered and managed, formally as well as informally. In fact, such beliefs and norms may well be embedded within the culture of everyday working practices and in people's typical, repeated, and legitimated responses to observed problems.

Culture both enables and reaffirms beliefs, norms, and responses. Culture is created as organizational members repeatedly behave and communicate in ways that are natural or taken for granted to them, and obvious and unquestioned by others around them. Warnings or potential signs of a deteriorating situation are discounted, somehow legitimated, or even integrated in current understandings of how things work or how they might fail. Beliefs and norms can thus grow increasingly at odds with the buildup of risk. An accident, said man-made disaster theory, disrupts those existing cultural beliefs and norms about risks and hazards, or even leads to their total collapse. The phrase 'safety culture' was never uttered by its originators. But culture was at the heart of man-made disaster theory, and it opened the way for more sociological and anthropological understandings of safety as driven not by technical, calculable hazards, but by people's collective assumptions, beliefs, and practices.

Not much later, high reliability theory (see Chapter 8) inverted the perspective and offered the ingredients that are necessary to build a strong safety culture:

- Personnel subjected to intensive training and socialization to make sure they can respond to hazards and contingencies swiftly and adequately;
- Decentralization that devolves decision-making about technologies and processes to the experts who work with them daily;
- Redundancy in personnel and parts to back up failures in primary systems or teams;
- Senior management commitment to safe operations, and setting and communication of safety objectives across the organization;
- Maintenance of relatively closed systems to minimize effects that the outside environment and actors can have on those objectives;
- Development of deep knowledge of operating technologies and processes through many ways of organizational learning;
- Informal networks of senior officials, technical specialists, and advisors can from an organization's collective mind to deal with unanticipated complexities.

The 'creation of a safety culture' was mentioned in some of the early high reliability theory work as a point in addition to the bullets above. Today, though, the creation of a safety culture would pretty much cover *all* the points above. From the 1970s onward, organizational safety theory was getting intertwined with culture: safety-scientific theory articulated more and more factors that could be linked to organization culture: top management commitment; safety as a managerial priority; status of safety officers; safety training; communication between workers and managers; environmental control and housekeeping; stable workforce; good industrial relations; and standard procedures. Safety theories did not come up with the resulting concepts themselves. They borrowed many of these ideas and inspirations from other disciplines such as management science, social psychology, industrial psychology, sociology, and anthropology. Together, these ideas could help explain how safety was created by, or depended on, a suite of social-organizational factors that had been out of scope for much of safety science until then.

10.1.4 Safety Climate

In 1980, Dov Zohar from the Technion Institute of Technology in Israel, published a paper in the *Journal of Applied Psychology* that many see as an important starting point. But it did not carry the title 'safety culture.' Instead, Zohar used 'safety climate.' As early as 1971, research on low-accident companies showed that their top management was personally involved in safety activities on a routine basis, and safety matters were given high priority in company meetings and production scheduling. Safety officers typically had a higher status in these companies, and safety training was taken seriously, both initially and recurrently. In these companies, leaders believed that accidents were symptoms of design faults in their system. Workforces in low-accident companies tended to be older and showed less turnover, enjoyed better industrial relations. Their companies promoted safety through guidance and counseling rather than enforcement and sanctions, and included individual praise or recognition for safe performance, even enlisting workers' families in safety promotions (Zohar, 1980).

Based on the industrial safety literature of the time, the Technion team developed organizational climate dimensions, each represented in turn by seven items that were short statements with five-point scales for evaluating subjects' agreement with them. All items were phrased positively so that full agreement resulted in a higher score in this dimension. This procedure resulted in a questionnaire of 49 items. The questionnaire was then given to a pilot sample of 120 production workers in four factories. These data were factor analyzed and the procedure resulted in eight factors that largely overlapped with the original ones derived from the literature, supporting the validity of the theoretical basis for the questionnaire. The eight factors were:

- Perceived importance of safety training programs
- Perceived management attitudes toward safety
- Perceived effects of safe conduct on promotion
- Perceived level of risk at work place
- Perceived effects of required work pace on safety
- Perceived status of safety officer
- Perceived effects of safe conduct on social status
- Perceived status of safety committee

Answers to questions on these factors reflected employees' perceptions about the relative importance of safety in their companies. It varied from highly positive to a neutral level. The Technion team showed that there was agreement among employees' perceptions regarding safety climate in their company and that the level of this climate was correlated with safety program effectiveness as determined by safety inspectors. The two dimensions of highest importance in determining the safety climate level were workers' perceptions of management attitudes about safety and their perceptions of the relevance of safety in general production processes (Zohar, 1980).

Climate was defined as a summary of perceptions that employees share about their work environments. These perceptions were assumed to have a psychological utility in serving as a frame of reference for guiding appropriate and adaptive task behaviors. According to psychological theorizing, employees develop these coherent and relatively stable sets of perceptions and expectations about the links between behaviors and outcomes, and then behave according to what they have learned or successfully repeated. The Zohar study seemed to confirm that safety climate could be regarded as a measurable characteristic of industrial organizations, and that safety climate was related to the general safety level in these organizations as judged by inspectors and measured in unsafe outcomes. (Note that this study did not show a causal relationship. Positive measures on the safety climate questionnaire and fewer negative events were shown to be related, or to covary.)

The relationship between safety culture and safety climate has been debated ever since Zohar. References to organizational climate occur as early as the 1930s, and the concept was turned into a 'measurable' characteristic of organizations in the 1960s (Cox & Flin, 1998). Following in the footsteps of Zohar, organizational climate researchers have focused on workforce perceptions of the social and managerial aspects of people's work environment, typically using surveys to gather their data. The connection between their definition of 'safety climate' and what the surveys told them was never seen as very contentious.

Today, perhaps to put an end to an otherwise endless debate, researchers seem to gravitate toward the idea that culture creates the conditions for climate. Culture, in other words, defines the setting within which climates appear (Hudson, 2006). Venerable culture researcher Hofstede proposed that organizational climate is within the remit of supervisors and lower- and middle management. Culture, in contrast, should be the concern of top management or the organization's senior leadership (Hofstede, 1991).

10.2   SAFETY CULTURE TODAY

10.2.1   What Is It Exactly?

So what is a safety culture exactly? There must be something intuitively appealing about the concept, because a number of studies in safety science do not define it at all. One review showed that "only eight (8) of twenty-seven (27) selected studies define safety culture" (Henriqson, Schuler, van Winsen, & Dekker, 2014, p. 468). In other words, 30% of scientific studies sampled here did not bother to define the very concept the studies were purporting to investigate. That is obviously problematic. That said, safety science has been there before (e.g., how many studies or investigations that talk about human error actually define what they mean by it?).

But if science does not define the concept it sets out to study, we will never know whether it was able to measure or investigate it, or say something meaningful about it, or whether the absence or presence of that concept correctly predicted an outcome. If we assume that readers all agree on a particular understanding of a concept, then we run the risk that we talk past each other. That is already happening. When safety culture is defined, there is such a rich diversity (and difference of depth) in the definitions that it is likely that researchers are not talking about the same thing at all (or that they would not know it if they were). The work on safety culture, in the words of Pidgeon (1998) has remained unsystematic, fragmented across, and driven by a multitude of disciplines and influences, and underspecified in theoretical terms. Here are two definitions, just to show the spread:

Safety culture is the best way we do things around here.

e.g., Hudson (2006, p. 11)

The safety culture of an organisation is the product of individual and group values, attitudes, perceptions, competencies, and patterns of behaviour that determine the commitment to, and the style and proficiency of, an organisation's safety management.

Cox and Flin (1998, p. 191)

Both definitions show the vast scope and ambition of the concept 'safety culture.' The first one is superficial and tries to be almost all-encompassing (and leaves questions like what is 'best' or what 'things' entirely up to the researcher to answer). The second definition is more specified than the first. One of the problems, of course, is that culture contains many of the less tangible aspects of organizational life—such as:

- practices
- rituals
- symbols
- informal relationships
- underlying assumptions
- norms
- attitudes
- beliefs
- values

If you were to take into account all of these aspects, you indeed end up with either an inordinately complex definition of safety culture, or with a very short and superficial one. The larger or longer the definition, the more it is likely to offer more precise ideas and hints of how to operationalize—or describe, if not measure—how and why things are done the way they are. The more this gets specified into measurable features, the more functionalist the definition of safety culture becomes (Glendon & Stanton, 2000). The distinction between functionalist and interpretivist points of view on culture is useful, and we will look at it in more detail below.

For now, you might already be persuaded that it is difficult to count and quantify everything that matters about the 'things' in the list above. Naturally, you can count how often an organization engages in, say, a ritual such as a 'safety moment.' But the sheer number of times it does so, says little about the substance or significance of that safety moment (or whether people in the organization actually believe in the value of it). These are things you need to investigate by qualitative rather than quantitative methods. You will see the distinction between quantitative and qualitative approaches to studying culture return in below, where the difference between climate and culture has slowly been overtaken by different interpretations (functionalist and interpretivist) of culture itself, driven by different research traditions.

Twenty years after Zohar, Guldenmund (2000) observed that much research had been done, but little consensus had been reached. This is confirmed by other researchers in the field. In fact, Guldenmund said, most writings on safety culture have not progressed beyond face validity. The concept seems to make sense at first appearance, but that is it. Few researchers tried what Zohar did, establishing the construct validity of safety climate or culture. Construct validity covers the operationalization of the concept (or construct, if you will). It demonstrates a relationship between the construct (e.g., people's attitudes, beliefs, norms) and context-specific, measurable features in the world. If this is not done, it leads to studies with serious flaws (Glendon & Stanton, 2000).

There is also the problem of predictive capacity, which would hopefully be connected to the definition of safety culture. Can safety culture, if measured in the way determined by the definition we have, predict how the organization is going to fare in reality? It turns out that the link is not strong at all (as you will see later in this chapter):

- Organizations with observed deficiencies of their safety culture (as measured by people's attitudes and beliefs, for instance) have not been proven to generate more accidents;
- Organizations with a great safety culture as measured by attitude surveys, on the other hand, have been involved in high-potential incidents and accidents.

Tests to show the causal power of safety culture would need to demonstrate not only covariation between the presumed cultural causes and their effects, but also a temporal precedence of the causes. That is, the conditions associated with safety-cultural deficiencies must have been in place before any accident (for how long, though? No definition of safety culture will help you answer that). Most importantly, a causal demonstration requires some kind of control, so as to rule out alternative explanations that could be responsible for the observed effect. This is almost impossible to do with safety culture. Think about it:

- It would mean that researchers require two comparable organizations, one with and one without the supposed cultural deficiencies, and then wait for an accident in the former, which then needs to be linked theoretically to the alleged deficiencies. How long will you have to wait? And will you get ethical clearance if you know enough bad things about the organization's safety culture to 'see the accident coming' and not say anything?;
- Conducting such research in retrospect does not really count. Because that means you select an organization which has already suffered an accident. And then you might compare it to a similar organization which has not. And then you try to trace back to supposed elements of their safety cultures. Doing it this way selects the samples on their dependent variable (the accident) rather than the independent one (safety culture). This introduces so many biases (hindsight and outcome biases being important ones) that few would believe the research results.

Without any proof that cultural features of an organization can predict anything, why do managers and others seem so seduced at all? It goes back, Guldenmund argues, to face validity. Safety culture is a great concept because, well, it is a great concept. How can we make more sense of such vulnerabilities and the continued seduction of the concept? The next section might shed some light on this, distinguishing a functionalist from an interpretivist take on culture.

10.2.2 A Functionalist Approach to Safety Culture

The functionalist approach is characterized by disciplines such as social psychology, management science, and engineering (Henriqson et al., 2014). These disciplines often have rather pragmatic goals. Studies on safety culture were initially dominated by social psychology, with analyses that take safety and culture as objectively quantifiable categories. The methods that were adopted by these disciplines are usually quantitatively oriented, including various forms of questionnaires and multivariate data analysis. Surveys are overwhelmingly the tool of choice. A special issue of *Safety Science* devoted to safety culture in 2000, for instance, featured seven empirical papers of which six had relied on surveys for gathering their data (Antonsen, 2009b). Surveys are considered to be well suited, or at least to not present any fundamental problems, for the measurement of attitudes, values, beliefs, and perceptions about organizational practices.

The functionalist approach defines and measures culture as something that an organization 'has.' For sure, it considers culture to be a complex and multidimensional object. But it can be measured. And understanding and measuring it can support management strategies and ideologies that can change and engineer that culture. One example of this approach is Reason's (1997) division of safety culture into five different parts, all of which can be 'worked on' by management and others in an organization. According to him, a safety culture is made up of:

- An *informed* culture: one in which those who manage and operate the system have current knowledge about the human, technical, organizational, and environmental factors that determine the safety of the system as a whole;
- A *reporting* culture: culture in which people are willing to report errors and near misses in the knowledge that their reporting will support the learning of the organization as a whole;
- A *just* culture, or a culture of no blame where an atmosphere of trust is present so that people can report safety issues without fear of retribution. This is a culture where people are encouraged or even rewarded for providing safety-related insights that would otherwise remain invisible to organizational decision makers. For Reason, a just culture is a culture with a clear line between acceptable and unacceptable behavior (but see Dekker, for some of the difficulties with this position and alternative ways of thinking about a just culture);
- A *flexible* culture that can take different forms depending on the needs of the situation, but that is nonetheless characterized by a flatter rather than conventional hierarchical decision structure;

- A *learning* culture that has the willingness and the competence to draw appropriate conclusions from safety information systems and other sources, and the will to implement major reforms when the need is indicated.

Dividing culture up into such seemingly manageable bits can be empowering to those in management and leadership roles. Various things can be done to enhance the organization's ability to identify hazards, manage risks, and withstand safety challenges. It has led to the uptake by various other domains, who were not originally involved in the study of safety culture. Have a look at the following example from healthcare:

> The Institute of Medicine (IOM) has recommended that healthcare organizations develop safety cultures to align delivery system processes with the workforce requirements to improve patient outcomes. Until health systems can provide safer care environments, patients remain at risk for suboptimal care and adverse outcomes. Health science researchers have begun to explore how safety cultures might act as an essential system feature to improve organizational outcomes. Since safety cultures are established through modification in employee safety perspective and work behavior, human resource (HR) professionals need to contribute to this developing organizational domain. The IOM indicates individual employee behaviors cumulatively provide the primary antecedent for organizational safety and quality outcomes.
>
> **Palmieri, Peterson, Pesta, Flit, and Saettone (2010, pp. 97–98)**

Functionalist studies are concerned with the practical significance of safety culture, and finding those levers of managerial influence and change that can affect behaviors down the line. The Agency for Healthcare Research and Quality, for example, proposed the following dimensions for a survey on patient safety culture in hospitals (see Palmieri et al., 2010):

1. Hospital management support for safety
2. Organizational learning
3. Teamwork within units
4. Supervisor/manager expectations and actions promote safety
5. Compliance with procedures
6. Staffing
7. Error feedback and communication
8. Overall perceptions of safety
9. Openness of communication
10. Nonpunitive response to error
11. Positive-reporting norms
12. Frequency of event reporting
13. Teamwork across units
14. Hospital handoffs and transitions

These things can all be measured, though some (e.g., frequency of event reporting) more obviously than others (teamwork across units). The numbers produced by cultural surveys are a logical endpoint of a functionalist analysis that focuses on (supposed) causes and consequences, an analysis that defines risk and safety instrumentally, in terms of minimizing errors and presumably measurable consequences. As is usual, though, the theoretical connection with 'culture' is not discussed in great detail, if at all: these dimensions are assumed to somehow represent or be linked to culture in ways that are so obvious that it does not need any further inquiry. With a functionalist approach, researchers take safety culture as an object for understanding and influencing how members of the organization signify risk and safety. What do they see, hear; what should they be reporting? What is there to learn from and improve?

But problems abound when safety culture is used and applied in this way. The first large multiple institution safety culture survey of the attitudes and experiences of workers in hospitals studied 15 hospitals with 6,312 employee responses (Singer et al., 2003). It found that cultures differed significantly not only between hospitals but also within. Even between clinical departments and job functions inside the institutions, large differences in survey responses were found. In addition, the Singer study flagged, survey scores like the ones they got, were seldom correlated with dependent outcome measures. In other words, it had not yet been established whether manipulation of the rather complex variable 'safety culture' indeed lead to changes in the dependent variable of patient (or other kinds of) safety outcomes.

Some have tried to establish this connection in empirically more creative ways. To find out what of those typical safety culture interventions worked and what did not, Hallowell and Gambatese (2009) analyzed the relative contributions of a number of safety interventions in the construction industry. They conducted a Delphi analysis, which is a way to glean expert forecasts through multiple rounds of questionnaires. It revealed some surprising things (or perhaps not). Some programs and initiatives that safety managers or entire companies cherish did not actually generate much safety yield. In descending order of importance for their effects on safety (with 1 the most, 13 the least), their results were as follows (Hallowell & Gambatese, 2009):

1. Upper management support. This involves the explicit consideration of safety as a primary goal of the firm. Upper management can demonstrate this commitment by, for example, participating in regular safety meetings, serving on committees, and providing funding for safety equipment and initiatives.
2. Subcontractor selection and management. This involves the consideration of safety performance during the selection of subcontractors. That is, only subcontractors with demonstrated ability to work safely should be considered during the bidding or negotiating process.
3. Employee involvement in safety and work evaluation. This may include activities such as performing job hazard analyses, participating in toolbox talks, or performing inspections.
4. Job hazard analyses. This includes reviewing the activities associated with a work process and identifying potential hazardous exposures that may lead to an injury.

5. Training and regular safety meetings. These aim to establish and communicate project-specific or work-specific safety goals, plans, and policies before the start of the project or work.
6. Frequent worksite inspections. These help identify uncontrolled hazardous exposures to workers, violations of safety standards or regulations, or behavior of workers.
7. Safety manager on site. The primary responsibility of a safety manger is to perform, direct, and monitor the implementation of safety program elements and serve as a resource for employees.
8. Substance abuse programs. These target the identification and prevention of substance abuse of the workforce. Testing is a crucial component of this safety program element. Methods of testing and consequences of failure may differ between organizations and industries.
9. Safety committees made up of supervisors, practitioners, workers, representatives of contractors, owner representatives, and safety consultants can be formed with the sole purpose of addressing the safety of work.
10. Safety orientation. This involves the organization- or site-specific, but not necessarily project-specific, orientation, and training of all new hires and contractors.
11. Written safety plan serves as the foundation for an effective safety program. The plan must include documentation of project- or work-specific safety objectives, goals, and methods for achieving success.
12. Record-keeping and incident/accident analyses involve documenting and reporting the specifics of incidents and accidents including information such as time, location, work-site conditions, or cause. It includes the analyses of accident data to reveal trends or weak spots.
13. Emergency response planning. This may be required by owners, insurers, or regulators, and involves the creation of a plan in case of a serious incident. Planning for emergencies can define the difference between an accident and a catastrophic event.

While these results were generated out of data from the construction industry, parallels with other industries and safety-critical settings are not hard to find. Frequent work inspections, for example, are common in aviation, where airlines themselves do line checks (where line-check captains fly with other crews to monitor and discuss their performance). This is true for a safety manager on site as well: one of the airlines I flew for had the safety manager sit in an office directly off the crew room at the airline's hub airport, where pilots and others could easily walk in and discuss anything related to safety. The airline also had a safety committee, on which I participated, for example, when a change was proposed to the pre-takeoff procedures to ensure the correct wing flap setting before departure. What is known as job hazard analyses or regular prework safety meetings in construction and other project work, is known as preflight briefings in aviation. These not only happen in the crew room to discuss the entire duty day, but in the cockpit before every critical phase of flight as well.

It is interesting, for any industry, to reflect on the order of importance of safety initiatives as identified in Hallowell and Gambatese's research. Does this order, and what it says about the role of the human factor, apply to them as well? The critical importance of upper management support has been identified in high reliability organization research across multiple industries, for example. Maintaining the goal of avoiding serious operational failures in those organizations has been said to nurture an organizational perspective in which short-term efficiency gains take a second seat to high reliability operations. This has been found in air traffic control, navy operations as well as electricity generation (LaPorte & Consolini, 1991). In industries such as those, there are obvious reasons why organizational leaders must place a high priority on safety:

> High reliability organizations require both significant levels of redundancy and constant operational training, and both of these factors cost a great deal of money... If political authorities and organizational leaders are not willing to devote considerable resources to safety, accidents will become more likely.
>
> **Sagan (1993, p. 18)**

Many organizations might say they have safety as their highest priority. Of course, the highest priority of any business is probably to stay in business. But to the extent that safety is declared a priority too, this needs to be backed up by resources and visible action. Whether high reliability organizations have really accomplished this or whether other factors account for their apparent success, is still a matter for debate (Sagan, 1994). The selection of subcontractors or partners who can demonstrate acceptable safety performance is also something that is widely shared beyond construction. In aviation, for example, airlines are not allowed to join a global alliance without undergoing a whole series of audits.

The problem with these initiatives, of course, is that it easily reproduces the almost fraudulent practices that we can also see in some organizations that have embraced 'Zero Harm' (Dekker, 2014b; Sheratt & Dainty, 2017). Where safety is mainly assessed by an absence of negatives, incentives exist to make negative events invisible. In industries such as construction and energy, it is unfortunately still canonical for contractors and subcontractors to demonstrate low numbers of negatives, otherwise they may not even be considered for tender at all. If the client is a government (city, state, federal), the demonstration of low numbers of incident- and injury rates is pretty much obligatory. Selecting a contractor or subcontractor on such low numbers may say more about their strenuous creativity in making the numbers look good, then about their safety or resilience. It moved David Capers, a long-time expert in energy processing, to quip that the numbers that matter are not LTIs (loss time injuries) or MTIs (medical treatment injuries), but an LGI—a Looking Good Index.

Let's turn to the bottom of the list. Safety inductions or orientation have developed, in many industries, a negative connotation. They can be seen as a waste of time and money. Some sites or companies have orientations that take days, not hours. And even then, they may be ill-connected to the work people will do or the risks they will face. Rather, they are seen as yet another exercise in liability management and reduction on part of the company (so that they can say, "see, told you so, warned you!" if things do go wrong). Many inductions or orientations are also top-down in an almost Tayloristic sense ("we are smart, you are dumb, so listen to us to know how to be safe").

Success stories do exist, however, and it is not difficult to see what it takes. One chemical production site, for example, made sure that its safety orientation was very short in the classroom but much longer on the site. It had developed a buddy system, in which new hires were mentored for weeks or more by somebody specifically assigned to them (and trained as a "buddy"). This allowed the experienced person and the new hire to actively engage with the work and jointly explore what was dodgy about it. Learning about safety was a coproduction, set in context—not a one-way dictation in a vacuum. The relationship between those already on site or in the company and those new to it was, as a result, more organic, more equal.

The low score of record-keeping and incident and accident analysis may come as a damper to those who make it their livelihood. Record-keeping is no aim in itself, of course, and they would not speak to safety people like some oracle. Records and data are only any good if you know what to ask of them. There is probably no organization in the world that has a perfect balance between all the data it gathers about things going wrong (or right) and the analytic yield it gets from it. In project-driven industries, like construction, little data is actually mined from previous projects to understand what accounted for the successes and the foundering. The data is probably there, but the questions that need to be asked of it before the next project are not consistently developed.

That written safety plans and emergency response planning have hit the bottom of the list may be no surprise to people familiar with these sorts of things in their own organizations. It is no surprise either to normal accident theory, which has concluded that organizations regularly ignore the bulk of their experience that shows these plans and documents to be inaccurate. This is why Clarke called them "fantasy documents" which are neither wholly believed, nor disbelieved. They are tested against reality only rarely, and seem to draw from a rather unrealistic or idealistic view of the organization or the environment in which it operates (i.e., every contingency is known and prepared for). The extent to which these plans and documents amount to fantasy was once again obvious from the Fukushima nuclear disaster in 2011 (already alluded to above), the largest since Chernobyl in 1986:

> Japan seemed underprepared for the disaster. At the plant itself, for instance, procedures and guidelines were woefully insufficient for a meltdown, forcing operators to respond on an almost entirely ad hoc basis. When the lights went out they had to borrow flashlights from nearby homes to study the plant's gauges. On-site dosimeters maxed out at levels that were far below those that could be expected from a catastrophe, unable to display readings any higher. And, as the plant's former safety manager would later testify, emergency plans '... had no mention of using sea-water to cool the core,' an oversight that caused unnecessary, and perhaps critical, delays. The bureaucratic infrastructure beyond the plant evinced similar shortcomings. Official announcements were often ill-considered and characterized by denial, secrecy, and refusal to

accept outside help... The pervasive idealization of formal risk assessments, which so many narratives of Fukushima reaffirm, both narrows the democratic discussions around nuclear power, and perverts the processes through which it is governed. The false surety it projects allows the cost–benefit projections that frame nuclear decision making to silently discount the evidence of past accidents in the tacit understanding that disasters are somehow aberrant and avoidable rather than endemic. The result, as outlined above, is a deep-rooted institutional reluctance to adequately plan for worst-case scenarios.

Downer (2013, pp. 2–3)

And finally, back toward the top of Hallowell & Gambatese's list. For employee involvement in safety and work assessments to score as high as it did is encouraging and consistent with the growing literature on high performance teams and resilient organizations. The themes that animate these literatures, after all, include participative leadership, coordinative relationships, and decision-making, common purpose, a preference for bottom-up rather than top-down initiatives, open communication and mutual trust, valuing of diversity of inputs, and openly dealing with conflict between viewpoints.

Functionalist safety culture interventions are often meant to generate a sense of commitment, motivation, and to enhance social stability, guide and shape behavior, and get coherent values about safety spread among the workforce. The functionalist logic assumes that there is a causal relationship between behavior, beliefs, and values: values are ascribed to be at the core of a cultural system. These then drive people's beliefs, which in turn are supposed to determine their behavior. Thus, organizations have to make workers internalize safety both as a value and a belief, in order to foment the necessary conditions for safe behavior. This logic legitimizes 'safety first' and 'zero accident' slogans and campaigns, since these are ways to disseminate values and beliefs. It perpetuates and reinforces the idea that culture is a viable characteristic, a quality that can be managed, changed, or manipulated (Haukelid, 2008). It is a view of culture with consequences:

- Safety culture implies a normative homogeneity of values, beliefs, and behaviors, something that organizations can demand by policy and incentive (see also under Section 3. below). Its homogenizing effects can stand in the way of diversity, creativity, and improvisation (or at least push them out of view). Yet it is that kind of diversity that is linked to resilient systems; systems that are able to adapt to change and absorb disruptions that they were not directly designed for.
- Safety culture initiatives can go hand-in-glove with neoliberal and behavioral safety initiatives toward worker responsibilization (Gray, 2009). This is what Foucault referred to as governmentality: a complex form of power that links individual conduct and administrative practices—in this case, extending responsibility for safety from the state to organizations, and from organizations to individuals, expecting self-responsibilization and self-discipline. People are a problem to control, and through subtle and less subtle processes and technologies, organizations exercise such

control. Workers are expected to commit to safety, adopt accountabilities and responsibilities, participate and comply, communicate openly, and actively look out for each other (Henriqson et al., 2014).

Safety, in this functionalist approach, is seen in part as a moral commitment (remember "hearts and minds" from not long ago, and the human factor as the "moral or mental shortcomings" of individual people from the turn of the last century). Safety becomes that moral, normative choice, a supposedly "free" choice, even though it can be in great conflict with other stated and unstated organizational goals on production and efficiency, or with the design provisions offered in the equipment people need to work with. The sorts of behavioral encouragements that come from, and with, a safety culture, however, are based on faith more than on evidence—much like zero vision. Their embrace seems driven by conviction and moral commitment.

POSTERS TO ENGAGE THE WORKFORCE

I was at a construction site not long ago, accompanying the site's production director. Posters with a photograph of him, wearing the requisite protective equipment, were plastered all over the place. On the photograph, he had a stern look, his arms folded across his chest. Next to his head were the company's slogans about its commitment to zero harm. I asked him what evidence the company had that any of this was working to prevent injury and incident. He had no such evidence, nor had his company, anywhere. The posters seemed based entirely on faith. His company was not alone of course. I recall working with an air traffic service provider, whose latest anti-incident campaign was based on posters throughout the control room that goaded air traffic controllers into greater "situation awareness."

Leaving the construction site, I asked the production director whether the company would do the same when facing a decision to engage in a project, in say, Tanzania. Whether they would print posters with him, or the CEO on it, and a slogan next to his head saying "Tanzania is a great place to invest in a project!" and then paste them in the lifts, and board room, and hallways and toilets. Would he truly have the hope that in a few weeks' time most people in the company would be committed to investing in a project in Tanzania? He looked at me blankly. Yet are we not asking people to embrace a zero vision of safety on the same sort of faith and absence of evidence?

This approach, as many safety practitioners would know, may potentially backfire. The credibility and sustained effect of safety value and poster campaigns is generally low. Slogans on such posters are typically too underspecified to be of practical value in day-to-day situations (Lee & Harrison, 2000). "Here we follow standard operational procedures" or "Safety is our number one priority" can quickly become ignored as fantastical, irrelevant, and ill-informed by those who do the organization's safety-critical work. Also, the attempt to rally a workforce behind a single, small set of values communicated through policies and posters can prevent workers from legitimately developing a 'dissenting view' about what it takes to get the job done. It might affirm managerial confidence in control and even maintain or radiate the belief that the organization is safe and that risk is managed well.

It would seem that in most of safety science, and in practical approaches to safety culture, the functionalist approach reigns—despite its epistemological and practical problems. What is attractive about it from a managerial but also researcher's point of view is that it renders human culture decomposable and redesignable. As Batteau summarized:

> safety culture results from a decomposition of culture ... a set of homilies toward prudence, following rules, and communicating openly. By decomposing habits of the heart into elements sized to a designer's checklist, one eliminates from view the ineffable impulses that distinguish cultures from commands. In a list focusing on specific items, such as reporting systems (check), open communication (check), non-punitive fact-finding (check) and adherence to standard operating procedures (double-check), there is no formatted space for a dialogue, [a dialogue or conversation that might construct other versions or perceptions of operational risks].

> **Batteau (2001, p. 203)**

Safety culture is not unique in this regard. It reflects something, says Gideon Kunda, that has been going on in management science for decades. Safety culture is just another instantiation of it. Culture, in this perspective, becomes a management tool (Kunda, 2006). It is a vehicle for the coding and dissemination of the dominant symbolic power or values in an organization. This means that the manipulation of culture—through all the functionalist means possible—is an alternative to the imposition or execution of rational-legal bureaucratic authority by managers. This sets culture up as a control mechanism. It is supposed to drive performance and action and shape a pragmatic, present-oriented, solutions-driven business environment. When culture is seen in this functionalist way, it is worlds apart from the emergent phenomena and subtle connections of human contexts that constitute it. Instead, culture has morphed into a branch of engineering. A functionalist take on safety culture, when you get down to it, uses engineering strategies and engineering discipline to solve what is seen as an engineering problem. "When social sciences support this strategy, they strike a pose of positivism, with any insights of interpretation, of social construction, of discourse, and of historical conflict carried along as contraband" (Batteau, 2001, p. 204).

10.2.3   An Interpretivist Approach to Safety Culture

It probably comes as no surprise that the functionalist approach has been criticized for invoking culture as an "iconic concept with little of the theoretical edifice sociologists and anthropologists have built" and for "instrumental and reductionist epistemologies antithetical to cultural analysis" (Silbey, 2009, p. 342). These critics suggest that culture as the functionalist sees it, is not culture at all. It is a set of behavioral and attitudinal measures, questions about beliefs and perspectives, homilies and managerial prescriptions, and policy guidelines that have little bearing on how people actually do work. And it is focused on manipulating and engineering the organization into a particular direction. The functionalist perspective is in a sense a holdover from the way culture was seen earlier in the 20th century: stable and internally harmonious, but still formable from the top-down:

> During the first half of the twentieth century, scholars taught that every culture was complete and harmonious, possessing an unchanging essence that defined it for all time. Each human group had its own world view and system of social, legal and political arrangements that ran as smoothly as the planets going around the sun. In this view, cultures left to their own devices did not change. They just kept going at the same pace and in the same direction. Only a force applied from outside could change them. Anthropologists, historians and politicians thus referred to culture as if the same beliefs, norms and values had characterized [it] from time immortal.
>
> Today, most scholars of culture have concluded that the opposite is true. Every culture has its typical beliefs, norms and values, but these are in constant flux. The culture may transform itself in response to changes in its environment or through interaction with neighboring cultures. But cultures also undergo transitions due to their own internal dynamics. Even a complete isolated culture existing in an ecologically stable environment cannot avoid change.
>
> **Harari (2011, pp. 181–184)**

The shift from a modernist mindset—where the social world is controllable, where good science can lay bare the objective order of things, where cultures can be taken apart, redesigned and formed, and where cultural progress can be made by actively intervening—to a postmodernist one is visible here. A postmodern understanding of culture is more skeptical of our ability to control that culture. Of course it is possible to influence a culture, but interpretivists would probably argue that functionalism overestimates the ability to control it, and underestimates the consequences of its attempts to do so. There is also skepticism of our ability to make 'progress' with our culture, if anything because we have so many views on what progress is, or because we keep changing our own view of what counts as progress. Just think about the difference between a purely materialist pursuit (which was thought of as progress in Western culture only a few decades ago) to a new focus on life-work balance, family time, health, and well-being. A postmodernist understanding of culture is also more interested in perspectives from the bottom-up: how does culture look from those who shape it and live in it every day? The control of a culture through hierarchical imposition of rules, managerial order, and centralized governance is thought to be difficult or impossible. Instead, culture is decentralized, all over the place, diverse and dynamic.

Thus, the interpretivist approach is driven by scientific disciplines such as anthropology and sociology, which are often defined as the 'home' disciplines of studies on culture. These disciplines usually adopt qualitative methods such as ethnography and ethnomethodology, including data collection techniques based on various forms of observations, interviews, discussions, and document analysis:

> In recent years, there have been some signs that qualitative research may be gaining greater prominence in safety culture research. Authors such as Richter and Koch and Haukelid have introduced ethnographically inspired perspectives to the study of safety culture. Also, some of the quantitatively oriented researchers are starting to question

the usefulness of survey methods. For instance, Guldenmund suspected that safety culture questionnaires 'invite respondents to espouse rationalizations, aspirations, cognitions or attitudes at best,' and that we 'simply do not know how to interpret the scales and factors resulting from this research.'

Antonsen (2009b, p. 244)

The interpretivist, or qualitative, approach defines culture as something that an organization 'does'. It considers culture as a bottom-up, complex, emergent phenomenon; greater than the sum of its parts; resistant to reductionist analysis, measurement, and engineering. For this reason, it cannot be trained and injected into individual minds. Culture is a medium by which individuals understand identity, values, beliefs, and behaviors (Haukelid, 2008). The interpretivist approach emphasizes the discovery of underlying structures of meaning within an organization (Wiegmann, Zhang, von Thaden, Sharma, & Mitchell, 2002). In doing so, researchers usually use safety culture as an 'ex-post-fact' object to explain accidents and the behavior of workers and organizations. The interpretivist logic often defines safety as a form of expertise connected to organizational practices. In this sense, safety culture is asserted to be a consequence or perhaps even more an emergent property, rather than a cause. It is continually redefined and negotiated in relation to a larger range of organizational factors, processes, goals, and environmental pressures that force adaptation, and produce interpretation and meaning making about danger and risk (Starbuck & Farjoun, 2005). The interpretivist approach has, predictably, been condemned for not offering pragmatic solutions for safety culture improvements (Table 10.1).

TABLE 10.1
The Interpretivist and Functionalist Views of Safety Culture

Interpretivist View	Functionalist View
At home in culture studies, like sociology, anthropology	At home in management studies, organizational psychology, engineering
Sees culture as something an organization *does*	Sees culture as something an organization *has*
Culture is complex and emerges from interactions *between* people	Culture can be reduced to the attitudes and behaviors *of individual* people
Culture can only be influenced, by what people anywhere in it do and how that interacts with others	Culture can be controlled. It can be imposed, changed, taken away, replaced, typically from the top-down
Studies culture with qualitative methods such as observations, interviews, discussions, document study	Studies culture with quantitative methods such as surveys, measurements, questionnaires
Takes the 'emic' or inside-out perspective	Takes the 'etic' or outside-in perspective
Assumes a diversity of perspectives and ideas about safety	Assumes a homogeneity of views and attitudes ('vision zero,' 'safety first')
Leads to little other than more studying of culture	Leads to safety campaigns, behavior modification programs, posters
Typically accused of being non-pragmatic, no control over culture	Typically accused of being overly pragmatic, myth of control

Some more interpretivist researchers see safety as a system property that emerges (like other system properties) from complex interactions between people and lots of other things both inside and outside the organization. It does not just derive solely from individual beliefs, behaviors, or performance (Palmieri et al., 2010). This means that identifying managerial levers for control is an almost hopeless endeavor. Managers, like other people inside and outside an organization, can certainly influence things: people's actions do so every day. But they ultimately control very little, because of the complexity of the organization, and because of the even more complex, emergent ways in which both safety and culture are produced. Both are much more than the measurable and manageable, surveyable properties of the functionalist approach.

The idea behind measuring safety culture is that it is some kind of objective, stable (and perhaps ideal) reality, a reality that can be measured and reflected, or represented, through method. But does this idea hold? Rochlin (1999, p. 1550), for example, proposed that safety is a "constructed human concept" and others in human factors have begun to probe how individual practitioners construct safety, by assessing what they understand risk to be, and how they perceive their ability of managing challenging situations. Note how this approach resists:

> ...conceptualization of safety culture as a separate entity or aspect of organizations. [Indeed] safety culture is an analytical concept, not an empirical entity. The term safety culture is viewed as nothing more than a conceptual label that denotes the relationship between culture and safety. This shifts the attention away from the concept of safety culture to the concept of culture.
>
> This is fruitful for at least two reasons: first, it bypasses the conceptual chaos of earlier safety culture research. Second, focusing on (organizational) culture introduces powerful theoretical perspectives from social sciences such as anthropology and sociology into the study of safety culture. Especially important in this respect is anthropologists' emphasis on analyzing work and other social activity as being situated in a wider context. This implies taking the actors' point of view and asking questions as to why actions and practices stand out as meaningful for the actors involved in concrete work processes. This conception of culture focuses more on how actors make sense of their work situation and, in the process, how they choose between alternative ways of performing work.
>
> This leads to a definition of culture as a frame of reference for meaning and action, which encompasses the skills, beliefs, basic assumptions, norms, customs and language that members of a group develop over time. This definition requires two immediate clarifications: first, culture can neither be seen as a static property of a group nor as something that is radically changed in every instance of interaction. Culture is socially constructed but will, nevertheless, provide guidance for future interaction. Second, defining culture as something that is shared is not tantamount to saying that culture is all about integration.
>
> **Antonsen (2009b, p. 243)**

A substantial part of people's construction of safety, says the interpretivist approach, is reflexive. Reflexive is a term that comes originally from grammar, meaning that what is said refers back to the subject of the clause (e.g., I hurt *myself*). In the social sciences, reflexive means that our understanding of something has to take into account the effect we ourselves have on what is being investigated. Safety, then, is also something that people relate to themselves. If safety is reflexive, or a 'reflexive project,' then people constantly assess their own competence or skill in maintaining safety across different situations.

In this, there may be a mismatch between risk salience (how critical a particular threat to safety was perceived to be by the practitioner) and frequency of encounters (how often these threats to safety are in fact met in practice). The safety threats deemed most salient were the ones least frequently dealt with (Orasanu, 2001). Safety is more akin to a reflexive project, sustained through a revisable narrative of self-identity that develops in the face of frequently and less frequently encountered risks. It is not something referential, not something that is objectively 'out there' as a common denominator, open to any type of approximation by those with the best methods.

The interpretivist approach is more socially and politically oriented than the functionalist one, and places emphasis on representation, perception, and interpretation rather than on functional or structural features (Rochlin, 1999). The managerially appealing numbers generated by culture surveys do not carry any reflexivity, and conveys none of the nuances of what it means to be there, doing the work, creating safety on the line. What it means to be there, however, says the interpretivist approach, ultimately determines safety as outcome. People's local actions and assessments are shaped by their own perspectives. These in turn are for sure embedded in histories, rituals, interactions, beliefs and myths, both of people's organization and organizational subculture and of them as individuals. This confirms the importance of culture, but also explains why good, objective, empirical indicators of social and organizational definitions of safety are difficult to obtain. Operators of reliable systems "were expressing their evaluation of a positive state mediated by human action, and that evaluation reflexively became part of the state of safety they were describing" (Rochlin, 1999, p. 1550).

The description of what safety means to an individual operator is itself part of that very safety: dynamic and subjective. "Safety is in some sense a story a group or organization tells about itself and its relation to its task environment" (Rochlin, 1999, p. 1555). That these stories matter and that they have consequences for what people in an organization do or do not do, or see or do not see, has become ever more visible. Recent work on appreciative inquiry, for example, suggests that narrative and relational ways of knowing are critical to understanding what happens inside an organization. Through such processes, people construct, or co-construct the realities that they subsequently act upon. Therefore, even the interpretivist approach can help turn culture into something with visible consequences, and something that can be influenced—through dialogue, narrative, and relationships.

10.3 PROBLEMS AND CRITIQUE

10.3.1 Cultures That Are 'Better' or 'Worse'

Recall the healthcare safety culture study conducted by Singer in 2003 (see above). It ran into a particular problem with the functionalist approach. The scores for their

TABLE 10.2

Three Stages of Safety Culture (Westrum, 1993)

Pathological	Bureaucratic	Generative
Information is hidden	Information may be ignored	Information is actively sought
Messengers are "shot"	Messengers are tolerated	Messengers are trained
Responsibilities are shirked	Responsibility is compartmented	Responsibilities are shared
Bridging is discouraged	Bridging is allowed but discouraged	Bridging is rewarded
Failure is covered up	Organization is just and merciful	Failure causes enquiry
New ideas are crushed	New ideas create problems	New ideas are welcomed

own study, and others, had not been normalized. Because what is the norm, or the standard? Where should respondents ideally be on these surveys to gage how well or how poorly the safety culture is in their organization, or their part of the organization? The functionalist approach, after all, implies manageability: safety culture is an object that can be manipulated and changed. But if we do not know what the baseline is, or what the norm should be, then how can we even know we have to advocate change, or where to push or nudge an organization?

In 1993, Ron Westrum came up with the beginning of a solution to this. Based on a series of case studies in different domains, he produced a graduated scale of safety cultures: pathological, bureaucratic, and generative (Westrum, 1993). These offer different 'norms' (though not quantitative) against which any survey results could be contrasted. The three stages of cultures he derived from his research implied that an organization's safety culture could have a kind of 'maturity' and that organizations could be on a 'journey' to higher or more enlightened stages of safety culture (Table 10.2).

This was later extended by Hudson (2006), who had seen more nuances in the way organizations can 'grow' in their safety culture or get stuck in a particular phase:

- A pathological safety culture is one where people ask 'who cares, as long as we're not caught.'
- A reactive safety culture says that safety is important, and shows that the organization does a lot every time there is an accident;
- A calculative safety culture is one that has all the systems in place to manage hazards;
- A proactive safety culture is continuously improving, driven by safety leadership and values;
- A generative safety culture is one in which 'safety is the way we do business around here.'

The sense that some cultures are 'better' than others is seen as nonsensical to some interpretivist researchers. In the extreme interpretivist view, cultures just 'are.' Whether they are 'better' or 'worse,' or more or less mature than other comparable cultures, is a question that can only be answered if we, perforce, impose certain expectations or standards. This has happened before in history, of course: a culture with a greater proportion of people with blond hair and blue eyes has on occasion been considered a 'better' culture, for instance (Harari, 2011).

The safety culture literature on cultural progress is similarly normative. It explicitly attempts to 'rank' organizations according to a culture 'ladder' toward some cultural ideal. Organizations (or their managers) will likely want to achieve better rankings, or see themselves in a kinder light, and might do whatever it takes to get themselves to look better. Safety consultants, in the meantime, and perhaps also safety culture researchers, might have a stake in the opportunities that safety culture offers: by telling organizations that they have come a long way, but still have a long way to go in their 'cultural journeys,' they could well create continued demand for their services.

The suggestion that there is a hierarchical cultural ordering, of cultural development, of an inferiority and superiority, can of course only work if we somehow agree on the features of culture that make it so. The problem is that science, of whatever kind, cannot do this. You only need to look at the enormous diversity of definitions and lack of consensus on how to operationalize safety culture. But does that mean that we must dismiss the idea that there are 'better' or 'worse' cultures? That would seem problematic too. You might see it is a moral relativism that is unacceptable.

Conclaves Filho and Waterson (2018) reviewed the published literature that made use of safety culture maturity models. Their review showed growth in the use of maturity models to assess safety culture, but also significant variation in the ways in which they are used and reported. Beyond the published literature, particularly inside of organizations themselves, the application (and variation) of safety culture maturity models could be even more widespread. As Andrew Hale commented in a speech at Delft University, this kind of application (and even the maturity scales themselves) is probably amateurish at best:

Maturity scales... represent the steps leading from the pathological state or face to the smiling, generative one, but we know little or nothing about whether it is indeed possible to mount that scale, and if so how. We have not defined whether we should be trying to shift companies all the way from one to another end of the scale, or that we would be happy if the bulk of companies made it to the halfway point and became 'calculative'. Longitudinal research studies of companies to plot such shifts and how to facilitate them are desperately needed. We might also need to expand our discipline base to do it.

Hale (2006), cited in Conclaves Filho and Waterson (2018, p. 208)

The review showed that assessments of the reliability and validity of the use of maturity models to assess safety culture tend to be the exception, rather than the rule (44% of the sample did not report any evidence of reliability or validity assessments). The authors (2018, p. 209) conclude that "our understanding of the scientific and practice-oriented aspects of maturity models remains relatively 'immature'."

Another idea conveyed by cultural stages or maturity levels is that there is a homogeneity or cohesiveness of an organization's culture. For example, 'the culture at this organization is bureaucratic.' Such a general impression may of course be consistent with many of the things the organization is, or is not, doing. But it is unlikely that the

FIGURE 10.1 'Safety culture' tries to do and say so much that it may end up doing very little. (Drawing by author.)

whole organization is bureaucratic, the whole time. Instead, parts of the organization may be pathological, or generative, or bureaucratic, or anything in between. And it is even more likely that how all these features of culture are locally distributed shift and move during a workday—in response to events, to people showing up on shifts or leaving, to interactions between people and between them, and other factors and influences.

In an elegant case study, Walker demonstrated how a small group of workmen was caught in a common dilemma regarding work safety. They had to work safely and maintain production targets, but they were configured in a pathological organization that did not meaningfully reward or encourage participation or communication. In their own group, however, they collaborated and communicated, and socially constructed danger, avoided injury, and promoted safety for themselves. Thus, they created a small but functioning counterculture that challenged or could at least insulate them against the safety climate contrived by managers (Walker, 2010).

Insights produced by such studies should indeed an important addition to the research on safety culture: the sense of disharmony, of the dynamics that create conflict and a lack of cohesiveness. Perhaps a culture is not the pristine absence of conflict, contradiction, or opposition. Perhaps a culture evolves precisely because it is able to deal with all that. Let's turn to this now.

10.3.2 Consistency and Agreement Versus Conflict and Contradiction

The measurement of safety culture by questionnaires or surveys that inquire about people's perceptions, beliefs, norms, and practices, also unwittingly introduces a sense of agreement, harmony, and cohesiveness. Likert scales might ask people to rank the extent to which they agree or disagree with a statement (e.g., "I will always tell my boss if I see a safety problem"). Likert scales are ordinal scales, which allow one answer. A Likert scale's possible answers have a logical order according to the amount of agreement or disagreement they represent. Though ordinal scales do not allow much fancy statistics, researchers typically use them to suggest a mean, median, and mode. Particularly the mode (the answer that occurs most often) is used as a guide to what most people in an organization believe about a particular statement.

Logically, this leads to a presentation of research results in which 'most people believe that...' or 'a majority of people slightly disagree with the statement that....' In other words, it leads to a suggestion of harmony, or of the regression of most people's opinion toward some kind of agreement, cohesiveness, and lack of conflict. Such research, by the very way does its empirical work, cannot reveal, let alone illuminate or explain, conflict and inconsistency, other than by the empty gesture of a larger spread of answers across a scale. The sense of monolithic, homogeneous culture, or internal harmony, then, may in part be a by-product of the way safety culture is defined and measured. The research often speaks of culture as consisting of beliefs or values that are shared. This reduces culture to agreement among individuals.

But culture does more than provide the means to agree. It can equally provide individuals with the means to understand and handle disagreement. If safety culture, to take another example, is the 'best way we do things around here,' then it is a safety culture only if everyone does those things the best way. This allows no inconsistencies between how people define that best way, and a strong safety culture allows no deviations from that best way.

Decades ago, man-made disaster theory identified an important feature of culture which has perhaps fallen by the wayside a bit since. This is the feature of inconsistency, contradiction, and conflict. As you will recall from theorizing throughout the past chapters, organizational goal conflicts and resource constraints have been accorded an overriding role in the creation of safety problems. Organizational culture can be seen as a set of ways to deal with these conflicts and inconsistencies.

This creates a paradox that not much safety culture research is capable of handling. The same organizational culture is capable of illuminating some hazards, but also neutralizing or deflecting attention away from others *at the same time*. Organizations, because they are made up of people in all kinds of hierarchical and other relationships, have a way of selective gathering of data and directing of attention. Hazards might be known or foreseen by some, but they do not come up to a level of decision makers who have the power to stop or redirect operations (you might recall how this featured in both space shuttle accidents, for example). Or the knowledge may be subtly squelched or forgotten, or subsumed by other concerns or pushed out of the way in the pursuit of other goals also important to the organization. It is this lack of harmony, and this fundamental conflict between multiple goals under the pressure of limited resources, that needs managing every minute by many different organizational members. That makes what they have together into a culture.

Man-made disaster theory was the first to articulate this with respect to safety. The assumptions and norms that comprise a culture of safety also have the capacity to create conditions where people are left with few clear-cut opportunities to challenge those norms, beliefs, and assumptions—if indeed they even recognize the need for such a challenge. Culture shapes people's perspective, which reveals *and* hides things. It makes organizational members see, and it makes them blind, at the same time.

Weick and colleagues noted that organizations and the people in them are defined in large part by what they ignore (Weick, Sutcliffe, & Obstfeld, 1999). This is why organizational cultures can act simultaneously as a precondition for safe operations *and* as an incubator for hazard and disaster (Turner, 1978). Vaughan summed it up for the Challenger disaster like this: NASA's culture "provided a way of seeing that was simultaneously a way of not seeing" (1996, p. 392). As worded by Harari:

> Unlike the laws of physics, which are free of inconsistencies, every man-made order is packed with internal contradictions. Cultures are constantly trying to reconcile these contradictions, and this process fuels change. ... Such contradictions are an inseparable part of every human culture. In fact, they are culture's engines, responsible for the creativity and dynamism of our species. ... If tensions, conflicts and irresolvable dilemmas are the space if every culture, a human being who belongs to any particular culture must hold contradictory beliefs and be riven by incompatible values. It's such an essential feature of any culture that it even has a name: cognitive dissonance. Cognitive dissonance is often considered a failure of the human psyche. In fact, it is a vital asset. Had people been unable to hold contradictory beliefs and values, it would probably have been impossible to establish and maintain any human culture.
>
> *(2011, pp. 181–184)*

If you want to understand culture, said Harari, do not look for a pristine set of values that every member of that culture holds dear. Rather, inquire into the catch-22s of that culture. Look for those places where rules are at war, where standards scuffle, and goals clash. Focus on the very spot where culture members teeter between imperatives. That is where you will understand them, and their culture, best. This is when they are being safe *and* being productive. Where they are on time *and* get there alive. Where they are following best engineering practice *and* getting the next contract by undercutting the competitor.

Culture includes things that are shared *and* contested, even inside a particular group or occupation. The legitimacy of the parties involved in a conflict, the value placed on the perspectives and knowledge they argue from—these all entail different kinds of cultural interpretations. The possible and acceptable manners in which resolutions might reasonably occur, and any notions of fairness invoked to resolve a conflict, do too. Similarity and difference are things that culture handles, whatever is at stake.

10.3.3   SAFETY CULTURE AND POWER

Once we acknowledge that cultures exist to somehow handle conflict and disharmony, it becomes necessary to address power (Dekker & Nyce, 2014). Issues of culture and power are so intertwined that safety culture research should incorporate perspectives of power and conflict. This is necessary in order to be able to give a realistic account of the dynamics of organizational life (Antonsen, 2009a).

Safety science's unrealistically harmonious image of organizational life may have deep epistemological roots. Safety science seems to constitute one of the last research literatures that strongly reflect Enlightenment ideals with its appeals to be both rational and pragmatic. According to these ideals, science, the highest expression of reason, can make the world a better place. After all, science can explain, predict, and ultimately help prevent that which we do not want—disease, disaster. Safety science aspires to be a normal or paradigmatic science, with systematic, unified production of evidence so that it can measurably affect things in the real world—including the production of generative safety cultures.

And so, safety culture research mostly continues to adopt a technical and problem-solving approach consistent with theories of organizational life dominated by rational choice and regulative management. In this, the environment is seen as a target of managerial control, exercised through rational practices of evidence gathering and decision-making. Power, as something that (some) people inside organizations have, or possess, of course plays a role in this. If understood that way, there are various kinds of power, which can influence safety culture all in their own ways (Antonsen, 2009a):

1. *Position power*: Organizational structures distribute different amounts of formal authority to the various layers in the hierarchy. Also, the division of labor in an organization inevitably creates power differences. Different tasks and activities, after all, are not all equally critical to organizational survival (recall the section on Prima Donnas in Chapter 2).
2. *Information and expertise*: This is the power of 'know-how' and 'know-what.' Possessing and controlling the release of knowledge or information crucial to the organization is an important source of power. Some positions imply access to important information which can be used strategically or tactically, as an exercise of power. Some positions are crucial for an organization's information or decision flow, and can hold things up at will. As another example, doctors are seen as powerful in hospitals because they have a monopoly on the medical expertise required for the organization's key activities (Bosk, 2003; von Thaden, Hoppes, Yongjuan, Johnson, & Schriver, 2006).
3. *Control of rewards and resources*: Control over resources that are either material (money, employment, data processing) or immaterial (recognition, political support) form perhaps a most visible and common form of power-as-possession. Monetary rewards are used to promote organizational goals, even if these can be in conflict (e.g., efficiency versus safety). Safety rewards that are reputational or economic (or both) are another example.
4. *Coercive power*: Coercive power is closely connected to actors' or groups' control over sanctions, as it rests on 'the ability to constrain, block, interfere or punish'. Industrial action, over a real or imagined safety concern, is an example of the use of this form of power.

5. *Alliances and networks*: This is sometimes known as the power of 'know-who.' Alliances and networks can allow actors to tap into other actors' sources of power, or together form coalitions that identify and exploit solutions that are mutually beneficial.
6. *Personal power*: Charisma, energy, political skills, and verbal facility are among the individual characteristics that constitute a source of power and the reputation that comes with it.

This is of course a limited consideration of power and its role in safety culture. Power is not just a possession that limits or constrains the will or possibilities that other people have. Power can also be seen as a possibility, literally as an empowerment, which can be exercised from the bottom-up as well as from the top-down (Bosk, 2003).

When seen this way, power is everywhere—embodied in discourse, knowledge, agency, structure, and procedure. It can work in a 'capillary' sense, pervading even the most seemingly innocuous interactions. Nurses, for example, are not just the passive recipients of power possessed by practitioners higher up in the hierarchy. They too can deploy an array of strategies through which power flows—by reporting patient symptoms in a particular way and not another, by modulating the tone or urgency of particular messages to physicians so as to compel them into a particular action, including calling for back-up (Dekker, 2012). They might threaten to use a hospital's incident reporting system to 'rat' on a doctor who does not do what the nurse believes is the right thing to do for that patient at that time. Power does not just repress or limit human agency. Rather, it enables and sets the stage for *all* human action.

10.3.4 METHODOLOGICAL INDIVIDUALISM

The functionalist study of safety culture is easily critiqued for its methodological individualism (Newlan, 1990). Methodological individualism, you might recall from Chapter 2, is the belief that social processes can be explained by reducing them to the level of an individual. The question of safety culture, according to this belief, can be made intelligible by referring, for example, to the attitudes, behaviors, perceptions, and relational styles of individual workers. It is generally blind to the interactions *between* people from which cultures emerge and develop.

Methodological individualism dominates industrial psychology and its links to safety. Remember that culture is sometimes even defined that way—"the way we do things around here"—as routinized behaviors by individuals that become legitimated through repetition. Culture, in studies typical of industrial psychology or management science, is reduced to worker attitudes and behaviors, because that is what these fields know how to measure. The problem is that this does not allow for properties of human experience and behavior that are emergent; the sorts of effects that are not visible at the level of one individual worker, but that come out of the interactions *between* humans.

And of course, if ever there was an emergent phenomenon, it is culture. A culture is not the simple sum of the behaviors and attitudes of the individuals that belong to it. Culture cannot be reduced to an individual: it arises precisely because individuals interact, and can only really be captured and described at the level of those interactions and the effects they produce (Guldenmund, 2007). Methodological individualism, however, does not allow us to define or study it that way. We need to ask ourselves, said Guldenmund

> ...to what extent individual measures can be used to say something about organisational levels higher than the individual one. Clearly, this is an issue of great importance because, very often, aggregated individual measures, from questionnaire surveys for instance, are used to say something about the full organisation or certain parts thereof. [As stated in a] study of US Navy enlisted personnel on various ships: 'aggregations of such data carry the potential for erroneous inference' especially so when 'perceptions are combined across groups of increasingly heterogeneous context or structure'... Aggregation to ship-wide or departmental-wide conditions did not appear warranted but aggregation to divisional or functional level like Navigation, Maintenance and Radio Communication did. Such studies show that seemingly obvious aggregational levels within organisations might not be so homogenous in practice.

> **Guldenmund (2000, p. 227)**

The risk that safety researchers see in this is not just methodological or epistemological. It is also ideological and practical. If safety culture is boiled down to the attitudes and behaviors of individual employees, and those become the target for any intervention, then it is no better than behavior-based safety (Hopkins, 2014). The seduction is there, of course. As Hudson said, "in a safety culture it becomes possible that people carry out what they know has to be done not because they *have* to, but because they *want* to" (2006, p. 10, italics in original).

This is where winning workers' 'hearts and minds' becomes a tempting managerial aspiration. The problem arises when targeting the 'hearts and minds' of employees becomes akin to believing that safety is a problem of 'moral and mental shortcomings' of workers. If their hearts are not in it, then they are morally deficient. If their 'minds' need focusing or working on, then they are mentally deficient. Recall from the beginning of the book that this was the way we thought about safety at the beginning of the 20th century (see Chapter 3). And so, say safety researchers:

> The fashion for 'safety culture', insofar as it relies on a campaign for hearts and minds, is at best wishful thinking and at worst a thinly disguised version of the blame-the-worker strategy which we've been combatting for years.

> **Hopkins (2014, p. 6)**

As with behavior-based safety, this approach is thought to start at the wrong end of the problem. And as with Taylorism, it seems to suggest that managers know what is best, and what has to be done. Workers need to comply, but not out of obligation or threat of sanction. They comply because they have been made to *want* to comply. An approach like this tries to change the way people at the sharp end think and feel about safety. It tries to change people's attitudes and behaviors, in other words, by targeting their attitudes and behaviors.

As you might recall from earlier chapters, safety research has gradually become more directed at the conditions under which people work, not at the people themselves. Rasmussen (1997) has captured this trend of safety science as well: away from a focus on prescriptive, normative models of behavior, and toward a growing interest in identifying and changing behavior-shaping features of the system. This involves studying and influencing the incentives, tools, tasks, expectations, goals, and resources of people's work. If safety culture is used to win the hearts and minds of workers, this research says, it seems to do little more than relabeling a discredited strategy that went out of fashion in safety science many decades ago.

10.3.5 Is Safety Culture Useful for Regulators or Investigators?

In 2012, the Organization for Economic Cooperation and Development published a report that promoted the use of 'safety culture' in regulating safety-critical industries like nuclear power (OECD, 2012). It cited a number of investigative reports that played up the role of the regulator in the oversight of safety culture, but concluded that there was not much guidance on *how* that should be achieved. What strategies and practical approaches should a regulator adopt for maintaining oversight of and influence over those facets of an organization's leadership and management that have a profound influence on safety culture? The report asked the question rather than answering it. What it also did not do was to ask the underlying question: Can a safety culture be regulated at all? Is the concept even useful in trying to regulate a safety-critical activity?

The results from regulatory experimenting with 'safety culture' are probably not so surprising given what you have read above. A regulator can specify the expectations it wants organizations to meet on a 'soft topic' like safety culture, yes (Le Coze & Wiig, 2013). But it needs to leave rather open, or unclear, how to translate these expectations into operations and control by the organization (Antonsen, 2009b). Others argue that culture can never be regulated, even though regulators can exert influence on organizational priorities and processes that might eventually impact safety culture (Grote & Weichbrodt, 2013).

What happens to the role of safety regulation if we take a more interpretivist approach? What possibilities are left then? Remember what was said about reflexivity above. The question the interpretivist approach asks is how we could capture what groups tell about themselves. Recent accidents provide some clues of where we might start looking. A main source of residual risk in seemingly safe systems is the drift into failure (recall this from Chapter 7). Pressures of scarcity and competition narrow an organization's focus on goals associated with production. With an accumulating base of empirical success (i.e., no accidents, even if safety is increasingly traded off against other goals such as maximizing profit or capacity utilization), the organization, through its members' multiple little and larger daily decisions, will begin to believe that past success is a guarantee of future safety and that historical success is a reason for confidence that the same behavior will lead to the same (successful) outcome the next time around.

The absence of failure, in other words, might be taken as evidence that hazards are not present and that countermeasures already in place are effective. Such a model of risk is embedded deeply in the reflexive stories of safety that Rochlin (1999) mentioned, and it can be made explicit only through qualitative investigations that probe the interpretative aspect of situated human assessments and actions. Surveys do little to elucidate any of this, according to the interpretivist approach. More qualitative studies could reveal how currently traded models of risk may increasingly be at odds with the actual nature and proximity of hazard, though it may of course be difficult to establish the objective presence of hazard.

Particular aspects of how organizational members tell or evaluate safety stories can serve as markers. Woods (2003, p. 5), for example, has called one of these markers "distancing through differencing." In this process, organizational members look at other failures and other organizations as not relevant to them and their situation. They discard other events because they appear at the surface to be dissimilar or distant. Discovering this through qualitative inquiry can help specify how people and organizations reflexively create their idea, their story of safety. Seemingly divergent events can represent similar underlying patterns in the drift toward hazard.

High reliability organizations characterize themselves through their preoccupation with failure: continually asking themselves how things can go wrong and could have gone wrong, rather than congratulating themselves on the fact that things went right. Distancing through differencing means underplaying this preoccupation. It is one way to prevent learning from events elsewhere, one way to throw up obstacles in the flow of safety-related information. Additional processes that can be discovered include to what extent an organization resists oversimplifying interpretations of operational data, whether it defers to expertise and expert judgment rather than managerial imperatives. Also, it could be interesting to probe to what extent problem-solving processes are fragmented across organizational departments, sections, or subcontractors.

The 1996 ValuJet accident, where flammable oxygen generators were placed in an aircraft cargo hold without shipping caps, subsequently burning down the aircraft, was related to a web of subcontractors that together made up the virtual airline of ValuJet. Hundreds of people within even one subcontractor logged work against the particular ValuJet aircraft, and this subcontractor was only one of many players in a network of organizations and companies tasked with different aspects of running (even constituting) the airline. Relevant maintenance parts (among them the shipping caps) were not available with the subcontractor, ideas of what to do with expired oxygen canisters were generated ad hoc in the absence of central guidance, and local understandings for why shipping caps may have been necessary were foggy at best. With work and responsibility distributed among so many participants, nobody may have been able anymore to see the big picture, including the regulator itself. Nobody may have been able to recognize the gradual erosion of safety constraints on the design and operation of the original system.

Investigating the role of safety culture in an accident, Barry Strauch found, was difficult for several reasons. These are variations on the themes that you will have already encountered in the various sections above (Strauch, 2015):

- Measuring traits of a safety culture through questionnaires after an accident throws up familiar problems of validity. These may even be worse because of the distortions in beliefs and perceptions, or people's willingness to disclose in the wake of an accident;

- Using ethnographic techniques to study culture with a more interpretivist approach is subject to severe resource constraints that operate on most investigations (particularly time, money, investigator expertise);
- Examining an organization's safety management system or risk identification and mitigation practices is not going to yield much, since these are only effective against the risks that were recognized in advance. There is not much use citing a company for not recognizing what it could not or what peers in the industry (nor even the regulator) did not recognize either.
- Also, companies that embrace all the desirable safety culture characteristics may still end up producing an accident (recall the banality of accidents thesis from Chapter 7). The presence or absence of a safety culture is not strongly linked to the outcome at all, in other words.

This does not mean, Strauch (2015) reminds us, that understanding the influence of organizational factors is irrelevant to investigations. It is hugely important, as has been demonstrated by the direction of safety science since WWII. But, he concludes, we cannot really examine safety culture in an accident investigation, and perhaps we should not even try. One of the reasons we should not even try is on display in the case below. The examination of safety culture in the wake of an accident can quickly become a reinvented search for 'human errors,' committed by people and teams further away from the sharp end, but just as deficient in their understanding of the world (as we now, in hindsight understand it) as the operators we once blamed for human error.

SPACE SHUTTLE COLUMBIA: "A BROKEN SAFETY CULTURE"

In the Space Shuttle Columbia investigation (CAIB, 2003), safety culture was defined as the collection of characteristics and attitudes in an organization, promoted by its leaders and internalized by its members, that makes safety an overriding priority. In a case of what Erik Hollnagel would call 'what you look for is what you find,' this 'overriding priority,' did not exist. The Columbia Accident Investigation Board examined how and why an array of processes, groups, and individuals in the Shuttle Program failed to appreciate the severity and implications of the foam strike on Space Shuttle Columbia. After all (at least based on hindsight), the Shuttle Program should have been able to detect the foam trend and more fully appreciate the danger it represented.

The flight readiness process, which involved every organization affiliated with a shuttle mission, missed the danger signals in the history of foam loss. Generally, the higher information is transmitted in a hierarchy, the more it got 'rolled-up,' abbreviated, and simplified. Sometimes information got lost altogether, as weak signals dropped from memos, problem identification systems, and formal presentations. The same conclusions, repeated over time, could result in problems eventually being deemed nonproblems. An extraordinary example of this phenomenon is how Shuttle Program managers assumed that the foam strike on a previous mission was not a warning sign.

During the subsequent Flight Readiness Review, the bipod foam strike to the previous mission was rationalized by simply restating earlier assessments of foam loss. The question of why bipod foam would detach and strike a solid rocket booster triggered no further analysis or heightened curiosity or questions. After the successful flight, once again the previous foam event was not discussed at the next Flight Readiness Review. The failure to mention what was really an outstanding technical anomaly, even if not technically a violation of NASA's own procedures, desensitized the Shuttle Program to the dangers of a foam strike. It demonstrated just how easily the flight preparation process can be compromised. In short, the dangers of foam strikes got "rolled-up," which resulted in a missed opportunity to make shuttle managers aware what the shuttle required, and did not yet have a fix for the problem.

Once the Columbia foam strike was discovered, the Mission Management Team Chairperson asked for the rationale the previous Flight Readiness Review used to launch in spite of the earlier foam strike. In her e-mail, she admitted that the analysis used to continue flying was, in a word, "lousy." This admission, that the rationale to fly was rubber-stamped, was deemed unsettling by the Investigation Board.

The Flight Readiness process, after all, is supposed to be shielded from outside influence, and is viewed as both rigorous and systematic. Yet the Shuttle Program is inevitably influenced by external factors, including schedule demands. Collectively, such factors shape how the Program establishes mission schedules and sets budget priorities, which affects safety oversight, workforce levels, facility maintenance, and contractor workloads. Ultimately, external expectations and pressures impact even data collection, trend analysis, information development, and the reporting and disposition of anomalies. These realities contradict NASA's optimistic belief that preflight reviews provide true safeguards against unacceptable hazards.

The premium placed on maintaining an operational schedule, combined with ever-decreasing resources, gradually led shuttle managers and engineers to miss signals of potential danger. Foam strikes on the shuttle's thermal protection system, no matter what the size of the debris, were "normalized" and accepted as not being a "safety-of-flight risk." Clearly, the risk of Thermal Protection damage due to such a strike needed to be better understood in quantifiable terms. External Tank foam loss should have been eliminated or mitigated with redundant layers of protection. If there was in fact a strong safety culture at NASA, safety experts would have had the authority to test the actual resilience of the leading edge reinforced carbon panels, as the Board has done.

The Debris Assessment Team's efforts to obtain additional imagery of Columbia were not successful. When managers in the Shuttle Program denied the team's request for imagery, the Debris Assessment Team was put in the untenable position of having to prove that a safety-of-flight issue existed without the very images that would permit such a determination. This is precisely the opposite of how an effective safety culture would act. Organizations that deal with high-risk operations must always have a healthy fear of failure: operations must be proved safe, rather than the other way around. NASA inverted this burden of proof.

Another crucial failure involved the Boeing engineers who conducted the crater analysis. The Debris Assessment Team relied on the inputs of these engineers along with many others to assess the potential damage caused by the foam strike. Prior to Space Shuttle Columbia, crater analysis was the responsibility of a team at Boeing's Huntington Beach facility in California, but this responsibility had recently been transferred to Boeing's Houston office. In October 2002, the Shuttle Program completed a risk assessment that predicted the move of Boeing functions from Huntington Beach to Houston would increase risk to Shuttle missions through the end of 2003, because of the small number of experienced engineers who were willing to relocate. To mitigate this risk, NASA and United Space Alliance developed a transition plan to run through January 2003.

The Board discovered that the implementation of the transition plan was incomplete and that training of replacement personnel was not uniform. Space Shuttle Columbia was the first mission during which Johnson-based Boeing engineers conducted analysis without guidance and oversight from engineers at Huntington Beach. Even though Space Shuttle Columbia's debris strike was 400 times larger than the objects that the crater program was designed to model, neither Johnson engineers nor Program managers appealed for assistance from the more experienced Huntington Beach engineers, who might have cautioned against using the crater program so far outside its validated limits. Nor did safety personnel provide any additional oversight. NASA failed to connect the dots: the engineers who misinterpreted the crater program, a tool already unsuited to the task at hand, were the very ones the Shuttle Program identified as engendering the most risk in their transition from Huntington Beach. The Board viewed this example as characteristic of the greater turbulence the Shuttle Program experienced in the decade before Columbia as a result of workforce reductions and management reforms.

In the Board's view, the decision to fly the previous mission without a compelling explanation for why bipod foam had separated on ascent during the preceding mission, combined with the low number of Mission Management Team meetings during Space Shuttle Columbia, indicates that the Shuttle Program had become overconfident. Over time, the organization determined it did not need daily meetings during a mission, despite regulations that state otherwise.

Status update meetings should provide an opportunity to raise concerns and hold discussions across structural and technical boundaries. The leader of such meetings must encourage participation and guarantee that problems are assessed and resolved fully. All voices must be heard, which can be difficult when facing a hierarchy. An employee's location in the hierarchy can encourage silence. Organizations interested in safety must take steps to guarantee that all relevant information is presented to decision makers. This did not happen in the meetings during the Columbia mission. For instance, e-mails from engineers at Johnson and Langley conveyed the depth of their concern about the foam strike, the questions they had about its implications, and the actions they wanted to take as a follow-up. However, these e-mails did not reach the Mission Management Team.

The failure to convey the urgency of engineering concerns was caused, at least in part, by organizational structure and spheres of authority. The Langley e-mails were circulated among coworkers at Johnson who explored the possible effects of the foam strike and its consequences for landing. Yet, they kept their concerns inside of local channels and did not forward them to the Mission Management Team. They were separated from the decision-making process by distance and rank (a problem that also showed up in the Challenger Launch decision).

Similarly, Mission Management Team participants felt pressured to remain quiet unless discussion turned to their particular area of technological or system expertise, and, even then, to be brief. The initial damage assessment briefing prepared for the Mission Evaluation Room was cut down considerably in order to make it 'fit' the schedule. Even so, it took 40 min. It was cut down further to a 3-min discussion topic at the Mission Management Team. Tapes of the Space Shuttle Columbia Mission Management Team sessions revealed a noticeable "rush" by the meeting's leader to the preconceived bottom line that there was "no safety-of-flight" issue. Program managers created huge barriers against dissenting opinions by stating preconceived conclusions based on subjective knowledge and experience, rather than on solid data. Managers demonstrated little concern for mission safety.

Organizations with strong safety cultures generally acknowledge that a leader's best response to unanimous consent is to play devil's advocate and encourage an exhaustive debate. Mission Management Team leaders failed to seek out such minority opinions. Imagine the difference if any Shuttle manager had instead asked, 'Prove to me that Columbia has not been harmed.'

Similarly, organizations committed to effective communication seek avenues through which unidentified concerns and dissenting insights can be raised, so that weak signals are not lost in background noise. Common methods of bringing minority opinions to the fore include hazard reports, suggestion programs, and empowering employees to call "time out." For these methods to be effective, they must mitigate the fear of retribution, and management and technical staff must pay attention. Shuttle Program hazard reporting was seldom used, safety time outs were at times disregarded, and informal efforts to gain support were squelched. The very fact that engineers felt inclined to conduct simulated blown tire landings at Ames "after hours," indicates their reluctance to bring the concern up in established channels.

The Board believed that NASA's safety organization, due to a lack of capability and resources independent of the Shuttle Program, was not an effective voice in discussing technical issues or mission operations pertaining to Space Shuttle Columbia. The safety personnel present in the Debris Assessment Team, Mission Evaluation Room, and on the Mission Management Team were largely silent during the events leading up to the loss of Columbia. That silence was not merely a failure of safety, but a failure of the entire organization, if not the safety culture.

What hope should we have of safety culture taking us beyond finding individuals to blame, and seeking more ways to regulate their behavior? Much of safety science has focused on human agency and its deficiencies, reflecting the rationalist assumptions of regulative management. Even in proto-safety culture research (e.g., Turner, 1978) it has found, for example, how people's erroneous assumptions let events go unnoticed or misunderstood, or how rigidities of human belief and perception can lead to a disregard of complaints and warning signals from outsiders. This produces judgment errors, cognitive lapses, deficient supervision, and communication difficulties that safety scientific orthodoxy sees as critical in creating a discrepancy between a safe system and actual system state (Reason, 1997). As you might recall from Chapter 7 (Weick & Sutcliffe, 2007):

> ...failure means that there was a lapse in detection. Someone somewhere did not anticipate what and how things could go wrong. Something was not caught as soon as it could have been caught.
>
> *(p. 93)*

This is another kind of ontological alchemy that turns judgmental attributions (cognitive lapse, judgment error) into remediable statements of fact. If disasters in systems are related to failures of intelligence, or not catching things as soon as possible, then the system's intelligence should be enhanced by increasing the organization's commitment, reach, and flexibility in its data infrastructure and interpretations of risk and safety. Many of safety science's solutions emerge from this, including (as you will recall from previous chapters):

- Senior management commitment to safety;
- Shared care and concern for hazards and a willingness to learn and understand how they impact people;
- Realistic and flexible norms and rules about hazards;
- Continual reflection on practice through monitoring, analysis, and feedback systems.

What is seldom addressed here is any mention of who does or decides what in sharing concern, in changing norms, in committing to certain priorities or principles, in learning and providing feedback. Or who, for that matter, got to call something a judgment error or cognitive lapse in the first place and on whose part. An analysis of Space Shuttle accidents by Feldman is a good example (Feldman, 2004). At first, the analysis and its findings remain consistent with the standard safety-scientific model and other published work on those accidents. Misunderstandings of flight risk were the systemic products of overconfidence in quantitative data, a marginalization of nonquantitative data, an insensitivity to uncertainty and loss of organization memory, of the illusion that engineering problems and solutions could be addressed independently from organizational goals. All this comfortably fits Turner's category of cognitive failures, or failures of organizational information processing that characterize the incubation period before disaster strikes (Turner, 1978).

In his conclusion, however, Feldman (2004) departed from any further consideration of institutions, power, or bureaucracy. Instead, he exhorted engineers to intervene, to be better aware of what they are doing, to speak up, to not be blinded by the situation(s) in which they are involved, to be more responsible: "Engineers need intense cultivation of their professional responsibilities within organizations," he argued (p. 713). Individuals need to work harder, be more conscientious, and virtuous to overcome the limitations of their institutions. To safety science in Anglo-Saxon traditions, such valorization of individual heroism in the face of institutional hysteresis may seem natural. As does the tendency to analyze "down and in" and trace organizational failure to a few who did not speak up. Even the use of "safety culture," though ostensibly a way to broaden out to more diffuse understandings of failure, can end up allocating responsibility to particular individuals or groups (Silbey, 2009):

> ...the endorsement of safety culture can be usefully understood as a way of encouraging and allocating responsibility ... Invoking culture as both the explanation and remedy for technological disasters obscures the different interests and power relations enacted in complex organizations. Although it need not, talk about culture often focuses attention primarily on the low-level workers who become responsible, in the last instance, for organizational consequences, including safety.

(p. 343)

Feldman continued that there should be consequences for those low-level workers who do not live up to their fiduciary responsibilities: "Engineering societies need to require engineers to act in accordance with the prevent-harm ethic. This requirement must include both training to inculcate the prevent-harm ethic and sanctions—up to losing one's license—when the ethic is violated" (p. 714).

10.3.6 Do Safety Culture Assessments Have Predictive Value?

One of the most important contributions that the concept of safety culture could make is telling managers and others something about the future. Descriptions of an organization's safety culture, after all, are considered to provide the grounds for proactive safety management. They can, or should, yield "predictive measures which may reduce the need to wait for the system to fail in order to identify weaknesses and to take remedial actions" (Flin, Mearns, O'Connor, & Bryden, 2000, p. 178).

Safety culture should indeed offer some of the most compelling leading (rather than lagging) indicators for those tasked with governing a safety-critical organization or operation. Can a safety culture assessment foresee accidents? Can a safety culture assessment foretell how likely it is that the organization is going to suffer a major calamity? Antonsen (2009b), in an empirical study of an oil rig, set out to answer that question:

> If survey methods are considered to be useful as safety assessments, the results of such assessments should provide some basis for making judgements about how safe or unsafe an organization is, as well as some sort of 'prediction' as to whether the organization is prone to having accidents in the future. The predictive value of culture surveys has not been the subject of much discussion or empirical investigation in the research literature.
>
> (p. 243)

The data used to test whether safety culture assessment has predictive value came from a Norwegian oil and gas installation, called *Snorre Alpha*. A safety culture survey was conducted there in 2003, and a very serious incident occurred on the installation in 2004. Did the survey foresee that incident? Did it point managerial attention beforehand to any of the aspects that showed up in the incident? Was there a link between proactive and reactive assessments of the cultural and climate conditions for safety onboard *Snorre Alpha*? Do safety culture surveys predict whether an organization is likely to experience a major accident? Two investigations into the incident, as well as the safety culture survey from the year before, were analyzed to find the answers. The description of the case below is from Antonsen (2009b, pp. 244–245).

Snorre is an oil field in the Tampen area in the northern part of the North Sea. The sea depth in the area is 300–350 m. The oil on Snorre is produced with pressure maintenance by water injection, gas injection, and water alternating gas. Snorre Alpha is a floating steel facility for drilling and processing. It provides accommodation for 220 people.

On Sunday November 28, 2004, an uncontrolled situation occurred during work in Well P-31A on the Snorre Alpha facility. The operation that was to be performed is called slot recovery. It consists of pulling pipes out of the well in preparation for the drilling of a sidetrack. During the course of the day, the situation developed into an uncontrolled gas blowout on the seabed, resulting in gas under and on the facility. Of the 216 people onboarding the platform at the time of the blowout, 181 were evacuated by helicopter to installations nearby. The 35 people left on the platform tried to gain control of the situation by pumping drilling mud into the well, a process known as 'bull heading' in offshore terminology. This work was complicated by the fact that there was gas under the installation, which made it impossible for any vessels to supply additional drilling mud (the proximity of a running combustion engine could have ignited the gas). During the night, the installation's supplies of mud ran out, and the only choice was manually to mix mud out of available drilling fluids. This approach was successful, and the well was stabilized on November 29.

The actual consequences of the incident are mainly related to disturbances in production and the economic losses connected with this. The production on Snorre Alpha and the neighboring platform Vigdis is about 200,000 barrels of oil per day. Production on both platforms was stopped by the incident and did not return to normal for several months afterwards. No one was killed or injured as a result of the incident. Under slightly different circumstances, however, the incident could have led to the loss of many lives. The gas could have ignited, causing a scenario similar to the Piper Alpha accident. Also, the quantities of gas flowing from beneath the platform could have resulted in stability problems, which, in turn, could have made the platform capsize. Both these scenarios imply the loss of many lives, environmental damage, and additional loss of material assets. Owing to the serious potential of the incident, the Norwegian Petroleum Safety Authority (PSA) has characterized the incident as one of the most serious ever to occur on the Norwegian shelf.

In the year before the incident, employees on Snorre Alpha had completed a self-administered questionnaire that consisted of 20 items. It was part of a company-wide behavioral program for improving operational safety. Interestingly, the survey was designed by researchers from the same institution who would later participate in one of the post-incident investigations. It contained the following questionnaire items (with a six-point Likert-type scale to indicate agreement) which are quite typical of surveys of this kind (Antonsen, 2009b, p. 246):

1. My supervisor sets a good example when it comes to safety at my workplace.
2. When somebody at my workplace brings forward safety-related information, the unit manager will make sure the problems are solved.
3. Management will follow up on actions from Health Safety & Environment (HSE)-inspections and meetings.
4. My supervisor appreciates my coming forward with safety-related issues.
5. My supervisor often calls on me at my working place to discuss safety.
6. In our organization it is common to intervene if someone works in a hazardous way.
7. We show care for each other in our daily work.
8. I always consider the risks involved before I carry out my work.
9. I continuously consider the risks involved while I carry out my work.
10. Improving safety has a high priority at my workplace.
11. The principle that 'we always have the time to work safely' is lived up to at my workplace.
12. At my workplace, work operations are always stopped if there are any doubts as to whether safety is ensured.
13. Our managers will take action if safety measures are not implemented within given deadlines.
14. At my workplace, operations that involve risk are carried out in compliance to rules and regulations.
15. Injuries and near misses are always reported in accordance with regulations.
16. At my workplace, deliberate breaches of rules and regulations will always be sanctioned.

17. When undesirable events happen at my workplace, measures will be taken to prevent similar incidents from happening in the future.
18. If I make a mistake, I can report it to management without fear of negative reactions.
19. I never refrain from reporting undesirable events out of fear of negative sanctions from my coworkers.
20. All in all, I think we have a good safety culture at my workplace.

Safety culture, according to the survey, was given a good score. Only two items scored below a mean of 4.5 (on a scale of six). Not all employees were convinced that they always had the time to work safely, or that their bosses would call on them at the workplace to discuss safety. Seven items had a score higher than five, among them items concerned with compliance and learning. Overall, the survey showed a highly positive safety culture. In fact, *Snorre Alpha* had a 'generative' organizational culture, had the survey results been superimposed on Westrum's (1993) ranking. What did the investigations show after the incident? Antonsen (2009b, p. 247) relates the following:

The PSA's official investigation of the incident identified a total of 28 aberrations from the regulations that governed technical, operational, and organizational safety barriers. In particular, there were severe deficiencies in the planning and accomplishment of the well operation. According to the report, these deficiencies related to the following issues:

There was a lack of compliance with procedures and governing documents. According to the PSA, the lack of compliance was evident in all phases of the operation, but in particular in the planning of the operation.

There was a lack of understanding of risk assessments, and deficiencies in carrying out risk assessments. The investigation showed that risk assessments were given low priority and that there was a lack of holistic understanding of the risks involved.

There was insufficient managerial involvement. In particular, the PSA emphasized the insufficient managing of resources in planning the operation, the failure to involve [external] competence necessary to reveal the deficiencies related to risk assessments and training in use of procedures.

There was a lack of control in the use of governing documents. The unit responsible for developing the procedures relevant for this operation had proved unable to uncover the deficiencies in an internal revision conducted only 4 months before the incident.

There were breaches of rules regarding safety barriers in sub-sea wells. The PSA noted that the operators chose to reopen a well which had been shut down due to lack of well integrity, knowing the complexity, and insufficient integrity of the well. Investigators attributed the identified aberrations to both individuals and groups in the mother company and the drilling contractor, at multiple organizational levels—both onshore and offshore. They also expressed concern that these organizational flaws were not identified before the accident. In the summary of the report, the PSA wondered about the fact that such extensive failures in established systems were not discovered, and questioned why these deficiencies were not detected and corrected at an earlier stage.

Further investigation by researchers delved into the history of Snorre Alpha to try to better understand its cultural and organizational features. They found that the installation had been governed by three different oil companies in the 4 years leading up to the incident, and that it had gone through an extended period of great organizational turmoil even though the people on the platform itself were pretty much the same throughout. At the time of the incident, however, it was neither socially nor culturally integrated into the mother organization, truly an 'island' in its North Sea operations. There was little calibration of views on safety by external perspectives, and few or no opportunities to intervene from the outside. One of the constants throughout the various organizational changes was a focus on efficiency. Success criteria were those associated with production, and delivery on budgets and plans. There was a high-risk tolerance. Long-term planning and maintenance suffered. Hard work and creative improvisation were central to *Snorre Alpha's* identity. In the end, and fascinatingly, these were also the features that got them *out of trouble* once the incident had started unfolding (Antonsen, 2009b).

There was a considerable gap between the situation depicted by the culture survey and the situation described by the two investigations, as shown in Table 10.3:

The lack of managerial involvement was picked up by both the survey and the post-incident investigations. Further investigation showed, however, that this was not unique to *Snorre Alpha*, but commonly reported across the company and its many installations. So what accounted for the difference in these findings? Three explanations were offered:

1. *Empirical*: the 'culture' in 2003 was different from that in 2004. So whatever measurement would have been applied in 2004 (even a survey) would have revealed these differences, whether an incident had occurred or not. The problem with this explanation is that Snorre Alpha was, as its own

---

TABLE 10.3

Comparison between 2003 Safety Culture Survey Results and 2004 Post-incident Investigation Results (Antonsen, 2009b)

Safety Culture Survey Pre-incident	Investigations Post-incident
Safety highly prioritized	Safety subordinate to meeting production targets
Risk assessments carried out before and during work operations	Lack of risk assessments, poor understanding of risk assessments
High degree of compliance to rules and procedures, breaches only when sanctioned by management	Severe breaches of procedures common, culture of noncompliance
Climate that welcomes communication of safety-relevant information	Various weaknesses in communication climate
Incidents and near misses reported, measures taken to prevent recurrence	Not all incidents and near misses reported; limited use of organization's and others' safety experience
Insufficient managerial involvement	Insufficient managerial involvement

island, very stable in its culture. It would be hard to argue that it had undergone rapid and transformational culture change that would be picked up by
such investigations.[1]

2. *Theoretical*: post-incident investigations suffer from hindsight- and outcome
biases, of course (Dekker, 2014a). A safety culture survey does not. Or does
it? The thing about the survey items typical of safety culture research is that
they represent, in fact, the accumulated experience of outcome-driven and
hindsight-based incident and accident investigations. In other words, there
should not be much theoretical divergence between what an investigation
discovers and a survey asks.

3. *Methodological*: did the survey actually ask the right questions, or is a survey capable at all of asking questions about culture in a way that reveals
what we need to know? Positively laden statements (which is how the
twenty items were phrased) may bias the person answering the survey. And
of course, surveys allow no opportunity for clarification either way. The
respondent has no way of assuring that she or he understands the question
the way the researcher meant it, and the researcher has no way of interro-
gating the respondent's interpretation of it. They simply have to take each
other's written word for it. Stating, for instance, that one never refrains from
reporting undesirable events out of fear of negative sanctions from coworkers leaves all kinds of things unspecified, such as what an undesirable event
is, what reporting (and to whom) actually means, what are seen as 'negative
sanctions' and who are included in the group of coworkers that one should
care about with respect to this item. People may have widely differing interpretations of all of these things, and none of that would be picked up by the
survey. That said, of course, it would be good to administer the very same
survey after the incident (despite its many imperfections), to see whether it
alone would have been able to reveal the difference between people's beliefs
and perceptions before and after.

Independent of what explanation may make most sense, research in healthcare also
shows a disconnection between how safety culture is measured by staff surveys, and
how well a hospital is actually doing in keeping its patients safe (Meddings et al.,
2016). Hospitals that had signed on to a national patient safety project were given
technical help—tools, training, new procedures, and other support—to reduce two
kinds of infections that patients can get during their hospital stay:

---

[1] A comparison like this, between an organization before and after an accident, is itself based on
assumptions about the nature of organizational life and how we can model and understand it. It is common, in safety science as elsewhere in the West in particular, to see organizations more as Newtonian
machines (just consider the typical organizational chart and its hierarchy and components) than as
living, organic, complex systems that grow, change, and give rise to emergent behaviors. From the
view of complexity, it is trivial that an organization is different before and after suffering an accident. Lots of things happen; new things emerge in the wake of such an event, including a heightened
awareness of how people think about themselves, their work, and their organization. These aspects of
complexity will be dealt with in detail in the next chapter.

- Central line associated blood stream infection (CLABSI) from devices used to deliver medicine into the bloodstream;
- Catheter-associated urinary tract infection (CAUTI) from devices used to collect urine.

Using data from hundreds of hospitals, the researchers showed that hospital units' safety culture scores did not correlate with how well the units did on preventing these two infections. As with *Snorre Alpha*, the expectation had been that units with higher safety culture scores would do better on infection prevention. They did not. In fact, in some hospitals where safety culture scores worsened, there were improvements on infection rates. There appeared to be no association between safety culture measurements and infection rates either way. One explanation was the relatively low overall survey response rate: less than half of surveyed staff actually sent in their answers. Their length (42 items) may have contributed to that. Another explanation was that critical aspects of patient care in these hospitals were not picked up by the survey, independent of how many people actually responded.

Of course, all methods of investigating culture have their limitations. They all make empirical, theoretical, and methodological sacrifices. The trust that both the research literature and managers have put in safety culture surveys, however, may need to be reassessed on the basis of these insights. After all, Antonsen (2009b) argued,

> ...the very utility of the concepts of safety climate and safety culture rests on an assumption that they have something to say about the safety level in an organization. After all, if assessments of safety culture and climate were not regarded as having anything to say about the conditions for safety in organizations, why should we bother conducting these assessments in the first place?
>
> (p. 243)

10.3.7 Safety Culture Says so Much, It Ends up Saying Very Little

Is 'safety culture' a concept that tries to say so much that it really ends up saying very little? This is always a risk with a 'container term' like it. The public launch of safety culture was not auspicious in this regard. In the INSAG-7 report of 1992, after all, the concept was pressed into service to cover so much. As you recall from the Pravda quote (above), it might have been used by a UN-affiliated body as code for "all that was wrong in the management of the [Soviet] national economy and had been so for many decades." *All that was wrong.* The same problem occurs when culture becomes defined as 'the way we do things around here.' If culture is the way we do things, then there are no boundaries to define, study, or change. This creates an attraction, for managers, policy makers, investigators, and others, as well as a frustration. Both could be more a product of marketing and convenience then of scientific rigor and substance. Cox and Flin (1998) sum up this sense quite nicely:

> It was the novelist Hanns Johst, not Henry Ford, who wrote: 'Whenever I hear the word 'culture'... I reach for my gun'. His words might well reflect some of the frustration

that managers and safety practitioners experience when they attempt to understand the current academic debate on the definition, measurement and utility of the concept of safety culture. They have to attempt some understanding because it has become the 'motherhood and apple-pie' of safety management that organizations should establish 'excellent safety cultures.' ... The ... concern is that a naïve belief in the concept has far out-stripped the evidence for its utility. The concept of safety culture has become something of a catch-all for social psychological and human factor issues. ... in the absence of sound theory, the notion of safety culture is in danger of becoming meaningless.

(pp. 189–190)

No one term or mechanism, no matter how abstract or seductive, can adequately explain the social order in which we live (Myers, Nyce, & Dekker, 2014). While this may seem obvious, it has become too easy for safety culture to encompass any and all aspects of social life that can somehow be linked to organizational safety. Perhaps that is too much to ask of *any* concept. Perhaps it is too broad. Perhaps it is too much to expect the concept to still do any verifiable and meaningful explanatory work.

One response to this dilemma, as you have seen above, is to try to narrow the operationalization of 'culture,' so that it is asked to do less. We do not have to expect so much of it then. But the more researchers do this, the more likely they are to be studying the mechanical makings of what others consider the organization's 'climate.' Or they end up asking people about things they *can* practically ask about, but which no longer have a theoretical connection to culture at all. Schein, another venerable writer on organizational culture, went the other way. Rather than trying to narrow culture down to a 'safety culture,' and then narrow 'safety culture' down into something operationalizable and measurable, he stuck with organizational culture. This, he said, is defined as:

a pattern of shared basic assumptions that the group learned as it solved its problems of external adaptation and internal integration, that has worked well enough to be considered valid and, therefore, to be taught to new members as the correct way to perceive, think and feel in relation to those problems.

Schein (1992, p. 12)

You can read a lot of what we have seen by now into this definition. You see that Schein leaves out behavior, as it in itself does not relate directly to culture but is itself a coproduct of many of the things he does specify. Learning happens as a result of encountering problems repeatedly. This happens partly because people have to adapt to external influences and pressures, and integrate multiple conflicting goals and pressures internally, even if this leaves fissures and disagreements, which in turn are also handled culturally. It also contains a sense of 'satisficing' solutions that have shown to work 'well enough' (which is not the same as 'perfect' or 'best') given the organization's multiple goals and resource limitations, are considered valid. They are a basis to inculcate newcomers and simultaneously constrain and enable what current organizational members see as common sense.

Organizations thus develop a 'culture' around how they handle the dilemmas and challenges of safety, for sure. But they also do this for how they handle competitive pressures and their finances and fiscal responsibilities internally and externally, around how they take on the role as responsible (or not) corporate citizen, around how they view and treat people in general, both inside and outside the organization. These are all linked, even if not necessarily consistent. It is unlikely that an organization would celebrate its open-plan office environment, flat wage structure and its open, collaborative decision-making processes, and then have a swiftly retributive, closed-door, hierarchical top-down response to an employee's rule breach. But *if* it does, then the organization will have evolved cultural scripts to handle and accommodate such paradoxes and inconsistencies as well. Schein considers organizational culture to be this overall, integrative concept. Cultures can be tied to the successes and failures of organizations—however defined: financial, business, commercial, competitive, safety, ecological, emancipatory—that much seems reasonable to conclude (Glendon & Stanton, 2000).

It would be premature to conclude that narrowing or nailing organizational culture down into the concern of a particular research community—which then becomes its prefix (as in: *safety* culture)—is theoretically nonsensical and not very useful. Perhaps the conversation for safety science to have is one that lies a step away from the fray, a level up, if you will. Whether we measure safety culture by the instrumental means of a survey or see safety culture as a reflexive project on the other, competing premises and practices reflect particular models of risk that safety science as a community produces. These models of risk are interesting not because of their differential abilities to access empirical truth (because that may all be relative), but because of what they say about safety science itself. It is not just the measuring or monitoring of safety culture that we should pursue, but also the monitoring of that monitoring. If this corner of science wants to help make progress on safety, one important step is to engage in such meta-monitoring, to become better aware of the models of risk embodied in our assumptions and approaches to safety.

STUDY QUESTIONS

1. In what sense are the origins of safety culture *political*? And what *theoretical* underpinnings already existed at the time to make the connection between 'safety' and 'culture' legitimate and possible?
2. Is the definition of safety culture as provided by INSAG in 1991 still applicable to organizations today? How much overlap is there between this definition and more contemporary ones?
3. Please characterize the functionalist approach to safety culture. How can slogans and posters make sense in this approach?
4. Please describe the interpretivist approach to safety culture. What perceived weaknesses in the functionalist approach does it try to deal with? And what are its own problems?
5. Is it possible to rate cultures as 'better' or 'worse'? And is there such a thing as cultural 'maturity' (and if so, relative to what standard of growth or maturation)?

6. What roles do conflict and power play in the creation of culture? Is safety culture research and measurement sensitive to these influences?
7. How can the unwitting notion of safety culture as a cohesive, harmonious, homogeneous characteristic of an organization be a by-product of the very way in which functionalist studies typically measure safety culture?
8. What is the problem of aggregation in the study of safety culture? There are plenty of studies that do not even acknowledge this problem, of course, but do you know of any studies that have dealt with it satisfactorily?
9. Describe the influence of methodological individualism in the measurement of safety culture.
10. In your experience, can safety culture be regulated? And does measuring it have predictive value on an organization's safety outcomes?

REFERENCES AND FURTHER READING

Antonsen, S. (2009a). Safety culture and the issue of power. *Safety Science, 47*(2), 183–191.
Antonsen, S. (2009b). Safety culture assessment: A mission impossible? *Journal of Contingencies and Crisis Management, 17*(4), 242–254.
Batteau, A. W. (2001). The anthropology of aviation and flight safety. *Human Organization, 60*(3), 201–211.
Bosk, C. (2003). *Forgive and remember: Managing medical failure.* Chicago, IL: University of Chicago Press.
CAIB. (2003). *Report volume 1, August 2003.* Washington, DC: Columbia Accident Investigation Board.
Conclaves Filho, A. P., & Waterson, P. (2018). Maturity models and safety culture: A critical review. *Safety Science, 105*(6), 192–211.
Cox, S., & Flin, R. (1998). Safety culture: Philosopher's stone or man of straw? *Work & Stress, 12*(3), 189–201.
Dekker, S. W. A. (2012). Complexity, signal detection, and the application of ergonomics: Reflections on a healthcare case study. *Applied Ergonomics, 43*, 468–472.
Dekker, S. W. A. (2014a). *The field guide to understanding 'human error'.* Farnham, UK: Ashgate Publishing Co.
Dekker, S. W. A. (2014b). The problems of zero vision in work safety. *Malaysian Labour Review, 8*(1), 25–36.
Dekker, S. W. A., & Nyce, J. M. (2014). There is safety in power, or power in safety. *Safety Science, 67*, 44–49.
Downer, J. (2013). Disowning Fukushima: Managing the credibility of nuclear reliability assessment in the wake of disaster. *Regulation & Governance, 7*(4), 1–25.
Feldman, S. P. (2004). The culture of objectivity: Quantification, uncertainty, and the evaluation of risk at NASA. *Human Relations, 57*(6), 691–718.
Flin, N., Mearns, K., O'Connor, P., & Bryden, R. (2000). Measuring safety climate: Identifying the common features. *Safety Science, 34*, 177–192.
Glendon, A. I., & Stanton, N. A. (2000). Perspectives on safety culture. *Safety Science, 34*, 193–214.
Grote, G., & Weichbrodt, J. (2013). Why regulators should stay away from safety culture and stick to rules instead. In C. Bieder & M. Bourrier (Eds.), *Trapping Safety into rules: How desirable or avoidable is proceduralization* (pp. 225–242). Farnham, UK: Ashgate Publishing Co.
Guldenmund, F. W. (2000). The nature of safety culture: A review of theory and research. *Safety Science, 34*, 215–257.

Guldenmund, F. W. (2007). The use of questionnaires in safety culture research: An evaluation. *Safety Science, 45*, 723–743.

Hale, A. R., Guldenmund, F., & Goossens, L. (2006). Auditing resilience in risk control and safety management systems. In E. Hollnagel, D. D. Woods, & N. G. Leveson (Eds.), *Resilience Engineering: Concepts and Precepts* (pp. 289–314). Aldershot, UK: Ashgate Publishing Co.

Hallowell, M. R., & Gambatese, J. A. (2009). Construction safety risk mitigation. *Journal of Construction Engineering and Management, 135*(12), 1316–1323.

Harari, Y. N. (2011). *Sapiens.* London, UK: Penguin.

Haukelid, K. (2008). Theories of (safety) culture: an anthropological approach. *Safety Science, 46*, 413–426.

Henriqson, E., Schuler, B., van Winsen, R. D., & Dekker, S. W. A. (2014). The constitution and effects of safety culture as an object in the discourse of accident prevention: A Foucauldian approach. *Safety Science, 70*, 465–476.

Hofstede, G. (1991). *Cultures and organizations: Software of the mind.* New York, NY: McGraw-Hill.

Hopkins, A. (2014). *Why 'safety cultures' don't work.* Paper presented at the 3rd Annual O!shore Safety Conference 2014 (Sept 29–Oct 1), Houston, TX.

Hudson, P. (2006). *Safety management and safety culture: The long, hard and winding road.* Leiden, The Netherlands: Leiden University.

INSAG. (1991). *INSAG-4 safety culture (Safety Series No. 75-INSAG-4).* Vienna, Austria: International Nuclear Safety Advisory Group.

INSAG. (1992). INSAG-7 the chernobyl accident: Updating of INSAG-1 (Safety Series No. 75-INSAG-7). Vienna, Austria: International Nuclear Safety Advisory Group.

Kunda, G. (2006). *Engineering culture: Control and commitment in a high-tech corporation.* Philadelphia, PA: Temple University Press.

LaPorte, T. R., & Consolini, P. M. (1991). Working in practice but not in theory: Theoretical challenges of "High-Reliability Organizations". *Journal of Public Administration Research and Theory, 1*(1), 19–48.

Le Coze, J. C., & Wiig, S. (2013). Beyond procedures: Can 'safety culture' be regulated? In C. Bieder & M. Bourrier (Eds.), *Trapping safety into rules: How desirable or avoidable is proceduralization?* (pp. 191–203). Farnham, UK: Ashgate Publishing Co.

Lee, T., & Harrison, K. (2000). Assessing safety culture in nuclear power stations. *Safety Science, 34*, 61–97.

Meddings, J., Reichert, H., Greene, M. T., Safdar, N., Krein, S. L., Olmsted, R. N., … Saint, S. (2016). Evaluation of the association between Hospital Survey on Patient Safety Culture (HSOPS) measures and catheter-associated infections: Results of two national collaboratives. *BMJ Quality and Safety, 26*(3), 226–235. doi:10.1136/bmjqs-2015-005012.

Myers, D. J., Nyce, J. M., & Dekker, S. W. A. (2014). Setting culture apart: Distinguishing culture from behavior and social structure in safety and injury research. *Accident Analysis and Prevention, 68*, 25–29.

Newlan, C. J. (1990). *Late capitalism and industrial psychology: A Marxian critique* (Master of Arts). San Jose State University, San Jose, CA. (1340534).

OECD. (2012). Oversight and influencing of licensee leadership and management for safety, including safety culture: Regulatory approaches and methods (NEA/CSNI(2012)13, JT03324289). Chester, UK.

Orasanu, J. M. (2001). *The role of risk assessment in flight safety: Strategies for enhancing pilot decision making.* Paper presented at the 4th International Workshop on Human Error, Safety and Systems Development Linköping, Sweden.

Palmieri, P. A., Peterson, L. T., Pesta, B. J., Flit, M. A., & Saettone, D. M. (2010). Safety culture as a contemporary healthcare construct: Theoretical review, research assessment and translation to human resource management. *Strategic Human Resource Management in Healthcare, 9,* 97–133.

Pidgeon, N. F. (1998). Safety culture: Key theoretical issues. *Work & Stress, 12*(3), 202–216.

Rasmussen, J. (1997). Risk management in a dynamic society: A modelling problem. *Safety Science, 27*(2–3), 183–213.

Reason, J. T. (1997). *Managing the risks of organizational accidents.* Aldershot, UK: Ashgate Publishing Co.

Rochlin, G. I. (1999). Safe operation as a social construct. *Ergonomics, 42*(11), 1549–1560.

Sagan, S. D. (1993). *The limits of safety: Organizations, accidents, and nuclear weapons.* Princeton, NJ: Princeton University Press.

Sagan, S. D. (1994). Toward a political theory of organizational reliability. *Journal of Contingencies and Crisis Management, 2*(4), 228–240.

Schein, E. (1992). *Organizational culture and leadership* (2nd ed.). San Francisco, CA: Jossey-Bass.

Sheratt, F., & Dainty, A. R. J. (2017). UK construction safety: A zero paradox. *Policy and Practice in Health and Safety, 15*(2), 1–9.

Silbey, S. (2009). Taming Prometheus: Talk about safety and culture. *Annual Review of Sociology, 35,* 341–369.

Singer, S., Gaba, D. M., Geppert, J. J., Sinaiko, A. D., Howard, S. K., & Park, K. C. (2003). The culture of safety: Results of an organization-wide survey in 15 California hospitals. *Quality and Safety in Health Care, 12*(2), 112–118.

Starbuck, W. H., & Farjoun, M. (2005). *Organization at the limit: Lessons from the Columbia disaster.* Malden, MA: Blackwell Publishing.

Strauch, B. (2015). Can we examine safety culture in accident investigations, or should we? *Safety Science, 77,* 102–111.

Turner, B. A. (1978). *Man-made disasters.* London, UK: Wykeham Publications.

van Erp, J. B. (2007). *Safety culture and the accident at Three Mile Island (Report XA0203452).* Argonne, IL: Argonne National Laboratory.

Vaughan, D. (1996). *The challenger launch decision: Risky technology, culture, and deviance at NASA.* Chicago, IL: University of Chicago Press.

von Thaden, T., Hoppes, M., Yongjuan, L., Johnson, N., & Schriver, A. (2006). *The perception of just culture across disciplines in healthcare.* Paper presented at the Human Factors and Ergonomics Society 50th Annual meeting, San Francisco, CA.

Walker, G. W. (2010). A safety counterculture challenge to a 'safety climate'. *Safety Science, 48*(3), 333–341.

Weick, K. E., & Sutcliffe, K. M. (2007). *Managing the unexpected: Resilient performance in an age of uncertainty* (2nd ed.). San Francisco, CA: Jossey-Bass.

Weick, K. E., Sutcliffe, K. M., & Obstfeld, D. (1999). Organizing for high reliability: Processes of collective mindfulness. *Research in Organizational Behavior, 21,* 81–124.

Westrum, R. (1993). Cultures with requisite imagination. In J. A. Wise, V. D. Hopkin, & P. Stager (Eds.), *Verification and validation of complex systems: Human factors issues* (pp. 401–416). Berlin, Germany: Springer-Verlag.

Wiegmann, D., Zhang, H., von Thaden, T., Sharma, G., & Mitchell, A. (2002). A synthesis of safety culture and safety climate research *(ARL-02-3/FAA-02-2).* Urbana-Champaign, IL: University of Illinois.

Zohar, D. (1980). Safety climate in industrial organizations: Theoretical and applied implications. *Journal of Applied Psychology, 65*(1), 96–102.

11 | The 2010s and Onward
Resilience Engineering

Johan Bergström and Sidney Dekker

KEY POINTS

- Resilience engineering is about identifying and then enhancing the positive capabilities of people and organizations that allow them to adapt effectively and safely under varying circumstances. Resilience is not about reducing negatives (incidents, errors, violations).
- Resilience engineering is based on the premise that we are not custodians of already safe systems. Complex systems do not allow us to draw up all the rules by which they run, and not all scenarios they can get into are foreseeable.
- Systems are not inherently safe: they need to operate under competitive pressure, having to meet multiple conflicting goals at the same time, and always with limited resources. People and organizations have to create safety under these dynamic circumstances.
- Resilience engineering wants to understand and enhance how people themselves build, or engineer, adaptive capacities into their system, so that systems keep functioning under varying circumstances and conditions of imperfect knowledge. How do they create safety—by developing capacities that help them anticipate and absorb pressures, variations, and disruptions?
- Resilience engineering is inspired by a range of fields beyond traditional safety disciplines, such as physical, organizational, psychological, and ecological sciences. The organic systems studied in these fields are effective (or not) at adjusting when they recognize a shortfall in their adaptive capacity—which is key to the creation of resilience (or its disruption).
- Although this is not unique to this theory, resilience engineering, too, appears vulnerable to three analytical traps: a reductionist, a moral, and a normative one.

11.1 THE NEED FOR RESILIENCE

11.1.1 RESILIENCE ENGINEERING AS THE ASSURANCE OF CAPACITY TO ADAPT

This chapter covers the background and drivers for resilience engineering as a new approach to safety science, and an overview of the people and ideas that have helped form it. Before we go there, let's look at a brief overview of what resilience engineering might be, and what it tries to do. Resilience engineering is a logical continuation of the idea that safety lies in the capacity of people, teams, and organizations to make things go right—even under varying circumstances. In his 2014 *Safety I and Safety II: The past and future of safety management*, Erik Hollnagel makes the argument that we should not (just) try to stop things from going wrong. Instead, we need to understand why most things go right, and then ensure that as much as possible indeed goes right. Many organizations have begun to recognize the organizational deficiencies, cultural problems, and ethical issues that lag indicators of negatives create for them. Most will be familiar with the following sorts of things:

- Numbers games and the hiding or renaming of injuries and incidents;
- Counterproductive and credibility-straining sloganeering ('Zero Harm!');
- Short-termism (driven by quarterly figures);
- Creative case management and a lack of compassion for those who do get hurt in the course of work (think of the cynical use of 'suitable duties' to keep someone off the injury stats or lost time books);
- The misdirection of accountability through sanctioning, dismissal, or exclusion of those who have been hurt in the past (just cancel the contractor's access card, for instance);
- The statistical insignificance of any change in typical lagging indicators because the number of negatives relative to hours worked or operated is so low;
- Organizational learning disabilities and cultures of risk secrecy;
- Worker cynicism, mistrust, and disenchantment;
- Cases of outright management fraud that have got managers dismissed or even in jail.

Resilience engineering instead suggests that we need to learn why things go right and find out what we can do to make it even more so. Safety is not about the *absence* of negatives; it is about the *presence* of capacities. The field of resilience engineering, formally founded at a meeting in Söderköping in Sweden over a decade ago (Dekker, 2006) was of course driven by this logic.

One way to illustrate this point is by way of a Gaussian, or normal curve. The curve shows that the number of the things that go wrong (the small shaded part on the left side of the curve) is tiny. On the right side of the curve are the unexpected surprises and heroic recoveries (a Hudson River landing, for instance) that fall far outside what people would normally experience or have to deal with (Reason, 2008). In between, the huge bulbous middle of the figure, sits the quotidian or daily creation of success. This is where good outcomes are achieved, despite the organizational, operational and financial obstacles, despite the rules, the bureaucracy, the common frustrations, and obstacles. This is where work can be hard, but is still successful.

The way to make even further progress on safety, suggests this figure, is not by trying to make the shaded part of things that go wrong even smaller, but by understanding what accounts for the big middle part where things go right, and then enhancing the capacities that make it so. That way, we do not make the shaded part smaller by making the shaded part smaller. We make the shaded part smaller by making the white part bigger. Research by René Amalberti suggests that it is indeed likely that this is *the* way to make progress on safety in already safe systems (Amalberti, 2001, 2006, 2013). In those systems, we have used up the recipes for how to prevent things from going wrong to the maximum already. We have many layers of protection in place. We have rules to the point of overregulation. We monitor, record, investigate, and standardize the designs people work with. The suggestion that comes from this is that ever more things targeted at the shaded part are not going to make it any smaller. The complexity of the system would not let us. And in fact, the more we do to make that part smaller with what we already know (more rules, more limits on people, more technology and barriers) may in fact contribute to novel pathways to breakdown, accidents, and failures (Figure 11.1).

Resilience engineering has been developing as a set of ideas that tries to understand why things go right, and to enhance the capacities that make it so. For example, looks for ways to:

- Enhance the ability of organizations to create processes that are robust yet flexible;
- Monitor and revise risk models;
- Use resources proactively in the face of disruptions or ongoing production and economic pressures.

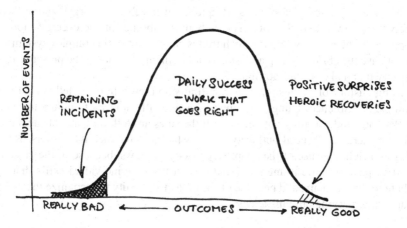

FIGURE 11.1 The way to make the shaded part (unwanted outcomes) on the left smaller is not by making it impossible for things to go wrong (as we have done almost everything in that regard already). We can make the shaded part smaller by making the white part bigger: focusing on why things go right and enhancing the capacities that make it so. (Drawing by author.)

This comes not only from the experiences of other approaches in safety science (many of which were focused on stopping things from going wrong), but also from a couple of intertwined theoretical positions. In resilience engineering:

- Failures are not seen as a breakdown or malfunctioning of normal system functions. Rather, they represent the converse of the adaptations necessary to cope with the real world complexity.
- Individuals and organizations must always adjust their performance to varying conditions.
- Because resources and time are finite it is inevitable that such adjustments are approximate.
- Success can be ascribed to the ability of groups, individuals, and organizations to anticipate the changing shape of risk before damage occurs; failure is the result of a temporary or permanent disruption of that capacity.

In resilience engineering, assuring safety does not mean tighter monitoring of performance, more counting of errors, or reducing violations. That, again, is all based on the assumption that safety should be defined as the absence of something because systems are already safe. The corollary of this assumption is that safety-critical systems need protection from unreliable humans—by more procedures, tighter monitoring, sanctioning of undesirable behaviors, and automation to get the human out of the loop altogether.

Resilience engineering reminds us that we are not custodians of already safe systems. Complex systems, operating under competitive pressure, always have to meet multiple opposing goals at the same time, and always with limited resources. It is only people who can reconcile these conflicting demands, who can hold together such inherently imperfect systems. People, at all levels of an organization, create safety through practice. So safety is not about the absence of bad events or about us stopping all the pathways along which things can go wrong (in complex systems, we do not have the ability to map and foresee all of them). It is about the presence of the capacity that makes things go right.

In his history of pirates and other sailing ship crews, Rediker tells of a typical example of resilience. Things go right not because all pathways to failure have been blocked out, and certainly not because all the rules are followed, and absolutely not because there is a central authority figure who can give the right orders. Rather, things go right because of the adaptive capacity of crews, because of the close and constant peer pressure to maintain and demonstrate the individual skills that contribute to that capacity, and because of the collective ability to recognize the need to adapt under varying circumstances:

Crews were extremely sophisticated in judging the quality of each man's contribution to the sailing of the ship. Everyone knew how to perform the basic tasks, and most men had been on other ships and had seen every chore, from the captain's duties down, executed by others. Consequently, even the lowest ordinary seaman considered himself a judge of his officers. Work was closely scrutinized since the collective well-being depended on it. There was considerable pressure to demonstrate one's skills, and when a man could do a job better than his superior, it was rarely a secret. When a captain was unskillful in his station, a crew might follow his incorrect orders with precision just to expose his ignorance... Seamen were usually able to counteract danger through their own knowledge of the labor process.

Rediker (1987, p. 95)

You might recall from other examples throughout this book, as well as other theoretical contributions, that when we see things go right under difficult circumstances, it is mostly because of people's adaptive capacity—their ability to recognize, absorb, and adapt to changes and disruptions—some of which may even fall outside of what the system has been trained or ordered or designed to do. This is why we call it resilience—the ability to accommodate change, conflict, disturbance, without breaking down, without catastrophic failure. In addition to Rediker's observations, you may have already picked some of the things that you see teams and organizations do that are good at this:

- They do not take past success as guarantee of future safety. Past results are not enough for them to be confident that their adaptive strategies will keep working.
- They keep a discussion of risk alive even when everything looks safe. That things look safe does not mean they are: the model of what is risky may have become old, wrong, so they keep updating it.
- They are able to bring in different and fresh perspectives on problems. They listen to minority viewpoints, invite doubt, stay curious and open-minded, complexly sensitized.
- And they inspire and reward in their people the courage to say "no" to trading chronic safety concerns for acute production pressures; the courage to put the foot down and invest in safety when everybody else says that they cannot. Because that is exactly the time when such investments may be necessary.

It is relatively easy to measure whether an organization or operation has become faster, better, or cheaper. But has it become less resilient as a result? In a complex, dynamic world, this question takes on a new urgency. Resilience might well be a coveted additional management variable to indicate whether safety-critical businesses can sustain their performance under uncertainty, production expectations, and resource pressures.

11.1.2 RESILIENCE AND COMPLEXITY

It is impossible to derive all the rules and write all the safety procedures for how a complex system should work. Al Gore explains our commitment to treating our world as something that can be reduced into components:

The cultural legacy that still influences the scientific method is reductionist—that is, by dividing and endlessly subdividing the objects of our research and analysis, we separate interconnected phenomena and processes to develop specialized expertise. But the focusing of attention on ever narrower slices *of* the whole often comes at the expense of attention *to* the whole, which can cause us to miss the significance of emergent phenomena that spring unpredictably from the interconnections and interactions among multiple processes and networks. That is one reason why linear projections of the future are so often wrong.

Gore (2013, p. xxi)

We make linear projections of the future when we write procedures and rules for people to follow in some future time and place. They are not only often wrong, as Gore explains, but actually hopeless. Attempts at fully specifying somebody's work in a complex system through rules quickly come to grief. Jeffrey Braithwaite, a researcher at Macquarie University, once asked how many rules apply to the work of a typical hospital ward nurse (Debono et al., 2012). He and his colleagues found some 600 rules that specify a ward nurse's daily work. But how many of these rules did the nurses actually know? On average, nurses were able to only recite between two and three of the 600 rules that supposedly specify their job. Yet they do their job effectively and safely every day or night—by moving nimbly inside the interacting ebbing and flowing work demands and the interconnected web of differing patient physiologies and anatomies, physician specialties, expectations, protocols, bureaucratic accountability requirements, department personalities, supervisor peculiarities, and professional cultures. Following all 600 rules would probably make it entirely impossible to even begin to do the job of nursing.

Then consider anesthesiology. There are over 4 million references to 'Operating Room standards of practice.' Rules and procedures that specify anesthetic work in the operating theater come from a variety of accreditors, regulators, institutions, bodies, reformers, or educators; some are in response to conflicting guidelines and standards issued by other professional bodies (such as the Association of Perioperative Nurses); and many are the accompaniment of new equipment, technologies, techniques, and drugs that get introduced to the OR. The result is such a plethora of rules that a zealous practitioner (studying 40 hours a week) would have to spend *2,000 years* to read it all (Johnstone, 2017).

This is the point: you cannot derive the full equation set for a complex system. You cannot specify its workings through a coherent, consistent set of rules. You would be writing thousands, millions of rules (which then take 2,000 years to read, as with anesthesia). And still they would not exhaustively specify how work actually gets done inside that complex system. Many of the rules would contradict each other, or not be relevant or at the right level for the context in which they supposedly apply, or would simply have become obsolete because the field has moved on since the rule was written. Complex systems can never be fully specified—not only because they are complex, but because they are dynamic. They change even while you are trying to write the rules or equation set by which it purportedly operates.

Complex systems tend to be naturally more resilient in dynamic environments than centrally governed systems. There is a good reason for that: a complex system can generate solutions to unexpected disturbances more effectively than a centrally controlled system. The latter, after all, either has only one mode of responding, or its responses come from only one source (an authority somewhere in the hierarchy). A complex system, on the other hand, can find ways to absorb and adapt to the disturbance (Cilliers, 1998; Heylighen, Cilliers, & Gershenson, 2007).

A 19th-century anarchist thinker Kropotkin offered two examples of complex, adaptive, distributed systems that worked well without a central authority (or in fact, worked well because they did not have a central authority) (Kropotkin, 1892). One was (and is) the global workings of mail: post offices, locally and nationally, manage to get a letter from pretty much anywhere in the world to anywhere else—at least most of the time. Any occasional delivery hiccups would not be remedied by a centralized global postal authority. Even prices for stamps are horizontally cross-coordinated in ways that leave every participant adequately remunerated, again without the involvement of a central arbiter. The same principles applied and still apply to international rail travel (which was available to Kropotkin in Europe). No central governor existed, nor exists today, to coordinate how trains cross all those borders, yet Kropotkin was able to buy a ticket in Russia to get himself to Bern in Switzerland, and that is still possible today. Local groups and associations combine in ways that do not require a centralized authority that coordinates or compels them into doing things. They locally interface and adjust where necessary. These examples show that:

- *Complex systems do not fare well under a central authority.* They could not, because the central authority that is truly in control would need to have a stable model of the whole complex system internalized. That would make the authority as complex as the system itself, which would mean that the system could actually not be complex.
- *Complex systems are open systems.* Organizations constantly and often deliberately interact with their environment, taking in newcomers and different ideas. The environment is not just some shell around the organization. In the words of Cilliers, the environment literally 'folded in,' touching people everywhere in the organization, since they all interact with the outside as well as with each other.
- *Complexity does not lead to anarchy and disorder.* Typical for interactions in a complex system is that they lead to repeated patterns, to other kinds of order, and to new ways of working. Complexity tends to produce and encourage horizontal, reciprocal self-organization.
- *Feedback loops are an emergent aspect of complex systems.* It is not as if the constituent subsystems of a complex system (such as various nations' postal services) are entirely free to do what they want. In fact, because of their coupling and reciprocity, they are really quite constrained by each other. No single service can suddenly jack up the price of international mail delivery, for instance, without lots of pushback from the others. A type of cyclical dependence and coupling between subsystems can give rise to that kind of negative feedback loop in which deviations and perturbations are suppressed. This self-correction or self-organization is characteristic of complex systems.

Complex-adaptive, multi-agent systems that have no central controller, that allow freedom of interaction between the people who make up that system, can produce new insights (Cilliers, 1998; Prigogine, 2003). In this way, resilience and innovative solutions can evolve from complex systems: they are capable of evolving and producing *new* things. The origin of their order, their organization, their behavior, and their apparent intelligence is not a structure or process imposed from the top-down. It coevolves in the interaction of many different agents, who come with diverse insights and ideas.

11.1.3 COMPLEX SYSTEMS OPERATE FAR FROM EQUILIBRIUM

Complex systems, even though seemingly stable, are not in equilibrium. Rather, complex systems are constantly adapting to balance multiple goal conflicts. Such complex systems are inherently vulnerable to "drift toward failure as defences erode in the face of production pressure" (Woods, 2003). Using these quoted words David Woods, in his 2003 *Testimony on the future of NASA to senate committee on commerce, science and transportation*, a committee led by senator John McCain, introduced the field *resilience engineering* as a school of thought dedicated to "help organizations maintain high safety despite production pressure" (p. 2). In his testimony, Woods connected the problems of complexity as they played out in the case of Columbia to potential solutions offered by focusing on system resilience. Let us summarize the testimony. Woods argues that the following five general patterns of failure can be constructed from the Columbia case:

1. Drift toward failure with goal conflicts;
2. Past success as reason for confidence;
3. Fragmented problem solving clouding the big picture;
4. Failure to revise assessment as new evidence accumulates;
5. Breakdowns at the boundaries of organizational units.

Woods argues that the patterns above are all signs of organizational problems and that NASA essentially needs to be redesigned. Woods argues that such a redesign should be based on a new approach, resilience engineering, which "is concerned with assessing organizational risk, that is the risk that holes in organizational decision making will produce unrecognized drift toward failure boundaries" (p. 8). As a footnote to this claim he gives various references to Rasmussen's work on the dynamics of safety and the notion of drift. The resilience engineering approach, Woods argues, is dedicated to monitoring management decision-making and invests in safety despite pressures on production and efficiency. It is open when it comes to the reporting on safety concerns, it is proactively picking up on signs of developing problems, monitoring the potential eroding of safety margins, and it is flexible and adaptive toward change, disruptions, and opportunities. Through such an approach, Woods envisioned that NASA:

> will create a new safety organization and culture that is skilled at the three basics of resilience engineering: (1) detecting signs of growing organizational risk, especially when production pressures are intense or increasing; (2) having the resources and authority to make extra investments in safety at precisely these times when it appears least affordable; (3) having a means to recognize when and where to make targeted

investments to control rising signs of organizational risk and re-balance the safety and production tradeoff.

(p. 9)

Most researchers today agree that organizations or socio-technical systems should be viewed as a less stable, less predictable, and more complex than what has previously been assumed in safety science. Rather than a Newtonian–Cartesian machine (e.g., made up of layers of defense that are in place to stop linear trajectories of harm), socio-technical systems are now increasingly seen as active, adaptive, self-organizing, resilient—even creative and innovative. Science, accordingly, has begun to adopt different perspectives that attempt to account better for instability, for complexity and nonlinearity, and for the role of perturbations, bifurcations and emergence. The language and analytics of systems thinking and complexity theory that have been emerging, direct inquiry and interpret safety-scientific findings in different ways.

11.1.4 RESILIENCE IN OTHER FIELDS

Though Woods introduced the notion of resilience to safety scientists and practitioners in the early 2000s, resilience had been an academic concept of several different scientific disciplines for much longer than that. In this section, we will briefly look at resilience from the perspective of:

- physics, mechanics, and material science
- societal functions and critical infrastructure
- psychology and health sciences
- ecosciences

In his 'etymological journey,' Alexander traces the scientific use of resilience to Sir Francis Bacon, who during the 17th century brought resilience (until then a Latin/Roman term) to English (Alexander, 2013). However, it was through the development of material physics in the 19th century that resilience got established as a scientific concept. In material science and mechanics, resilience represents the capacity of a material (or shape/construction of a material, e.g., a spring) to absorb energy when elastically stressed and then regain its original form when the stress is released. This would be what Woods describes as resilience as 'rebound' or as 'robustness' (Woods, 2015). This sense of resilience is principally preserved in contemporary engineering applications, e.g., of resilience as the capacity of infrastructure networks to maintain and regain their functioning during and following a disturbance (Francis & Bekera, 2014; Ouyang, 2014; Pant, Barker, & Zobel, 2013).

The notion of resilience was first introduced into politics in relation to critical infrastructures and the maintenance of important societal functions. The oil crisis in the 1970s was one driver behind this: for the first time the U.S. Government argued for the need of a decentralized energy supply (Walker & Cooper, 2011). The broad use of the stress-strain model, and its ability to be both intuitively and metaphorically grasped, can partly explain the migration of resilience into policy discourse. When the communications director of the Swedish Civil Contingencies Agency (MSB) presented the concept of societal resilience in 2013, he did so by using a rubber ball which he squashed. He then released the pressure and watched the ball regain its original shape. He explained that this is the kind of functioning the agency wanted society to have.

A different literature, in psychology and health sciences, uses a crisis management- or trauma-related version of resilience. In these academic fields resilience represents the ability of a psychological subject to 'thrive despite adversity.' The origin of this branch of research is typically credited to Garmezy, an epidemiologist, and researcher of schizophrenia. After World War II, but mainly in the 1970s and 1980s, Garmezy studied children who suffered from risks of developing psychopathologies (Garmezy, 1971, 1974; Garmezy & Streitman, 1974). Garmezy and his colleagues were interested in the factors leading some children to thrive despite being faced with evident risk factors and stressors (Garmezy, 1987; Masten & Garmezy, 1985).

Resilience research within psychology and health sciences takes those perceived as vulnerable as their target of analysis. Such vulnerable people can, apart from children, typically also be war veterans or certain groups in developing countries. Researchers tend to see trauma and growth as two sides of the same coin. That is how resilience researchers argue for identifying factors that lead to 'post traumatic growth' rather than 'post traumatic stress.' This field has developed broadly and now includes research of what are typically seen as resilient capacities needs to be put into a context of history and culture (Kolar, 2011; Ungar, 2005). Resilience in this field has also been seen as a capacity of vulnerable subjects to resist challenges to their person. Though it has probably mainly solidified the link between resilience and trauma (e.g., through disaster or crisis), an important contribution from this psychological use of resilience is the focus on uncertainty: the uncertainty of crisis and the importance of building a generic capacity to cope with (any) future stress. This aspect of resilience was later adopted by eco-science.

A fourth body of resilience knowledge, from which the resilience engineering community has been developed, is Holling's ecosystems theory. Here, the notion of resilience is closely coupled with the challenge of complexity. In a seminal paper, Holling (1973) introduced resilience as:

> a measure of the persistence of systems and of their ability to absorb change and disturbance and still maintain the same relationships between populations or state variables, [contrasting it with] stability, which represents the ability of a system to return to an equilibrium state after a temporary disturbance.

> (p. 14)

It is clear that this early ecosystems view of resilience is formulated as a contrast to the engineering notion of resilience (Holling calls this stability. This should not be confused by the notion of engineering in resilience engineering). Instead of focusing on the return from stress, Holling argues that complex (eco)systems typically embody a capacity to maintain their functioning even when facing severe stress. They even develop new coping mechanisms out of suffering this stress (i.e., they do not just return to a previously held pre-stress state but adapt and evolve). A key to resilience in Holling's school of thinking is diversity; a system's ability to maintain its functioning through a broad repertoire of adaptive capacities rather than through one best method. Back in 1973, Holling argued against streamlining to achieve the highest possible level of system functioning. After all, this creates a vulnerability in a complex world where not only the threats are constantly changing and impossible to foresee, but where unintended consequences of the best method identified are hard to predict. Through this line of reasoning Holling came to be influential in the 1970s discussion on sustainable development (Walker & Cooper, 2011).

Central to Holling's resilience theory, as well as to complexity theory, are the connections between system levels. What are the relationships between a tree, a group of trees, and a forest? Is the forest resilient when all the trees are resilient? Or does the forest lose resilience when one tree is not resilient? Note how this is very different from thinking in terms of, for example, 'the chain being as weak as its weakest link.' The chain metaphor is linear and componential (one link breaks, the whole chain goes). Resilience, in contrast, does not simply come from the static strength or weakness of a constituent part. It is a much more complex, dynamic function that arises in parts, between parts, and between them and their environment. These kinds of questions are the focus of Gunderson's and Holling's research (e.g., Gunderson & Holling, 2002). It has been developed toward a focus on how different types of socio-ecological systems with different kinds of capital (e.g., social capital, economical capital, natural capital) can help to absorb different kinds of stress (Gunderson, 2010).

The research fields mentioned above have together created an interdisciplinary and partially overlapping knowledge base which has provided an analytical language and starting point for resilience engineering research:

- Physics and material science contributed with a view of resilience as a material or object being able to 'bounce back' after stress was applied and then released;
- Critical infrastructures adopted these metaphorically attractive models of regaining a system's functioning following a disruptive event;
- Psychological and health sciences see resilience as a capacity of subjects facing adversity;
- Ecosciences see resilience as a capacity of systems and communities facing unpredictable, uncertain, and dynamic threats (such as climate change or antagonistic threats) as well as the capacities needed to cope with such threats.

The more recent development of social ecology (mainly represented by *The Resilience Alliance's* own scientific journal *Ecology and Society*) represents a further development of resilience research—this time focusing on the connections between humans, social communities, and their ecological settings.

11.2 RESILIENCE ENGINEERING AS A NEW DISCIPLINE IN SAFETY SCIENCE

In October 2004, 1 year after his testimony to the U.S. Senate, David Woods gathered two dozen other 'complexity safety' scholars, including Erik Hollnagel, Richard Cook, Nancy Leveson, Sidney Dekker, René Amalberti, John Wreathall, Andrew Hale, Ron Westrum, and Jean Pariès. They were to have a first symposium dedicated to the topic of resilience engineering in Söderköping, Sweden. The briefing was simple and basically included 'come and present new and innovative ideas.' According to Hollnagel, who hosted the symposium, "almost everyone complied":

> The symposium itself was organized as a loosely structured set of discussions with a common theme, best characterized as long discussions interrupted by short presentations—prepared as well as *ad hoc*. The objective of the symposium was to provide an opportunity for experts to meet and debate the present and future of resilience engineering as well as to provide a tentative definition of organizational resilience.
>
> **Hollnagel, Woods, & Leveson (2006, p. xii)**

The symposium resulted in the first volume in an ongoing series of edited books on the topic of resilience engineering. The perhaps most informative read for any student interested in the core ideas embedded in the formation of resilience engineering as a school of thought is Woods' and Hollnagel's prologue to the first edited volume (Hollnagel et al., 2006). In no more than six pages Woods and Hollnagel make the argument for the need for resilience engineering:

> Resilience engineering is a paradigm for safety management that focuses on how to help people cope with complexity under pressure to achieve success. It strongly contrasts with what is typical today—a paradigm of tabulating error as if it were a thing, followed by interventions to reduce this count. A resilient organisation treats safety as a core value, not a commodity that can be counted. Indeed safety shows itself only by the events that do not happen! Rather than view past success as a reason to ramp down investments, such organisations continue to invest in anticipating the changing potential for failure because they appreciate that their knowledge of the gaps is imperfect and that their environment constantly changes. One measure of resilience is therefore the ability to create foresight—to anticipate the changing shape of risk, before failure and harm occurs.
>
> *(p. 6)*

In this short paragraph, Woods and Hollnagel manage to express a dedication, a frustration, a new view, a path, and even a hint on how to measure resilience. Resilience Engineers, they argue, are dedicated to helping people coping with complexity (the classic problem for Rasmussian scholars as will further be discussed below). Just as Woods argued in his testimony to the U.S. Senate, complexity is a source of risk. However, instead of drawing the Perrowian conclusion that such complexity then needs to be avoided, or its associated risks accepted, resilience engineering represents the optimist stance. Its agenda is to develop ways to control or manage a system's adaptive capacities based on empirical evidence.

The frustration expressed in Woods and Hollnagel's prologue is with the strong paradigm of reductionist safety management through classic quality control. This sees risk as rooted in errors, and safety management as an exercise in error tabulation and reduction. While Woods, Hollnagel and their colleagues had shared this frustration since at least the aftermath of Three Mile Island, they now introduced the notion of resilience as a new competing paradigm; setting out a new view for what safety *is*. This view, they argue in the quote above, sees safety as one of several core values rather than a system property; a value that is constantly negotiated and in conflict with others. The path for safety management then is to invest in continuous proactive anticipation and foresight; despite complexity denying any complete picture to ever be generated and even when there are no lagging indicators of risky processes or behavior. Organizational resilience, they further argue, should be measured as an organization's ability to create such foresight.

The founders of resilience engineering suggested that this is a new paradigm in safety science (we will get back to this point below). Still in the prologue to the first edited volume:

> Resilience engineering tries to take a major step forward, not by adding one more concept to the existing vocabulary, but by proposing a completely new vocabulary, and therefore also a completely new way of thinking about safety. With the risk of appearing overly pretentious, it may be compared to a paradigm shift in the Kuhnian sense.
>
> *(p. 2)*

Since its first symposium, resilience engineering has indeed grown and got established as a field of thought in safety science, with its own symposium and dedicated series of edited books, among them:

- *Resilience Engineering: Concepts and Precepts*
- *Resilience Engineering Perspectives (volume one and two)*
- *Resilience Engineering in Practice (volume one and two)*
- *Resilient Health Care (volume one and two)*
- *Governance and Control of Financial Systems: A Resilience Engineering Perspective*

In 2015, the academic journal *Reliability Engineering and System Safety* published the first special issue on the topic of resilience engineering. This further marks its status as a now established (rather than new) research discipline in safety science. In their editorial Nemeth and Herrera (2015) conclude that the following three core values of resilience engineering have emerged during its first 10 years as a research discipline:

1. *Finding resilience*: Nemeth and Herrera argue that observations of resilience is the core value which has developed most fully in the research community during the time between 2005 and 2015. Reflecting on the field they argue that "Resilience engineering can identify critical performance factors of adaptation, and verify their existence. This detection of patterns among and across cases can be used to build a body of properties and principles" (p. 2).

2. *Assessing resilience*: When it comes to assessing resilience, Nemeth and Herrera argue that resilience engineering has not come far during its first 10 years. Rather than reflecting on *how* the field has adopted this core value, they pose a number of questions for scholars to explore:

> How does RE assess, and re-assess, the various states of adaptation in a system? How can that adaptation be represented in ways that can be measured? What variables matter in the assessment of adaptation? How can they be used to create a landscape that describes the terrain of adaptive systems? How can "benchmarks" of resilient performance be verified, particularly in situations that have no precedent? How does RE objectively determine and model the ways in which a system is adaptive? In a similar way, how does RE determine and model how a system is vulnerable? How can RE ensure that results from one application area can be transferred successfully to other application areas? What leading indicators can be validated so that operators and supervisors can use them to anticipate the need for change? What means can RE make available to enable operators and supervisors to see how resources would need to be allocated, and the trade-offs that would be involved?
>
> *(p. 2)*

While assessing resilience might still be a rather unexplored topic, Nemeth and Herrera argue that it is a central step in order to support the third core value of the field, namely to engineer resilience.

3. *Creating resilience*: Resilience engineering is an optimistic school of thought, dedicated to finding applications that will help (high-risk) organizations *become* resilient. This is the reason why the word 'engineering' has been kept in the defining topic of study (not to all researchers' liking). Nemeth and Herrera argue that when it comes to creating resilience, the resilience engineering community has undergone some changes:

> Early discussions of RE tended to focus on the preparation for, and response to, large scale adverse events. More recent attention has included system improvement, such as better efficiency, through adaptive change. In any case, practitioners will need to prove how RE improves system design by offering defensible evidence of its value
>
> *(p. 2)*

WHY DO THINGS GO RIGHT?

A few years back, we were working, together with some students, with a large health authority which employed some 25,000 people. The patient safety statistics were dire, if typical: 1 in 13 (7%) of the patients who walked (or were carried) through the doors to receive care were hurt in the process of receiving that care. These numbers were not unique, of course. They were also problematic. Because what exactly is 'nosocomial harm,' harm that originates in a hospital? What is 'medical error' and when is it putatively responsible for the adverse event that happened to the patient? Indeed, when exactly does the patient become that '1' out of 13? These are important (and huge) epistemological and ontological questions.

But they are not the point here. When we asked the health authority what they typically found in one case that went wrong—the one that turned into an 'adverse event,' the one that inflicted harm on the patient—here is what they came up with. After all, they had plenty of data to go on: 1 out of 13 in a large healthcare system can add up to a sizable number of patients per day. So in the patterns that all this data yielded, they consistently found:

- Workarounds
- Shortcuts
- Violations
- Guidelines not followed
- Errors and miscalculations
- Unfindable people or medical instruments
- Unreliable measurements
- User-unfriendly technologies
- Organizational frustrations
- Supervisory shortcomings

It seemed a pretty intuitive and straightforward list. It was also a list that firmly belonged to a particular era in our evolving understanding of safety: that of the person as the weakest link, of the 'human factor' as a set of mental and moral deficiencies that only great systems and stringent supervision can meaningfully guard against. In that sort of logic, we have got great systems and solid procedures—it is just those people who are unreliable or noncompliant:

- People are the problem to control
- We need to find out what people did wrong
- We write or enforce more rules
- We tell everyone to try harder
- We get rid of bad apples

Many organizational strategies, to the extent that you can call them that, were indeed organized around these very premises. Poster campaigns that reminded people of particular risks they needed to be aware of, for instance. Or strict surveillance and compliance monitoring with respect to certain 'zero-tolerance' or 'red-rule' activities (e.g., hand hygiene, drug administration protocols). Or a 'just culture' process that got those lower on the medical competence hierarchy more frequently 'just-cultured' (code for suspended, demoted, dismissed, fired) than those with more power in the system, or some miserably measly attention to supervisor leadership training.

We were, of course, interested to know the extent to which these investments in reducing the '1 in 13' had paid off. They had not. The health authority was still stuck at 1 in 13.

This is when we asked the Erik Hollnagel question. We asked: 'What about the other twelve? Do you even know why they go right? Have you ever asked yourself that question?' The answer we got was 'no.' All the resources that the health authority had were directed toward investigating and understanding the ones that went wrong. There was organizational, reputational, and political pressure to do so, for sure. And the resources to investigate the instances of harm were too meager to begin with. So this is all they could do. We then offered to do it for them. And so, in an acutely unscientific but highly opportunistic way, we spent time in the hospitals of the authority to find out what happened when things went well, when there was no evidence of adverse events or patient harm.

When we got back together after a period of weeks, we compared notes. At first we could not believe it, thinking that what we had found was just a fluke, an irregular and rare irritant in data that should otherwise have been telling us something quite different. But it turned out that everybody had found that in the 12 cases that go right, that do not result in an adverse event or patient harm, there were:

- Workarounds
- Shortcuts
- Violations
- Guidelines not followed
- Errors and miscalculations
- Unfindable people or medical instruments
- Unreliable measurements
- User-unfriendly technologies
- Organizational frustrations
- Supervisory shortcomings

It did not seem to make a difference! These things showed up all the time, whether the outcome was good or bad. It should not come as a surprise. Vaughan reminds us of this when she alludes to 'the banality of accidents': the interior life of organizations is always messy, only partially well-coordinated and full of adaptations, nuances, sacrifices, and work that is done in ways that is quite different from any idealized image of it. When you lift the lid on that grubby organizational life, there is often no discernible difference between the organization that is about to have an accident or adverse event, and the one that would not, or the one that just had one (Vaughan, 1999).

This means that focusing on people as a problem to control—increasing surveillance, compliance, and sanctioning—does little to reduce the number of negatives. As related in *The Safety Anarchist* (2018), we analyzed 30 adverse events in 380 consecutive cardiac surgery procedures with colleagues in Boston and Chicago (Raman et al., 2016). Despite 100% compliance with the preoperative surgical checklist, 30 adverse events occurred that were specific to the nuances of cardiac surgery and the complexities associated with the procedure, patient physiology, and anatomy. Perhaps other adversities were prevented by completely compliant checklist behavior, even in these 30 cases. But we will never know.

You can also see this in measures of safety culture, which typically include rule monitoring and compliance. They actually do not predict safety outcomes. Recall the study by Norwegian colleagues from the previous chapter, conducted in oil production. It followed a safety culture survey which had inquired whether operations involving risk were carried out in compliance with rules and regulations (Antonsen, 2009). The survey had also asked whether deliberate breaches of rules and regulations were consistently met with sanctions. The answer to both questions had been a resounding 'yes.' Safety on the installation equaled compliance. Ironically, that was a year before that same rig suffered a significant, high-potential incident. Perceptions of compliance may have been great, but a subsequent investigation showed Vaughan's 'messy interior'; the rig's technical, operational, and organizational planning were in disarray, the governing documentation was out of control, and rules were breached in opening a sub-sea well. Not that these negatives were necessarily predictive of the incident (indeed, we need to be wary of hindsight-driven reverse causality): the messy interior would have been present without an incident happening too.

More research in healthcare shows a disconnection between rule compliance as evidenced in surveys, and how well a hospital is actually doing in keeping its patients safe (Meddings et al., 2016). Hospitals that had signed on to a national patient safety project were given technical help—tools, training, new procedures, and other support—to reduce two kinds of infections that patients can get during their hospital stay:

- Central-line-associated blood stream infection (CLABSI) from devices used to deliver medicine into the bloodstream;
- Catheter-associated urinary tract infection (CAUTI) from devices used to collect urine.

Using data from hundreds of hospitals, researchers showed that hospital units' compliance scores did not correlate with how well the units did on preventing these two infections. As with that oil rig, the expectation had been that units with higher scores would do better on infection prevention. They did not. In fact, some hospitals where scores worsened showed improvements on infection rates. There appeared to be no association between compliance measurements and infection rates either way.

But if these things do not make a difference between what goes right and what goes wrong, then what does? We were still left with a relatively stable piece of data: 1 in 13 went wrong, and kept going wrong. What explained the difference if it was not the absence of negative things (violations, shortcuts, workarounds, and so forth)? This is not just an academic question. If you are a manager (or clinician, or especially patient) in this sort of system, you would like to know. You would love to get your hands on the levers and push or nudge the system toward more good outcomes and further away from those few bad ones.

So we looked at our notes again because there was more in there. And we started holding them up against the literature that we knew, and some that we did not yet know. In the 12 cases that went well, we found more of the following than in the one that did not go so well:

- Diversity of opinion and the possibility to voice dissent. Diversity comes in a variety of ways, but professional diversity (e.g., compared to gender and racial diversity) is the most important one in this context. Yet whether the team is professionally diverse or not, voicing dissent can be difficult. It is much easier to shut up than to speak up (Weber, MacGregor, Provan, & Rae, 2018). I was reminded of Ray Dalio, CEO of a large investment fund, who has actually fired people for not disagreeing with him. He said to his employees: "You are not entitled to hold a dissenting opinion...which you don't voice" (Grant, 2016, p. 190).

- Keeping a discussion on risk alive and not taking past success as a guarantee for safety. In complex systems, past results are no assurances for the same outcome today, because things may have subtly shifted and changed. Even in repetitive work (landing a big jet, conducting the fourth bypass surgery of the day), repetition does not mean replicability or reliability: the need to be poised to adapt is ever-present (Woods, 2006). Making this explicit in briefings, toolboxes, or other pre-job conversations that address the subtleties and choreographies of the present task, will help things go right.

- Deference to expertise. Deference to expertise is generally deemed critical for maintaining safety. Signals of potential danger, after all, and of a gradual drift into failure, can be missed by those who are not familiar with the messy details of practice. Asking the one who does the job at the sharp end, rather than the one who sits at the blunt end somewhere, is a recommendation that comes from high reliability theory as well (Weick & Sutcliffe, 2007). Expertise does not mean only frontline people. The size and complexity of some operations can require a collation of engineering, operational and organizational expertise, but high reliability organizations push decision-making down and around, creating a recognizable pattern of decisions 'migrating' to expertise.

- Ability to say stop. As Barton and Sutcliffe found in an analysis of wildland firefighting, "a key difference between incidents that ended badly and those that did not was the extent to which individuals voiced their concerns about the early warning signs" (2009, p. 1339). Amy Edmondson at Harvard calls for the presence of 'psychological safety' as a crucial capacity in teams that allow members to safely speak up and voice concerns. In her work on medical teams, too, the presence of such capacities were much more predictive of good outcomes than the absence of noncompliance or other negative indicators (Edmondson, 1999).

- Broken down barriers between hierarchies and departments. A point frequently made in the organizational literature, and also in the sociological postmortems of big accidents—from Barry Turner in the 1970's to Diane Vaughan recently—is also one of Deming's reminders, as well as one from the literature on fundamental surprises (Lanir, 1986): the totality of intelligence required to foresee bad things is often present in an organization, but scattered across various units or silos (Woods, Dekker, Cook, Johannesen, & Sarter, 2010). Get people to talk to each other: research, operations, production, safety, personnel—break down the barriers between them.

- Do not wait for audits or inspections to improve. If the team or organization waited for an audit or an inspection to discover failed parts or processes, they were way behind the curve. After all, you cannot inspect safety or quality into a process: the people who do the process create safety—every day (Deming, 1982). Subtle, uncelebrated expressions of expertise are rife (the paper cup on the flap handle of a big jet; the wire tie around the fence so the train driver knows where to stop to tip the mine tailings; draft beer handles on identical controls in a nuclear power plant control room, so as to know which is which; the home-tinkered redesigned crash cart in a hospital ward). These are among the kinds of improvements and ways in which workers 'finish the design' of their systems so that error traps are eliminated and things go well rather than badly.

- Pride of workmanship, another of Deming's points, is linked to the willingness and ability to improve without being prodded by audits or inspections. Teams that take evident pride in the products of their work (and the workmanship that makes it so) tended to end up with more good results. What can an organization do to support this? They can start by enabling their workers to do what they want to do and need to do, by removing unnecessary constraints and decluttering the bureaucracy surrounding their daily life on the job.

How much 'more' of this did we find in the 12 cases (1 of 13) that went well? That is impossible to answer. As said, the 'study'—such as it was—was scrambled and unscientific, an opportunistic deep-dive into a complex organization with all the unprepared person-power we could throw at it during a few hectic weeks. So you should see the list above not as conclusions, but as a set of hypotheses. Are these starting points for you and your organization to identify some of the capacities that make things go right? And if so, how would you enhance those capacities? What can you do to make them even better, more omnipresent, and more resilient?

11.3   RESILIENCE IDEAS OF RASMUSSEN, WOODS, AND HOLLNAGEL

To understand resilience engineering, it makes sense to go back to the 1980s and the collaborative work of Jens Rasmussen, David Woods, and Erik Hollnagel in their development of Cognitive Systems Engineering (recall this from Chapter 5) and their focus on socio-technical systems.

11.3.1   TRACING RESILIENCE ENGINEERING TO THE RISØ COMMUNITY IN THE 1980S

In recent years, there has been an increased academic interest in studying the history, and founding ideas, of different schools of safety science. One of the young scholars involved in constructing this 'safety history' is Le Coze; who has done a thorough job in outlining the legacy of Jens Rasmussen (Le Coze, 2015). He traced resilience engineering to the ecological view on errors found in Rasmussen's work at the Risø National Laboratory in Denmark (Rasmussen, 1990). In what follows we will, with inspiration from the work of Le Coze, trace the following three fundamental focus areas of resilience engineering scholars to the work of Jens Rasmussen:

1. the underlying problem of coping with complexity;
2. the ecological view on error;
3. and the problem of dynamic migration toward the boundary of functionally acceptable (safe) behavior.

In resilience engineering, as outlined above, the main challenge that requires system resilience is complexity. Complexity (rather than any underlying energy to be contained, or the error-proneness of workers) is described as the source of unpredicted and dynamic emergence of risk. This requires constant adaptations as a coping strategy (Bergström, van Winsen, & Henriqson, 2015). However, rather than the Perrowian structural view on complexity, resilience engineering approaches the notion of complexity from a functional view, one which can be traced to Rasmussen's work following the Three Mile Island disaster. Le Coze sees Rasmussen's interest in things such as cybernetics (the science of control in engineered and living systems) coming from his education as an engineer rather than a psychologist (such as Reason) or sociologist (such as Vaughan, or Perrow).

One important trace of the link between Rasmussen and resilience engineering is Woods' seminal chapter on coping with complexity (Woods, 1988). In the chapter, which is a much easier read than Rasmussen's paper with the same title (Rasmussen & Lind, 1981), Woods reflects on his time working together with Rasmussen in the research center Risø and sets the agenda for future research into coping with increased system complexity. In a highly cybernetic/Rasmussian manner, Woods introduces complexity as emerging from the interactions between agents, the world, and the representations between them. However, what should be more interesting for scholars of resilience engineering is how Woods identifies ways to cope with complexity (recall that this is in 1988):

> For dynamic worlds, skill becomes the ability to adapt routines in the face of changing circumstances in the pursuit of a goal ... One element of adaptability is the ability to anticipate or predict the behaviour of the world. A second element is the ability to detect and respond to the next event.

(p. 134)

You might recognize three out of four of what Hollnagel later termed the four abilities (later also functions or corner stones) of resilience (without reference to the Rasmussian heritage). These are the ability to:

1. respond;
2. monitor (Woods used 'detect');
3. anticipate;
4. and learn (not mentioned in Woods' paper).

In the chapter Woods also introduces trade-offs, later popularized by Hollnagel in his 'ETTO' (or Efficiency-Thoroughness Trade-Off) principle, as a necessary means to cope with complexity:

> Competing goals can take two forms. One is a choice involving goal sacrifices where the trade-off is between repairing or replacing one means to achieve the primary goal (can I repair it? how much time is available to attempt repair?), and an alternative means that entails a post-condition sacrifice of other goals in order to meet the primary goal.

(p. 135)

In his safety-historical studies, Le Coze traces today's interest in resilience to Rasmussen's 'Micro period (1970/1980)' in which the ecological, or naturalistic, view of human errors was developed, seeing errors as "an intrinsic feature of individual's experiences while exploring the boundaries of acceptable practices. This idea developed into a more positive view of operators with the topic of resilience" (Le Coze, 2015, p. 125). As described above, resilience engineering is a departure from the belief that safety can be managed through the elimination of negatives, such as errors. Rasmussen was of course a foundational thinker in a community that viewed human error as at best an attribution (Rasmussen, 1980; Woods et al., 2010).

This school of error studies, in turn fundamental to the resilience engineering community, goes back to Rasmussen's work in the 1980s:

> to optimize performance, to develop smooth and efficient skills, it is very important to have opportunities to perform trial and error experiments, and human errors can in a way be considered as unsuccessful experiments with unacceptable consequences. Typically, they are only classified as human errors because they are performed in an 'unkind' environment. An unkind work environment is then defined by the fact that it is not possible for a man to observe and reverse the effects of inappropriate variations in performance before they lead to unacceptable consequences. When the effect of human variability is observable and reversible, the definition of error is related to a reference or norm in terms of the successful outcome of the activity.
>
> **Rasmussen (1982, p. 313)**

The analytical focus on the dynamic process of migrating toward the boundary of functionally acceptable performance (the safety boundary) goes back to Rasmussen's so-called 'Macro period (1990/2000).' During this time, Rasmussen too (like scholars such as Turner, Snook, and Vaughan, see Chapter 7) started to analyze how system features, including the design logic of defenses in-depth, tend to give little feedback on local erosions of layers of defense. This in turn might spur a gradual drift toward the boundary of safe performance. Rasmussen called this the 'defence in depth fallacy' (Rasmussen, Nixon, & Warner, 1990).

The dynamics of safety management are the core of Rasmussen's summary of his own career. He sets out a research path for the future in his seminal paper *Risk Management in a dynamic society: a modelling problem* (Rasmussen, 1997). A core idea of the paper, rooted in cybernetics and Rasmussen's cognitive engineering work, is the model of bounded discretion. The performance envelope drawn by Rasmussen shows simultaneously that:

- work in high-risk systems will always remain characterized by having only limited degrees of freedom in an environment characterized by the need to make constant trade-offs between multiple and often opposing goal pressures;
- freedom within a frame (or, in the words of Rasmussen, *bounded discretion*) is the only way in which complex systems can remain adaptive and resilient in a changing world.

Looking at the early writings defining their field, Rasmussen's agenda seems to remain a central inspiration for resilience engineering (Dekker, 2006). Resilience engineering is indeed a continuation, along multiple lines, of Rasmussen's intellectual work. It represents, in a sense, a collective attempt to pursue the research agenda he suggested in the late 1990s. To a great extent, this has been driven by two of Rasmussen's Risø collaborators; David Woods and Erik Hollnagel. Since the founding workshop, Woods and Hollnagel have continued to define complementary directions for the research agenda. They work with different kinds of high-risk systems to test their ideas and make their points. Below we will look at what distinguishes the two.

11.3.2 WOODS: THE ADAPTIVE UNIVERSE

As a scholar based in the tradition of Cognitive Engineering, Woods has remained dedicated to studying and theorizing the messy details of work in organizations such as NASA, healthcare, aviation, nuclear energy, and more recently the IT sector. Woods' contributions to resilience engineering are primarily to develop and apply theories and models of resilience based on the notion of adaptability (Woods, 2006). Woods outlines the following four models of resilience (and rather clearly argues that resilience engineering should adopt the third and fourth; which is why we will outline them in more detail below) (Woods, 2015):

1. *Resilience as rebound*: This is resilience in terms of returning to a previous equilibrium; as a process of recovering from a traumatic or disrupting event. Woods merely uses this theory as a rhetorical starting point to argue how resilience engineering is different. Resilience as rebound, Woods argues, has limitations, such as its preoccupation with a prior state or previous conditions. It is also stymied by a fundamental paradox of considering disrupting events as fundamental surprises, yet still being left with learning only from past events. But following a disruptive event, there is no stable past to return to because the event itself changes the system in various ways. Also, adaptation is more continuous than a simple response to major disrupting events.
2. *Resilience as robustness*: In this model, resilience is seen as the ability of a system to absorb (rather than recover from) disrupting events. Woods traces this view to control engineering and argues that it is a strategy which works for well-modeled disturbances, but that studying resilience is to study responses to the non-modeled disturbances.

   In other words, resilience comes to the fore when the set disturbances is *not well* modelled and when this set is changing. And, ironically, the set of poorly modelled variations and disturbances changes, based on a record of past success which triggers adaptive responses by other nearby units in the layered networks of interdependent systems ... confounding resilience and robustness turns out to be erroneous

   (p. 6)

3. *Resilience as graceful extensibility*: In this third model of resilience, Woods draws the agenda for resilience engineering not as a study of how systems or actors bounce back but *"how do systems stretch to handle surprises?"* (p. 7). Woods is rhetorically playing with the language of 'graceful degradation' as a classic design requirement for highly technological systems and infrastructures, but argues that while graceful degradation only refers to breakdowns while graceful extensibility is used:

   ...because adaptation at the boundaries can be very positive and lead to success, not simply less negative capability. Systems with high graceful extensibility have capabilities to anticipate bottlenecks ahead, to learn about the changing shape of disturbances and possess the readiness-to-respond to adjust responses to fit the challenges.

   (p. 7)

Woods argues that the study of resilience as graceful extensibility opens up new possibilities to learn from past events. After all, adaptations in the past can provide data on the system's potential for future adaptation. Further, Woods argues that studies of graceful extensibility can provide a means to study consequences of having gone through multiple stretches of adaptive capacities over time. He proposes how the exhausting of adaptive capacities can be labeled 'decompensation' while the positive pattern of maintaining graceful extensibility is rather labeled 'anticipation' (which is a link both to Woods' paper on coping with complexity as well as the cornerstones of Hollnagel). In lectures and keynotes on the notion of graceful extensibility, Woods typically introduces Rasmussen's classic graphical model of the performance envelope to show how the margins can be extended (as well as how there will be challenges lurking outside the space of performance experience).

4. *Resilience as sustained adaptability*: In his fourth model of resilience, Woods comes back to the notion of engineering adaptive capacity into the system architecture of high-risk organizations so as to sustain the capacity to perform over a longer time. This model of resilience, Woods argues, asks the following three questions:

   1. What governance or architectural characteristics explain the difference between networks that produce sustained adaptability and those that fail to sustain adaptability?
   2. What design principles and techniques would allow one to engineer a network that can produce sustained adaptability?
   3. How would one know if one succeeded in their engineering (how can one confidently assess whether a system has the ability to sustain adaptability over time, like evolvability from a biological perspective and like a new kind of stability from a control engineering perspective)? (p. 8)

   Woods paints the enquiry of resilience as sustained adaptability with broad brushes. The agenda, he argues, should identify basic architectural principles of sustained adaptability, identify fundamental system constraints and trade-offs, define resilience control mechanisms, and understand interactions across system scales.

As his domain of studying graceful extensibility and sustained adaptability, Woods has increasingly worked with the IT and software 'Development and Operations' communities.

11.3.3 HOLLNAGEL: CORNERSTONES, FUNCTIONAL RESONANCE, AND TRADE-OFFS

Hollnagel, who together with Woods and Rasmussen defined the field of Cognitive Systems Engineering in the 1980s (see Chapter 5), has been as instrumental in the formation and development or resilience engineering as Woods. Hollnagel has played a significant role in making sure that the early symposium contributions were published and has been behind the formation of professional networks such as the Resilience Engineering Association, the Resilience in Healthcare Net, and the FRAMily.

While Woods has focused on expanding a theoretical understanding of resilience and adaptability, Hollnagel's theoretical contribution has rather been to formulate some rather broad and simplified models targeted at the community dedicated to application (industry representatives and applied scientists).

Hollnagel introduces his theory of resilience as four abilities—later to be called corner stones or functions (Hollnagel, Nemeth, & Dekker, 2008). The rationale for introducing the four abilities is interesting. Hollnagel argues that they are the answer to the question whether resilience engineering really brings something new. His answer is:

- *yes* "because resilience engineering does offer a different approach to system safety"
- and *no* "because resilience engineering does not require that methods and techniques that have been developed across industries over several decades must be discarded" (p. xi).

Hollnagel sums up his argument as follows: "resilience engineering differs more in the perspective it provides on safety, than in the methods and practical approaches that are used to address real-life problems" (p. xi). The difference in perspective, Hollnagel argues, is that while traditional safety management sees failure as the result of failing components or erratic behavior, resilience engineering sees both failures and successes as the products of normal performance variability. Since "performance variability is both normal and necessary, safety must be achieved by controlling performance variability rather than by constraining it" (p. xii). This quality, Hollnagel argues, can be summarized in terms of four essential abilities:

1. The ability to respond to the actual;
2. The ability to flexibly monitor the critical;
3. The ability to anticipate the potential;
4. The ability to learn from the factual.

Unlike Woods' theoretical definitions of resilience as graceful extensibility and sustained adaptability, Hollnagel gives only one reference to the literature when outlining these abilities: the three types of threats proposed by Westrum in the first edited volume on resilience engineering (Westrum, 2006):

- the regular threat;
- the irregular threat;
- the unexampled event.

Hollnagel argues that the ability to respond is the ability to deal with the regular threat, while the ability to anticipate deals with the irregular threat. As we concluded above, Woods introduced the first three of Hollnagel's abilities already in his 1988 chapter reflecting on his time together with Rasmussen and Hollnagel at Risø, which again marks Rasmussen's legacy in resilience engineering.

In the following edited volume, Hollnagel comes back with a more elaborate description of what is then described as 'The Four Cornerstones of Resilience Engineering' (Hollnagel, Nemeth, & Dekker, 2009). The chapter is a much more elaborate description of the four. It is still not particularly data- or literature driven, but references are introduced to exemplify certain points. In this chapter, there is also a hint at the four cornerstones representing a functional model (again the Rasmussian approach to coping with complexity): the corner stones are pictured as hexagons (the functional representation in the functional resonance analysis method (FRAM) method, discussed below). This was developed even further for the next volume in which the four abilities/cornerstones have become the four 'capabilities.' The Resilience Engineering Grid, Hollnagel argues, should be seen as a set of questions to use to support resilience management. Hollnagel proposes a method in which a system/organization's resilience is assessed based on each of the capabilities to then form an aggregated star diagram of the system's resilience. Finally, Hollnagel visualizes the interdependence of the capabilities.

The tool used to visualize the interdependencies between the four cornerstones is the FRAM (Hollnagel, 2012), developed by Hollnagel as a method to analyze performance variability in terms of resonance between system functions. This methodological development overlaps greatly with Hollnagel's engagement in resilience engineering, since FRAM offers an analytical approach to the very modeling of dynamic interactions and relations that Rasmussen earlier defined as the challenge to risk management in a dynamic society (Rasmussen, 1997). With both safety and risk seen as the products of performance variability, and resilience engineering the field dedicated to the study of such performance variability, FRAM offers a tool to do just that (de Carvalho, 2011; Patriarca, Bergström, & Di Gravio, 2017; Woljter, 2009). The network dedicated to developing and applying FRAM (the so-called FRAMily) also overlaps with the network that makes up the Resilience Engineering Association.

When again revisiting the four cornerstones in the most recent Ashgate volume on resilience engineering (Hollnagel, 2014) (not considering the ones specifically focusing on healthcare), Hollnagel has yet another take. In the final chapter of the book, Hollnagel defines the need for "engineering a culture of resilience." To safety science historians this should read like a play with Reason's highly influential approach to engineering a safety culture (Reason, 1997), and the dedication to social engineering seems similar. Engineering the four cornerstones (even though Hollnagel now again calls them abilities) as the way to take an organization from being "dysfunctional" to become "resilient" becomes what Hollnagel defines as the "road to resilience" (p. 190).

In a further attempt to develop the theory of how safety and risk are the products of the same kind of processes (performance variability), Hollnagel has also repopularized the focus on trade-offs by introducing the so-called ETTO principle (Hollnagel, 2009). To Hollnagel, it seems as if resilience engineering offers a broad focus on organizational performance. He elaborates the principles of such performance in terms of ETTO, Safety I versus Safety II, and the tools to evaluate and improve such performance through the Resilience Engineering Grid and FRAM. It is indeed a comprehensive production of principles and methodological support and both the four cornerstones and FRAM have gained lots of attraction by the applied community surrounding resilience engineering.

11.4 DIMENSIONS OF RESILIENCE ENGINEERING

Given that it is a young contribution to the foundations of safety science, where is resilience engineering today? The following is a meta-analysis of a core dataset of 180 papers, chapters, and proceedings in terms of how closely they relate to each other (if two publications are cited together they get a relation, if they are often cited together they get an even closer relation). The analysis forms groups of publications (so-called factors) and also plots them on two different dimensions (scales) for the researchers to then interpret the meaning of both the factors and the scales. The analysis concludes that there are five main groups of resilience engineering studies (referred to below). Further, the analysis generated a plotted picture of all the 180 publications on a two-dimensional plane (so-called 'multi-dimensional scaling'):

- analysis of the research topic (the *x*-axis in Figure 11.2)
- analysis of the research field (the *y*-axis in Figure 11.2)

The map does not show the resilience engineering literature in its entirety (the analysis started with an initial dataset of 472 contributions), but only the contributions that are cited by others together with other contributions in the set. The map combines the factor analysis (the colored fields) and the multidimensional scaling (the two axes) to form a rich and informative picture of the state of the field.

The *x*-axis shows the span of studies from theoretical development to practical applications of resilience theory. Reflecting on the introduction to Woods and Hollnagel above, it is interesting to note that Hollnagel's publications are typically in the middle of this axis while Woods' contributions are positioned further to the left (theoretical development). This should not be surprising given Hollnagel's attempts to define and refine the four cornerstones of resilience to provide a practical tool for applications of resilience analysis. To the far right on the *x*-axis are studies dedicated

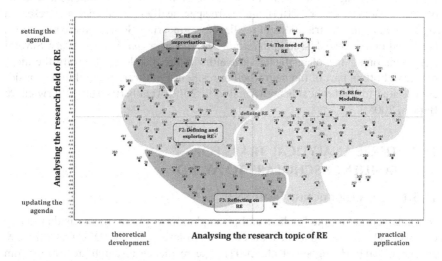

FIGURE 11.2   A conceptual map of the resilience engineering literature. (Drawing by authors.)

to resilience assessment and application. Two 'factors' span across the entire x-axis (actually over both): exploring and modeling. In the factor dedicated to explore resilience there is also a temporal dimension in that early contributions start to be positioned in the middle of the map (Hollnagel et al., 2006) and as time has passed more and more contributions have focused on additional theoretical development, typically focusing on adaptive capacities.

Moving into the dominant factor of modeling resilience, we again find some important chapters from early resilience work (Hollnagel, 2006; Wreathall, 2006), and Hale and Hayer (2006). Further toward practical applications, this factor contains papers dedicated to applying system dynamics models to the study of resilience (de Carvalho, 2011; Salzano, Di Nardob, Gallob, Oropallob, & Santillob, 2014), different kinds of network analysis techniques (Barker, Ramirez-Marquez, & Rocco, 2013; Pasman, Knegtering, & Rogers, 2013), and the development of resilience assessment and auditing programs (Costella, Saurin, & de Macedo Guimarães, 2009; Huber, Gomes, & de Carvalho, 2012; Shirali, Mohammadfam, Motamedzade, Ebrahimipour, & Moghimbeigi, 2012; Shirali, Motamedzade, Mohammadfam, Ebrahimipour, & Moghimbeigi, 2012).

The y-axis in the analysis differentiates different kinds of reflective approaches to the notion of resilience. The studies at the top of the map typically argue for (or against) the need for the resilience engineering approach and distinguish it from 'traditional' safety management and science approaches. Here is where a paper such as the classic one by Leveson (2004), making the argument for a new systems approach to safety management, is located. Here we also find early attempts to define resilience (Hale & Heijer, 2006) and the argument for the need for a new '(fifth) age of safety science: the adaptive age' (Borys, Else, & Leggett, 2009). Not all contributions in this factor, however, argue for the need for resilience. Hopkins (2013) reflects on such an argument, but finds that resilience engineering merely tries to redefine high reliability theory. Toward the bottom part of the map, contributions are tending toward critical reflections, even on the entire ontological status of safety science (Haavik, 2014), and more specifically on resilience as a research agenda and on the contributions of resilience engineering. Some of these reflections are found in the special issue in *Reliability Engineering and System Safety*. Several reflections on studies of resilience in healthcare are also located toward the more reflective bottom part of the map whether resilient healthcare is at all to be considered related to a safer healthcare (Amalberti, 2013; Amalberti, Auroy, Berwick, & Barach, 2005).

11.5 THREE ANALYTICAL TRAPS FOR RESILIENCE SCHOLARS TO AVOID

11.5.1 THE REDUCTIONIST TRAP

As explained above, the fundamental principle of complexity theory is that macro behavior of a system cannot be reduced to micro-level behavior (Dekker, Cilliers, & Hofmeyr, 2011; Heylighen et al., 2007). The resilience research stemming from ecology and socio-ecology shares this notion of emergence as a fundamental system principle (Gunderson & Holling, 2002; Holling, 1973; Walker & Cooper, 2011). However, there seems to be a risk that the resilience engineering community adopts a view in which resilience is constructed in a reductionist manner in which the target of operational resilience becomes the sharp-end (micro-level) operators (Bergström et al., 2015; Mendonça & Wallace, 2004); a risk that also has been pointed out for the application of resilience to wider societal systems. Let us look at what such an analysis might look like, by a quick sharp-end-focused resilience analysis of the case below.

POT LIDS AND BICYCLE BELLS

In 2010, several news media outlets reported on an event happening at the hospital Östra Sjukhuset in Gothenburg, Sweden. The emergency ward was crowded with patients. Patients were put on hold on beds in corridors and at one point there were no alarms left to hand out to the patients. The solution invented by the nursing staff was to start hand out pot lids and cutlery to the patients, and to go and buy bicycle alarm bells to maintain the function of allowing patients to call for the attention of staff, if needed. This happened repeatedly and at one point the media got to hear about the story and it became public.

The story above is a story of resilience. It is a story of how to maintain system functioning under great pressure. It is a story of inventing procedures and of being innovative in the act of balancing multiple conflicting goals. It is a story of human adaptive capacities in times of great performance variability. If you look solely at the sharp-end micro-level behavior of the system, you might even conclude that this is how organizational resilience is maintained. However, from a complexity perspective, is it really a story of a resilient system? Rather, like Rediker's pirates at the beginning of the chapter, this is a story of resilience *despite* the system—not resilience thanks to the system. It is a story of resilient behavior in a highly brittle system. And what is more, there is a risk that from higher system levels (if it was not for the media reporting on the case) the only thing visible is that the system works. This paradox of how local level resilience preserves macro-level brittleness (particularly in healthcare) has been observed by resilience scholars (Nemeth et al., 2007; Perry & Wears, 2012; Smith, Davis Giardina, Murphy, Laxmisan, & Singh, 2013; Sujan, Spurgeon, & Cooke, 2015):

> This kind of critique, that resilience policy and analysis run the risk of locating system resilience at the micro level of the system, is one frequently introduced in Political Science, critiquing national and international resilience policy as a neoliberal drive to responsibilize individual citizens (especially the poor) for having a preparedness for, and adapt to, the dynamic threats (natural and antagonist) of the modern world (the geological age called the Anthroposcene). Such critique is typically well formulated and harsh, such as in the following quotes by Joseph.

Joseph (2013a, p. 43)

So although resilience appears at first sight as a systems theory, its main effect is to emphasize the need for adaptability at the unit level. The idea behind this is shared with a range of other contemporary social ideas such as reflexive modernity, risk society, network society, and information age, all of which posit the idea that we must change our behavior and adapt to things beyond our control.

Perhaps resilience becomes appealing to organizations because, with its inherent skepticism toward centralized system control, it offers a promise that the adaptive capacities of the wider system are located at the sharp-end level of the system (Walker & Cooper, 2011). Resilience becomes the promise that no matter how challenging the goal conflicts, or how dynamic and surprising the threats, the adaptive capacities of the sharp end will (and should) guarantee the functioning of the wider system. Such a reductionist trap is not only a misreading of the complexity roots of resilience theory in order to make the transition from a descriptive to a normative discourse. It is also a discursive use of resilience with moral effects. These show up in how organizations delegate responsibility for continued system functioning and resilience to people at the sharp end or frontline.

11.5.2 THE MORAL TRAP

The moral trap of resilience analysis is closely connected to the reductionist trap. But it goes one step further in its responsibilization of (typically sharp end) actors for the resilient behavior of the system as a whole. Below, we exemplify this trap with the recent (2015) introduction of resilience into the EASA (European Aviation Safety Agency) regulatory framework for Crew Resource Management (CRM) training. Introduced after a number of high-visibility aviation disasters in the 1970s and 1980s, CRM training seeks to improve cockpit collaboration and make the best use of all possible resources for problem-solving (Helmreich, Merritt, & Wilhelm, 1999). The program, promising improved information management, communication, leadership, situational awareness, and decision-making, has since its introduction to aviation been exported to numerous other high-risk industries. Early on, the resilience engineering community made references to teamwork training and improvements, a process of which we have also taken part, so it should perhaps be seen as a great success that some 10 years after introducing the notion of resilience to a safety science community, it was enacted into the regulatory framework for aviation team training. But is it really?

NEW CRM TRAINING: BE MENTALLY FLEXIBLE AND ADAPT

Here is an example of a regulator adopting resilience engineering into the CRM regulatory framework (GM5 ORO.FC.115, CRM training, EASA Annex II to ED Decision2015-022-R):

RESILIENCE DEVELOPMENT

a. The main aspects of resilience development can be described as the ability to:

1. learn ('knowing what has happened');
2. monitor ('knowing what to look for');
3. anticipate ('finding out and knowing what to expect'); and
4. respond ('knowing what to do and being capable of doing it').

b. Operational safety is a continuous process of evaluation and adjustment to existing and future conditions. In this context, and following the description in (a), resilience development involves an ongoing and adaptable process including situation assessment, self-review, decision and action. Training in resilience development enables crew members to draw the right conclusions from both positive and negative experiences. Based on those experiences, crew members are better prepared to maintain or create safety margins by adapting to dynamic complex situations.

c. The training topics in (f)(3) of AMC1 ORO.FC.115 are to be understood as follows:

1. Mental flexibility

   i. The phrase 'understand that mental flexibility is necessary to recognise critical changes' means that crew members are prepared to respond to situations for which there is no set procedure.

   ii. The phrase 'reflect on their judgement and adjust it to the unique situation' means that crew members learn to review their judgment based on the unique characteristics of the given circumstances.

   iii. The phrase 'avoid fixed prejudices and over-reliance on standard solutions' means that crew members learn to update solutions and standard response sets, which have been formed on prior knowledge.

   iv. The phrase 'remain open to changing assumptions and perceptions' means that crew members constantly monitor the situation, and are prepared to adjust their understanding of the evolving conditions.

2. Performance adaptation

   i. The phrase 'mitigate frozen behaviours, overreactions and inappropriate hesitation' means that crew members correct improper actions with a balanced response.

   ii. The phrase 'adjust actions to current conditions' means that crew members' responses are in accordance with the actual situation.

The regulatory formulations in the example above make clear that operational resilience is located at the level of pilot mental processes (mental flexibility) and behavior (performance adaptation). It also proposes that this operational resilience can be improved through training. There is an interesting use of normative language which is open to interpretation. Examples include, in the same sentence, the use of words such as 'overreactions,' 'inappropriate hesitation,' 'improper action,' and 'balanced response.' This is a behaviorist approach to resilience with little support in academic traditions, except for the recognition that adaptive strategies (work as done) often take place beyond written procedures (work as imagined) as discussed previously (Patterson, Cook, Woods, & Render, 2006).

There is no doubt that introducing the notion of resilience into the EASA training regulation has taken tremendous efforts and has all been done with the best intentions. After all, it serves to inform the operational community of contemporary research that emphasizes how safety is not the absence of negatives (e.g., error, violation, complacency) but the presence of active strategies to dynamically adapt to complex risks. However, introducing resilience into the training curriculum will likely have effects beyond the theoretical ideas of resilience back in 2005. These are effects that hold the risk of further responsibilizing sharp-end operators for the safety and risks of airline operations (a kind of reinvention of human error in terms of 'overreactions,' 'inappropriate hesitation,' 'improper action' and a lack of 'balanced response' (Dekker, 2001)

CRM training has the unfortunate history of seeing its training categories turned into moral categories of accountability. It has been shown repeatedly how categories are traditionally used in CRM training, including situational awareness, complacency, decision-making, and even just CRM itself as a broad category, have been used to describe the failures (loss of situational awareness, poor decision-making, breakdown of CRM) of pilots in causal constructions of accidents (Cook & Nemeth, 2010; Dekker, 2014; Dekker, Nyce, & Myers, 2013; Hollnagel & Amalberti, 2001). There is a risk that the notion of resilience might become an additional category for holding operators morally accountable for their actions.

As said above, the language by which resilience is introduced to the EASA regulations is highly normative. It adds new categories of how pilots might exceed moral boundaries of their profession, including overreliance on standard procedures, overreactions, inappropriate hesitation, improper action, and unbalanced response. Rather than empowering pilots to influence their discretionary space, resilience provides means to further responsibilize pilots for the behavior of the system in which they are embedded. The language responsibilizes pilots not only for the performance of the system, but in a sense even for their beliefs in resilience theory.

11.5.3 The Normative Trap

The notion of resilience has established itself as a normative claim, one which discusses resilience in a value-laden, and as seen above even in a moralizing, way. In this discourse, resilience is typically something positive and desirable. To remain vulnerable (typically seen as the opposite of resilience) can consequently be seen as a moral failure. However, there are also other, and much less normative, meanings of resilience available in the wider academic literature. In its ecological meaning resilience is not necessarily a desirable system feature. How does that fit with resilience engineering? At the very least, it raises new questions, e.g., regarding the connections between system levels.

There is another assumption in resilience engineering research, which has not yet been made explicit, and that is that all systems are worth preserving. In 1984, of course, Perrow concluded that some systems should not be preserved (such as nuclear power generation) because their potential for normal accidents (based on their tight coupling and interactive complexity) is simply too great (Perrow, 1984). Jerneck and Olsson (2008) argue that both the notion of sustainability and the notion of resilience are based on the "implicit normative assumption of preservation of the system and thus resistance to change" (p. 175). They argue for introducing the notion of 'transition' to highlight the value of abandoning destructive system states. This discussion is so far absent from the organizational discussion on safety. Resilience engineering embraces the idea that the most desirable (and safe) system is a system where safety (and production) is continually ensured at the level of local adaptive capacities and strategies.

Yet another question, relating to the desirability of resilience, is whether this optimism toward the human ability to prepare for and adapt to dynamic and unpredictable threats also embodies pessimism toward the ability to instead eliminate or control the threat in the first place. This question is raised by a few critical Foucauldians, such as Joseph (2013a,b), Evans, and Reid (2013, 2014, 2015), as well as by Morel, Amalberti, and Chauvin (2008, 2009) who, in their studies of deep sea fishing, make the argument that resilience is not the quest to stay safe as much as the acceptance of danger:

> Although the best safety response would be to stop fishing in borderline conditions, the resilient response is to go on, and develop survival skills, according to the situation. This willingness to take risks is actually based on genuine craft-style knowledge of resilience, centered on a familiarity with the environment and the ability to anticipate the changes both of this environment and of one's own skill, thus achieving permanent and favourable adequacy.
>
> **Morel et al. (2008, p. 13)**

Evans and Reid make a similar point, which not only questions the normative 'good' of resilience, but also shows its connection to a moralizing biopolitics:

> To increase its resilience ... the subject must disavow any belief in the possibility to secure itself and accept, instead, an understanding of life as a permanent process of continual adaptation to threats and dangers which are said to be outside its control. As such, the resilient subject is a subject which must permanently struggle to accommodate itself to the world, and not a subject which can conceive of changing the world, its structure and conditions of possibility. However, it is a subject which accepts the dangerousness of the world it lives in as a condition for partaking of that world and which accepts the necessity of the injunction to change itself in correspondence with threats now presupposed as endemic.
>
> **Evans and Reid (2013, p. 85)**

If we apply this critique to organizational safety policy, then resilience implies a need for (sharp-end) people to be willing to accept and adapt to dangers originating beyond their control. Referring to great complexity, dynamics, and uncertainty, the discourse seems to silence—or even give up on—the Perrowian question whether we should be accepting such risks in the first place.

Given this, and the other traps listed above, the critical, and highly ethical, question for safety science to raise here is whether this is fair? Is this normative resilience movement creating discursive and moralizing effect beyond, or even contrary to, the ones intended? And if they are, then how can the safety science community find a relevant role for resilience research given its interest in complexity, variability, adaptive capacities, system dynamics, and goal conflicts? To keep progressing safety science, we will have to keep asking probing questions about the connections between organizational levels, organizational ethics, and power.

STUDY QUESTIONS

1. What are the reasons for proposing that we need to learn why things go right and find out what we can do to make it even more so; that safety is not about the *absence* of negatives but about the *presence* of capacities?
2. Where do you recognize the legacy of a reductionist approach to safety? Name at least three supposed safety 'solutions' or practices that typify this approach.
3. Why do complex systems tend to be naturally more resilient in dynamic environments than centrally governed systems?
4. What does it mean that complex systems operate far from equilibrium?
5. In what ways is resilience engineering a 'new' approach in safety science? And in which ways isn't it new? Name a few ways for each.
6. What does the 'defence in depth fallacy' have to do with a system's vulnerability to drifting into failure?
7. Explain the differences between resilience as rebound, resilience as robustness, resilience as graceful extensibility, and resilience as sustained adaptability.
8. In what sense do resilience engineering and Safety Culture share a susceptibility to the reductionist trap?
9. How could resilience engineering escape the normative trap, or rather avoid it altogether?

REFERENCES AND FURTHER READING

Alexander, D. E. (2013). Resilience and disaster risk reduction: An etymological journey. *Natural Hazards and Earth System Science, 13*(11), 2707–2716. doi:10.5194/nhess-13-27072013

Amalberti, R. (2001). The paradoxes of almost totally safe transportation systems. *Safety Science, 37*(2–3), 109–126.

Amalberti, R. (2006). Optimum system safety and optimum system resilience: Agonistic or antagonistic concepts. In E. Hollnagel, D. D. Woods, & N. G. Leveson (Eds.), *Resilience engineering: Concepts and precepts* (pp. 253–274). Aldershot, UK: Ashgate Publishing Co.

Amalberti, R. (2013). *Navigating safety: Necessary compromises and trade-offs—theory and practice*. Heidelberg, Germany: Springer.

Amalberti, R., Auroy, Y., Berwick, D., & Barach, P. (2005). Five system barriers to achieving ultrasafe healthcare. *Annals of Internal Medicine, 142*(9), 756–764.

Antonsen, S. (2009). Safety culture assessment: A mission impossible? *Journal of Contingencies and Crisis Management, 17*(4), 242–254.

Barker, K., Ramirez-Marquez, J. E., & Rocco, C. M. (2013). Resilience-based network component importance measures. *Reliability Engineering & System Safety, 117*, 89–97. doi:10.1016/j.ress.2013.03.012

Barton, M. A., & Sutcliffe, K. M. (2009). Overcoming dysfunctional momentum: Organizational safety as a social achievement. *Human Relations, 62*(9), 1327–1356.

Bergström, J., van Winsen, R., & Henriqson, E. (2015). On the rationale of resilience in the domain of safety: A literature review. *Reliability Engineering & System Safety, 141*, 131–141. doi:10.1016/j.ress.2015.03.008

Borys, D., Else, D., & Leggett, S. (2009). The fifth age of safety: The adaptive age. *Journal of Health and Safety Research and Practice, 1*(1), 19–27.

Cilliers, P. (1998). *Complexity and postmodernism: Understanding complex systems*. London, UK: Routledge.

Cook, R. I., & Nemeth, C. P. (2010). "Those found responsible have been sacked": Some observations on the usefulness of error. *Cognition, Technology and Work, 12*, 87–93.

Costella, M. F., Saurin, T. A., & de Macedo Guimarães, L. B. (2009). A method for assessing health and safety management systems from the resilience engineering perspective. *Safety Science, 47*(8), 1056–1067. doi:10.1016/j.ssci.2008.11.006

de Carvalho, P. V. R. (2011). The use of Functional Resonance Analysis Method (FRAM) in a mid-air collision to understand some characteristics of the air traffic management system resilience. *Reliability Engineering & System Safety, 96*(11), 1482–1498. doi:10.1016/j.ress.2011.05.009

Debono, D. S., Greenfield, D., Travaglia, J. F., Long, J. C., Black, D., Johnson, J., & Braithwaite, J. (2012). Nurses' workarounds in acute healthcare settings: A scoping review. *BMC Health Services Research, 13*, 175–183.

Dekker, S. W. A. (2001). The re-invention of human error. *Human Factors and Aerospace Safety, 1*(3), 247–266.

Dekker, S. W. A. (2006). Resilience engineering: Chronicling the emergence of confused consensus. In E. Hollnagel, D. D. Woods, & N. G. Leveson (Eds.), *Resilience engineering: Concepts and precepts* (pp. 77–92). Aldershot, UK: Ashgate Publishing Co.

Dekker, S. W. A. (2014). The psychology of accident investigation: Epistemological, preventive, moral and existential meaning-making. *Theoretical Issues in Ergonomics Science, 16*(3), 202–213.

Dekker, S. W. A., Cilliers, P., & Hofmeyr, J. (2011). The complexity of failure: Implications of complexity theory for safety investigations. *Safety Science, 49*(6), 939–945.

Dekker, S. W. A., Nyce, J. M., & Myers, D. J. (2013). The little engine who could not: 'Rehabilitating' the individual in safety research. *Cognition, Technology and Work, 15*(3), 277–282.

Deming, W. E. (1982). *Out of the crisis*. Cambridge, MA: MIT Press.

Edmondson, A. (1999). Psychological safety and learning behavior in work teams. *Administrative Science Quarterly, 44*(2), 350–383.

Evans, B., & Reid, J. (2013). Dangerously exposed: The life and death of the resilient subject. *Resilience, 1*(2), 83–98. doi:10.1080/21693293.2013.770703

Evans, B., & Reid, J. (2014). *Resilient life: The art of living dangerously.* Cambridge, UK: Polity Press.

Evans, B., & Reid, J. (2015). Exhausted by resilience: Response to the commentaries. *Resilience, 3*(2), 154–159. doi:10.1080/21693293.2015.1022991

Francis, R., & Bekera, B. (2014). A metric and frameworks for resilience analysis of engineered and infrastructure systems. *Reliability Engineering & System Safety, 121,* 90–103. doi:10.1016/j.ress.2013.07.004

Garmezy, N. (1971). Vulnerability research and the issue of primary prevention. *American Journal of orthopsychiatry, 41*(1), 101.

Garmezy, N. (1974). Children at risk: The search for the antecedents of schizophrenia: II. Ongoing research programs, issues, and intervention. *Schizophrenia Bulletin, 1*(9), 55.

Garmezy, N. (1987). Stress, competence, and development: continuities in the study of schizophrenic adults, children vulnerable to psychopathology, and the search for stress-resistant children. *American Journal of Orthopsychiatry, 57*(2), 159.

Garmezy, N., & Streitman, S. (1974). Children at risk: The search for the antecedents of schizophrenia: I. Conceptual models and research methods. *Schizophrenia Bulletin, 1*(8), 14.

Gore, A. (2013). *The future: Six drivers of global change.* New York, NY: Random House.

Grant, A. (2016). *Originals: How non-conformists change the world.* London, UK: W. H. Allen.

Gunderson, L. (2010). Ecological and human community resilience in response to natural disasters. *Ecology & Society, 15*(2), 18.

Gunderson, L., & Holling, C. S. (Eds.). (2002). *Panarchy: Understanding transformations in human and natural systems.* Washington, DC: Island Press.

Haavik, T. K. (2014). On the ontology of safety. *Safety Science, 67,* 37–43.

Hale, A., & Heijer, T. (2006). Defining resilience. In E. Hollnagel, D. Woods, & N. Leveson (Eds.), *Resilience engineering: Concepts and precepts* (pp. 35–40). Aldershot, UK: Ashgate Publishing Co.

Helmreich, R. L., Merritt, A. C., & Wilhelm, J. A. (1999). The evolution of crew resource management training in commercial aviation. *International Journal of Aviation Psychology, 9*(1), 19–32.

Heylighen, F., Cilliers, P., & Gershenson, C. (2007). Complexity and philosophy. In J. Bogg & R. Geyer (Eds.), *Complexity, science and society* (pp. 117–134). Oxford, UK: Radcliffe Publishing.

Holling, C. S. (1973). Resilience and stability of ecological systems. *Annual Review of Ecology and Systematics, 4,* 1–23.

Hollnagel, E. (2006). Resilience—the challenge of the unstable. In E. Hollnagel, D. Woods, & N. Leveson (Eds.), *Resilience engineering, concepts and precepts* (pp. 9–17). Aldershot, UK: Ashgate Publishing Co.

Hollnagel, E. (2009). *The ETTO principle: Efficiency-thoroughness trade-off. Why things that go right sometimes go wrong.* Aldershot, UK: Ashgate Publishing Co.

Hollnagel, E. (2012). *FRAM the functional resonance analysis method: Modelling complex socio-technical systems.* Boca Raton, FL: CRC Press.

Hollnagel, E. (2014). Becoming resilient. In C. P. Nemeth & E. Hollnagel (Eds.), *Resilience engineering in practice: Becoming resilient* (Vol. 2), pp. 179–192. Farnham, UK: Ashgate Publishing Co.

Hollnagel, E., & Amalberti, R. (2001). *The emperor's new clothes: Or whatever happened to 'human error'?* Paper presented at the 4th international workshop on human error, safety and systems development, Linköping, Sweden.

Hollnagel, E., Nemeth, C. P., & Dekker, S. W. A. (2008). *Resilience engineering: Remaining sensitive to the possibility of failure.* Aldershot, UK: Ashgate Publishing Co.

Hollnagel, E., Nemeth, C. P., & Dekker, S. W. A. (2009). *Resilience engineering: Preparation and restoration.* Aldershot, UK: Ashgate Publishing Co.

Hollnagel, E., Woods, D. D., & Leveson, N. G. (2006). *Resilience engineering: Concepts and precepts.* Aldershot, UK: Ashgate Publishing Co.

Hopkins, A. (2013). Issues in safety science. *Safety Science, 67,* 6–14. doi:10.1016/j.ssci.2013.01.007

Huber, G. J., Gomes, J. O., & de Carvalho, P. V. R. (2012). A program to support the construction and evaluation of resilience indicators. *Work, 41*(Supplement 1), 2810–2816. doi:10.3233/WOR-2012-0528-2810

Jerneck, A., & Olsson, L. (2008). Adaptation and the poor: Development, resilience and transition. *Climate Policy, 8*(2), 170–182.

Johnstone, R. E. (2017). Glut of anesthesia guidelines a disservice, except for lawyers. *Anesthesiology News, 42*(3), 1–6.

Joseph, J. (2013a). Resilience as embedded neoliberalism: A governmentality approach. *Resilience, 1*(1), 38–52. doi:10.1080/21693293.2013.765741

Joseph, J. (2013b). Resilience in UK and french security strategy: An anglo-saxon bias? *Politics, 33*(4), 253–264. doi:10.1111/1467-9256.12010

Kolar, K. (2011). Resilience: Revisiting the concept and its utility for social research. *International Journal of Mental Health and Addiction, 9*(4), 421–433. doi:10.1007/s11469-011-9329-2

Kropotkin, P. (1892). *La conquete du pain (The conquest of bread).* Paris, France: Tresse & Stock.

Lanir, Z. (1986). *Fundamental surprise.* Eugene, OR: Decision Research.

Le Coze, J. C. (2015). Reflecting on Jens Rasmussen's legacy: A strong program for a hard problem. *Safety Science, 71,* 123–141.

Leveson, N. (2004). A new accident model for engineering safer systems. *Safety Science, 42*(4), 237–270. doi:10.1016/s0925-7535(03)00047-x

Masten, A. S., & Garmezy, N. (Eds.). (1985). *Advances in clinical child psychology.* Heidelberg, Germany: Springer.

Meddings, J., Reichert, H., Greene, M. T., Safdar, N., Krein, S. L., Olmsted, R. N., … Saint, S. (2016). Evaluation of the association between Hospital Survey on Patient Safety Culture (HSOPS) measures and catheter-associated infections: Results of two national collaboratives. *BMJ Quality and Safety, 26*(3), 226–235. doi:10.1136/bmjqs-2015-005012.

Mendonça, D., & Wallace, W. A. (2004). Studying organizationally-situated improvisation in response to extreme events. *International Journal of Mass Emergencies and Disasters, 22*(2), 5–30.

Morel, G., Amalberti, R., & Chauvin, C. (2008). Articulating the differences between safety and resilience: The decision-making process of professional sea-fishing skippers. *Human Factors: The Journal of the Human Factors and Ergonomics Society, 50*(1), 1–16. doi:10.1518/001872008X250683

Morel, G., Amalberti, R., & Chauvin, C. (2009). How good micro/macro ergonomics may improve resilience, but not necessarily safety. *Safety Science, 47*(2), 285–294. doi:10.1016/j.ssci.2008.03.002

Nemeth, C. P., & Herrera, I. A. (2015). Building change: Resilience engineering after ten years. *Reliability Engineering and System Safety, 141,* 1–4.

Nemeth, C. P., Nunnally, M., O'Connor, M. F., Brandwijk, M., Kowalsky, J., & Cook, R. I. (2007). Regularly irregular: how groups reconcile cross-cutting agendas and demand in healthcare. *Cognition, Technology & Work, 9*(3), 139–148. doi:10.1007/s10111-006-0058-4

Ouyang, M. (2014). Review on modeling and simulation of interdependent critical infrastructure systems. *Reliability Engineering & System Safety, 121,* 43–60. doi:10.1016/j.ress.2013.06.040

Pant, R., Barker, K., & Zobel, C. W. (2013). Static and dynamic metrics of economic resilience for interdependent infrastructure and industry sectors. *Reliability Engineering & System Safety.* doi:10.1016/j.ress.2013.09.007

Pasman, H., Knegtering, B., & Rogers, W. (2013). A holistic approach to control process safety risks: Possible ways forward. *Reliability Engineering & System Safety, 117,* 21–29. doi:10.1016/j.ress.2013.03.010

Patriarca, R., Bergström, J., & Di Gravio, G. (2017). Defining the functional resonance analysis space: Combining abstraction hierarchy and FRAM. *Reliability Engineering & System Safety, 165,* 34–46. doi:10.1016/j.ress.2017.03.032

Patterson, E. S., Cook, R. I., Woods, D. D., & Render, M. L. (2006). Gaps and resilience. In S. Bogner (Ed.), *Human error in medicine* (2nd ed., pp. 66–87). Hillsdale, NJ: Lawrence Erlbaum Associates.

Perrow, C. (1984). *Normal accidents: Living with high-risk technologies.* New York, NY: Basic Books.

Perry, S. J., & Wears, R. L. (2012). Underground adaptations: Case studies from health care. *Cognition, Technology & Work, 14*(3), 253–260. doi:10.1007/s10111-011-0207-2

Prigogine, I. (2003). *Is future given?* London, UK: World Scientific Publishing Co.

Raman, J., Leveson, N. G., Samost, A. L., Dobrilovic, N., Oldham, M., Dekker, S. W. A., & Finkelstein, S. (2016). When a checklist is not enough: How to improve them and what else is needed. *The Journal of Thoracic and Cardiovascular Surgery, 152*(2), 585–592.

Rasmussen, J. (1980). What can be learned from human error reports? In K. Duncan, M. Gruneberg & D. Wallis (Eds.), *Changes in working life* (pp. 97–113). London, UK: John Wiley & Sons.

Rasmussen, J. (1982). Human errors. A taxonomy for describing human malfunction in industrial installations. *Journal of Occupational Accidents, 4*(2–4), 311–333.

Rasmussen, J. (1990). The role of error in organizing behavior. *Ergonomics, 33*(10–11), 1185–1199.

Rasmussen, J. (1997). Risk management in a dynamic society: A modelling problem. *Safety Science, 27*(2–3), 183–213.

Rasmussen, J., & Lind, M. (1981). *Coping with complexity (Risø-M-2293).* Paper presented at the European Conference on Human Decision and Manual Control, Delft, The Netherlands.

Rasmussen, J., Nixon, P., & Warner, F. (1990). Human error and the problem of causality in analysis of accidents [and discussion]. *Philosophical Transactions of the Royal Society of London. Series B, Biological Sciences, 327*(1241), 449–462.

Reason, J. T. (1997). *Managing the risks of organizational accidents.* Aldershot, UK: Ashgate Publishing Co.

Reason, J. T. (2008). *The human contribution: Unsafe acts, accidents and heroic recoveries.* Farnham, UK: Ashgate Publishing Co.

Rediker, M. (1987). *Between the devil and the deep blue sea: Merchant seamen, pirates, and the Anglo-American maritime world, 1700-1750.* Cambridge, UK: Cambridge University Press.

Salzano, E., Di Nardob, M., Gallob, M., Oropallob, E., & Santillob, L. C. (2014). The application of system dynamics to industrial plants in the perspective of process resilience engineering. *Chemical Engineering Transactions, 36,* 457–462.

Shirali, G., Mohammadfam, I., Motamedzade, M., Ebrahimipour, V., & Moghimbeigi, A. (2012). Assessing resilience engineering based on safety culture and managerial factors. *Process Safety Progress, 31*(1), 17–18. doi:10.1002/prs.10485

Shirali, G., Motamedzade, M., Mohammadfam, I., Ebrahimipour, V., & Moghimbeigi, A. (2012). Challenges in building resilience engineering (RE) and adaptive capacity: A field study in a chemical plant. *Process Safety and Environmental Protection, 90*(2), 83–90. doi:10.1016/j.psep.2011.08.003

Smith, M. W., Davis Giardina, T., Murphy, D. R., Laxmisan, A., & Singh, H. (2013). Resilient actions in the diagnostic process and system performance. *BMJ Quality & Safety, 22*(12), 1006–1013. doi:10.1136/bmjqs-2012-001661

Sujan, M., Spurgeon, P., & Cooke, M. (2015). The role of dynamic trade-offs in creating safety—A qualitative study of handover across care boundaries in emergency care. *Reliability Engineering & System Safety, 141*, 54–62. doi:10.1016/j.ress.2015.03.006

Ungar, M. (2005). *Handbook for working with children and youth: Pathways to resilience across cultures and contexts.* Thousand Oaks, CA: Sage Publications.

Vaughan, D. (1999). The dark side of organizations: Mistake, misconduct, and disaster. *Annual Review of Sociology, 25*(1), 271–305.

Walker, J., & Cooper, M. (2011). Genealogies of resilience: From systems ecology to the political economy of crisis adaptation. *Security Dialogue, 42*(2), 143–160. doi:10.1177/0967010611399616

Weber, D. E., MacGregor, S. C., Provan, D. J., & Rae, A. R. (2018). "We can stop work, but then nothing gets done." Factors that support and hinder a workforce to discontinue work for safety. *Safety Science, 108*, 149–160.

Weick, K. E., & Sutcliffe, K. M. (2007). *Managing the unexpected: Resilient performance in an age of uncertainty* (2nd ed.). San Francisco, CA: Jossey-Bass.

Westrum, R. (2006). A typology of resilience situations. In E. Hollnagel, D. Woods, & N. Leveson (Eds.), *Resilience engineering—concepts and precepts* (pp. 43–53). Aldershot, UK: Ashgate Publishing Co.

Woljter, R. (2009). *Functional modeling of constraint management in aviation safety and command and control* (Dissertation No. 1247). (Ph. D), Linköping University, Linköping.

Woods, D. D. (1988). Coping with complexity: The psychology of human behavior in complex systems. In L. P. Goodstein, H. B. Andersen, & S. E. Olsen (Eds.), *Tasks, errors, and mental models* (pp. 128–148). New York, NY: Taylor & Francis Group.

Woods, D. D. (2003). *Creating foresight: How resilience engineering can transform NASA's approach to risky decision making.* Washington, DC: US Senate Testimony for the Committee on Commerce, Science and Transportation, John McCain, chair.

Woods, D. D. (2006). Essential characteristics of resilience. In E. Hollnagel, D. D. Woods, & N. G. Leveson (Eds.), *Resilience engineering: Concepts and precepts* (pp. 21–34). Aldershot, UK: Ashgate Publishing Co.

Woods, D. D. (2015). Four concepts for resilience and the implications for the future of resilience engineering. *Reliability Engineering and System Safety, 141*(1), 5–9.

Woods, D. D., Dekker, S. W. A., Cook, R. I., Johannesen, L. J., & Sarter, N. B. (2010). *Behind human error.* Aldershot, UK: Ashgate Publishing Co.

Wreathall, J. (2006). Properties of resilient organizations: An initial view. In E. Hollnagel, D. D. Woods, & N. G. Leveson (Eds.), *Resilience engineering: Concepts and precepts* (pp. 275–286). Aldershot, UK: Ashgate Publishing Co.

Postscript

The safety-scientific approaches of a century have set a pattern. From an innovation that typically targets the system in which people work, almost every approach seems to end up reverting, one way or another, to the people who work in that system. Indeed, at the heart of this pattern is a dialectic—a constantly recurring discussion about the truth of either of two options. Should we look for safety improvements, risk management, and hazard containment in the system or in the person? In the organization or in the individual? In technology or in people? Upstream or downstream? Distal or proximal? Blunt end or sharp end? Do we fix work or do we fix the worker?

Virtually every approach dealt with in this book developed—or developed out of—a discovery that there is something about the way we organize work; that there is something about how we build our systems that we can do better. Because the way we have done it up to that moment structurally creates problems, risks, error traps, and inefficiencies. The problems identified were in the system, and the solutions were targeted at it. As you will have seen throughout the chapters:

- Taylor's and the Gilbreths' war on waste and inefficiency analytically deconstructed and then 'scientifically' reconstructed the management, organization, and order of work. Prodding individuals to work harder, or smarter, or cheaper, or better, or safer was of no use. The tasks they were told to do had to be strictly systematized first, and the tools with which they were told to work needed to be methodically engineered.
- Epidemiological data started pointing to a differential probability of suffering harm and accidents, which eventually led to the accident-prone thesis. The original impetus was systematic and scientific, however, and its interventions were targeted at the level of the system in which people worked, matching skills with demands.
- Heinrich was one of the first to investigate systematically ways to stop hazard trajectories from causing accidents and injury. On practical grounds, he advocated selecting remedies as early as possible—upstream, in the selection of work and workers, in the design of work and workplaces. He placed a strong emphasis on improving the physical conditions and physical safeguards of work.
- Human factors was born out of the realization that the human was the recipient of error-prone and error-intolerant system designs. Operational people inherited safety problems (from their technologies, tools, organizations, working environments, and conditions) rather than creating safety problems. Solutions were targeted at the technology, the system, which were systematically adapted to human strengths and limitations.

- System safety promoted the notion that safety needs get built into the system from the very beginning. And once a system is in operation, system safety needed to specify the requirements for effective and safe management of the system. It should not be left to frontline heroics to make things work in practice, or to recover from built-in error traps. These could, and should, be designed and managed out—upstream.
- Man-made disaster theory understood accidents and disasters squarely as administrative or organizational phenomena. The intentions and actions that incubated accidents and disasters were bred—by very normal, every-day processes and bureaucratic workings—at the organizational blunt end. Prevention efforts should be targeted there—upstream.
- Normal accident theory proposed how structural features of a system—its interactive complexity and tight coupling—not only create fundamental paradoxes in any attempts to safely manage it, but set that system up for a predictable kind of breakdown: the normal accident. The target for safety improvements is not in the people managing or operating it, but in the system itself and in the political arena that allows it to operate at all.
- Reason showed that the further people were removed from the frontline activities of their system, the greater the potential danger they posed to it. Only responding to frontline errors or violations would not lead to improvements. Attempts to discover and neutralize 'resident pathogens' in the system, the organization was going to have a greater beneficial effect on system safety than chasing localized acts by those on the sharp end.
- Safety culture gave organizations an aspiration: it got leaders and others to think about what they wanted to *have* rather than what they needed to *avoid*. The target was more than just upstream, it encompassed the entire organization and its culture. Researchers and practitioners became concerned with specifying what is necessary inside an organization to enhance safety.
- Resilience engineering represents a valiant and honest attempt to specify the possibilities for creating safety in complex, dynamic systems. In these systems, novel phenomena emerge, and much more variation occurs than could ever be procedurally specified, or designed or trained for. Engineering resilience into them is a matter, in part, of recognizing and matching the requisite variety of the system.

The innovations that have come from these approaches gradually congealed into their respective sets of technocratic projects, of programmatic operations and mechanistic schemes. The approaches all successively became professionalized, spawning time-and-motion researchers, psychotechnologists, behavioral safety consultants, human factors engineers, system safety experts, safety professionals, and practitioners trained in methods such as STAMP and FRAM.

It seems, though, that as a particular approach professionalizes, it moralizes. The steely emptiness of its methods and techniques alone may not be felt as sufficient to keep propelling it forward. More is necessary to keep mobilizing the commitment to it, or the passion for it, or perhaps simply to sustain its economic attraction and viability. This is where a safety-scientific approach can become a moral ideal and moralizing project, an object lesson in what to do and what to avoid, a catechism of what to believe and what to discard as false. And almost invariably, in this transformation from professionalization to moralization, the approach—that started with the system, the technology, the organization, upstream—lands back on the individual people in that system. And—in any other words—it appeals to them to try yet a little harder. The original impetus targets the system; the ultimate moral appeal is directed at the individual. As you will have seen throughout the chapters of this book:

- Taylor's call to become a 'high-class man' could be answered only by an individual submitting completely to the emptiness of the tasks designed and 'scientifically managed' by others. This kind of work is now known to have been so mind-numbing or soul-destroying that Henry Ford had to 'bribe' his workers with a $5-a-day wage.
- Accident-prone theory descended into the morally dubious (and scientifically untenable) separation of workers who were 'fit' from those who were not, justifying why we could legitimately give up hope on some people, and aligning itself with theories and practices that quantified who was an idiot, a moron, an imbecile.
- Heinrich placed increasing emphasis on eliminating unsafe acts by workers. He advocated creating an environment where even small undesirable acts were not tolerated. It spawned behavior-based safety. This squarely targeted workers and at times ignored the work or the environment of that work. It can deteriorate into retributive rituals in response to not meeting low-injury targets and arbitrary firings or exclusions on account of 'safety violations.'
- From a focus on solving safety problems by targeting the technology and the system, human factors has been enjoined to revisit the individual. It has been co-opted into methods and theories that squarely target human shortcomings, such as line management deficiencies, supervisory shortcomings, fallible decisions, unsafe acts, and violations. Even the dead can now be blamed for their own complacency and 'loss of situation awareness.'
- When pushed to explain how disaster incubation happened, man-made disasters had little else to go on than collective human deficiencies— erroneous assumptions, not noticing information or misunderstanding signals, communication failures, not heeding warnings, and a human reluctance to imagine the worst.
- The structural analysis of normal accident theory was counterbalanced with high reliability research that showed the extent to which people actually go (and go successfully) to manage even seemingly unmanageable complexity. But in this too, hid a moral judgment. The recent focus on 'mindfulness' allows this same approach to say that failure equals a lapse in detection, that someone somewhere did not anticipate or catch what went wrong, and that they should have paid attention earlier.

- From an enthusiastic embrace of systems thinking, Swiss cheese became entwined in a new effort to focus on the individual. In our search for distal causes, we had thrown the net too widely; we let the pendulum swing too far toward the system to be morally justifiable. The human could occasionally be a hero, but was best treated as a hazard, a system component whose unsafe acts are implicated in the majority of catastrophic breakdowns.
- Safety culture research, when operationalized, stopped being about culture and upstream, organizational factors fairly swiftly. It instead became reduced to the attitudes and behaviors of individuals in the organization. Safety culture could be improved by targeting those in ways that are similar to what behavior-based safety might propose (hearts-and-minds campaigns, posters, incentive programs—all focusing on fixing the worker, not the work or the environment).
- From an explicit commitment to understanding some of the complexity and variability of the systems in which people work, and literally helping to engineer resilience into these systems, resilience engineering has been turned around and used to blame people for not having enough resilience. Its approach has been deployed to force individuals to accept and adapt to dangers that brew and grow beyond their control.

The pattern—of discovery and innovation and system improvement, of technicalization and professionalization, and finally of moralization and moral emphasis on the individual—repeats itself through the decades of the past century. It is visible in the approaches of this day, including resilience engineering. Are we forever beholden to the dialectic between system and person, between work and the worker, between upstream and downstream, organization and individual, blunt end and sharp end? As I argued a few years ago, perhaps we should liberate ourselves from the dialectic by speaking not of people versus systems, but of people *in* systems (Dekker, 2012). We are all an indelible part of the complex systems we inhabit, after all. Borrowing Paul Cilliers' term, we are 'folded into' those complex systems, and they are folded into our lives and our consciousness in ways that we do not, and cannot, even recognize.

Of course, we should look at the system in which we and other people work, and improve it to the best of our ability. At the same time, we know that all safety-critical work is ultimately channeled through relationships between human beings, or through direct contact of some people with the risky technology. A system creates all kinds of opportunities for action. And it also constrains people in many ways. Beyond these opportunities and constraints, we could argue that there remains a discretionary space, a space that can be filled only by an individual caregiving or technology-operating human. This is a final space in which a system really does leave people freedom of choice (to launch or not, to go to open surgery or not, to fire or not, to continue an approach or not). It is a space filled with ambiguity, uncertainty, and moral choices.

This is why people matter. They will forever be behind the design and operation of these systems, if not actually actively operating them. That is why safety science will forever be, in an important sense, a social science. Rather than individuals versus systems, we should continually strive to understand the relationships and roles of individuals *in* systems. Maybe our moral commitment is to look for ways in which we can empower each other to make things go right in the systems that create safety risks, but that do so much else for us too.

REFERENCE

Dekker, S. W. A. (2012). *Just culture: Balancing safety and accountability* (2nd ed.). Farnham, UK: Ashgate Publishing Co.

Index

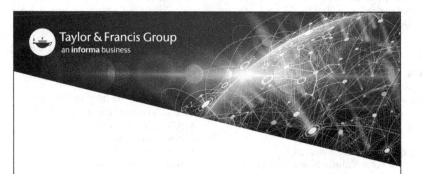

Printed in the United States
by Baker & Taylor Publisher Services